**Prentice Hall Series
in Geographic
Information Science**

**KEITH C. CLARKE,**
*Series Editor*

Arnold, *Interpretation of Airphotos and Remotely Sensed Imagery*

Avery/Berlin, *Fundamentals of Remote Sensing and Airphoto Interpretation, 5th edition.*

Clarke, *Analytical and Computer Cartography, 2nd edition.*

Clarke, *Getting Started With Geographic Information Systems, 3rd edition.*

Clarke, Parks and Crane (eds), *Geographic Information Systems and Environmental Modeling*

Foresman, *The History of Geographic Information Systems*

Jensen, *Introductory Digital Image Processing: A Remote Sensing Perspective, 2nd Edition*

Peterson, *Interactive and Animated Cartography*

Slocum, *Thematic Cartography and Visualization*

Vincent, *Fundamentals of Geological and Environmental Remote Sensing*

**Prentice Hall Series in
Geographic Information Science**

# Geographic Information Systems and Environmental Modeling

## Keith C. Clarke
*University of California, Santa Barbara*

## Bradley O. Parks
*Cooperative Institute for Research in Environmental Sciences,
University of Colorado, Boulder*

## Michael P. Crane (Eds)
*United States Geological Survey, EROS Data Center, Sioux
Falls*

Prentice Hall, Upper Saddle River, New Jersey 07458

*Library of Congress Cataloging-in-Publication Data*

Geographic information systems and environmental modeling / Keith C. Clarke, Brad E. Parks, and Michael P. Crane (eds.)
    p. cm.—(Prentice Hall series in geographic information science)
  "Fourth International Conference on Integrating Geographic Information Systems and Environmental Modeling, held in Banff, Canada, in September 2000"—Pref.
  Includes bibliographical references.
  ISBN 0-13-040817-4
    1. Geographic information systems—Congresses. 2. Ecology—Simulation methods—Congresses. I. Clarke, Keith C., II. Parks, Brad E., III. Crane, Michael P. IV. International Conference on Integrating Geographic Information Systems and Environmental Modeling (4th : Banff, Alta. 2000) V. Series.

G70.212 .G44547 2001
363.7'02'011—dc21

                                   20010148516

Executive Editor: Dan Kaveney
Assistant Editor: Amanda Griffith
Editorial Assistant: Margaret Ziegler
Production Editor: Kim Dellas
Marketing Manager: Christine Henry
Art Director: Jayne Conte
Cover Design: Bruce Kenselaar
Cover Art: Cordon art B. V.
Cover Image Description: M. C. Escher (1898–1970), "Puddle"
Manufacturing Manager: Trudy Pisciotti
Manufacturing Buyer: Michael Bell

 © 2002 by Prentice-Hall, Inc.
Upper Saddle River, NJ 07458

ISBN 0-13-040817-4

Printed in the United States of America
10 9 8 7 6 5 4 3 2 1

Prentice-Hall International (UK) Limited, *London*
Prentice-Hall of Australia Pty. Limited, *Sydney*
Prentice-Hall Canada Inc., *Toronto*
Prentice-Hall Hispanoamericana, S.A., *Mexico*
Prentice-Hall of India Private Limited, *New Delhi*
Prentice-Hall of Japan, Inc., *Tokyo*
Pearson Education Asia Pte. Ltd.
Editora Prentice-Hall do Brasil, Ltda., *Rio de Janeiro*

*To the participants of GIS/EM4*

# Contents

# *Preface*

This book serves three purposes. First, it is one of the five permanent records that are outcomes of the Fourth International Conference on Integrating Geographic Information Systems (GIS) and Environmental Modeling (GIS/EM4), held in Banff, Canada in September 2000. The other four permanent records are the research record of the meeting: special issues of the *Journal of Environmental Management* (Clarke et al., 2000a) and *Transactions in GIS* (Clarke et al, 2000b), plus the CD-ROM Proceedings for the meeting (Parks et al., 2001), and the permanent conference Web site (http://www.colorado.edu/research/cires/banff/). This book is different from the research results from the meeting in a fundamental way, since it is designed with the second purpose in mind, that of education. The book is intended for use in the advanced classroom or seminar to introduce students to the power that GIS offers to computer-based models of the environment. This education purpose is critical, because the editors of this volume believe that the decade of work on integrating GIS and environmental modeling has now yielded fruit, and the benefits will be reaped by a new generation of authors with much work ahead of them. Third, the book is an opportunity for a set of scholars, most of whom attended GIS/EM4, to establish where the state of the art lies at the turn on the century, after a productive first decade. In retrospect, it seems astonishing that when GIS/EM1 was held, there was no World Wide Web, GIS was still largely command-line driven stand-alone computer programs, and practical data interoperability was merely a pipedream.

The key word in the conference name, not reflected in this book's abbreviated title, is *integration*. The selection of sites in the Rocky Mountain region for the GIS/EM conferences over the years, Boulder, Breckenridge, Santa Fe, and Banff, was deliberate. It symbolized the divide between the two adjacent fields of environmental modeling and GIS. Environmental models, regardless of disciplinary specialization, shared a mathematical and statistical core, an algorithmic and a methodological periphery, and common concerns of application, model effectiveness, calibration and validation, computational implementation, and utility. Geographic information systems combined methods and functions yet faced immense practical and theoretical barriers as they gained use as ubiquitous tools for the management, display, and analysis of spatial data. At the surface there was a common boundary at the divide based on needs: for data, for visualization of pattern, and for objective scientific problem solving. GIS brought the bright light of complex reality to the elegant but blemished face of simplified mathematical models. It was an uneasy meeting at first, but throughout the 1990s the crispness of the divide line became fuzzy, and the separate sciences became integrated in their quest to explain the world and its systems, rather than advocate a single discipline as holding the solutions. The GIS/EM4 visual used with permission for the conference and on the cover of this book, Escher's "Puddle," is an excellent metaphor for this transformed science. In a single glimpse, Escher integrates clouds and weather, trees and soil, rain and surface water. And yet this integrated view shows these physical phenomena by indirect reflection, instead directing the viewer toward human impacts upon these "natural" environmental elements. Footsteps, vehicle tracks, and puddles hint at an unseen environmental modifier, homo sapiens. The scene makes sense only in unity, viewed through the puddle, just as the environment yields to simulation only by integrating knowledge.

The image of trees seen via a puddle in tire tracks is inverted and geometrically distorted, of course. Equally transformed is the environment, seen in isolation through the eyes of a single scientific discipline. In Chapter 12, the editors explain the logic behind the organization of GIS/EM4. The challenge, of course, is for scientists trained in depth as biologists, ecologists, geographers, geologists, and so forth, to think outside the box. Team research, involving disciplinary specialists and integrated science, is much harder than working in isolation. Universities recognize only departments, funding agencies have divisions, and disciplines have favorite journals, cultures, and traditions. It was this challenge, to think by synthesis and analysis, that was given to the authors of the chapters of this book. The editors met together for two years prior to the Banff conference, and slowly the chapter topics emerged. Next came a template, or generic look for a single chapter. Authors were selected by invitation, most self-identified. After generating a first draft, the pool of authors and others were used to peer-review the work. This continued through a review by the editors and a final production edit with the publishers. The final book was laid out by the first editor, and the last chapter was added to unify the book and to report the results of the conference. We hope that the result is a lasting tribute to the authors and the many people who have contributed to the quality of what you now see before you.

In Chapter 1, Nigel Waters presents a disciplinary perspective on the history of GIS/EM. Any history must be disciplinary, of course, because there is no history of integrated science. Waters examines the geographic tradition of modeling, its roots in the quantitative revolution, and gives a perspective on the four GIS/EM conferences and their output. The chapter's Canadian flavor is a tribute to the Canadian venue for GIS/EM4. In Chapter 2, Helen Couclelis explores modeling frameworks, paradigms and approaches. Environmental modeling has often been bottom-up in the past, without a major consideration of theory of approach. The chapter discusses the complex systems paradigm, popular in recent approaches to environmental systems modeling and especially appropriate when GIS is involved. The impact of the computer itself as a contributor toward models is discussed, termed *geocomputation*. The computer is seen as a valuable test-tube in which experiments are feasible and useful, fortunately with benign environmental consequences. Chapter 3 examines the consequences of intergating GIS and environmental models for human decision making. The authors use a single decision support system that they have developed to demonstrate that when models and GIS are linked, the implications can follow all the way through the science to the application, including involving users in informed decision making by interacting with the GIS and models themselves. This theme has permeated the GIS/EM sequence, yet here the disciplinary divide is crossed to the extent that new inroads in social science seem possible and useful.

The role of data in GIS and environmental modeling has been of long concern. Chapter 4 reviews the state of the art from the point of view of data, especially data suitable for broad-scale modeling and GIS. Various data types and the issues of their use are examined, and new data streams with high potential for future use are highlighted. Included is a guide to the availability of the data over the World Wide Web. One wonders if in the future models themselves, or at least their predictions and forecasts, could be as easily accessed as the data. If so, then a high degree of interaction and integration could be accomplished with the Web as the vehicle.

Models, of course, are merely intellectual exercises in description without calibration and validation. In Chapter 5, Mauro Giudici examines these critical steps in modeling from the viewpoint of model theory and structure. The process of model development is presented in the light of the calibration and validation stages, and some examples of environmental model calibration are presented. In Chapter 6, Andrew Rogowski and Jennifer Goyne emphasize the fourth or time dimension in modeling. GIS allows us to collect multitemporal data for modeling, and new approaches allow us to use them to calibrate and build truly dynamic models. These models require new theory, and the relevant concepts are shown using two case studies in hydrology.

In the Escher print, humans made the puddle that forms the window on the world. In Chapter 7, Morgan Grove and his colleagues look at the problems inherent in building models that cross the human–physical divide. Complexity again comes into play here, and the chapter examines future directions that clearly are paths GIS/EM needs to follow if integrated science is to be fully productive. Chapter 8 shows the other side of the coin, with Helena Mitasova and Lubos Mitas giving methodololgical and practical examples of the importance of process modeling for physical systems. Separate discussions of human and physical systems beg a discussion of issues of integration. Steven Frysinger examines in Chapter 9 what concerns arise from integration, including model interactions, uncertainty propagation, and user interface issues. The chapter also covers practical tools and approaches that allow integration to take place in modeling systems.

In Chapter 10, Brendan Mackey and Shawn Leffan give some examples from down under. The editors anticipate that educational use of this book will emphasize the case study approach. A series of case studies are presented via a typology of models in such a way that the reader can ask questions of communality, method, and direction. The models discussed involve applications of GIS to predictive vegetation mapping, forest ecosystem processes, assessment of wild river status, integrated modeling of global change, and integrated land use assessment as a GIS/environmental modeling learning tool.

GIS has always provided for modeling a direct link to visualization tools, whether for simple mapping or three-dimensional and animated display. In Chapter 11, Reed Copsey uses the Environmental Visualization System software's capabilties to demonstrate and explore the methods, capabilities, and problems specific to visualization of environmental data. While many GIS packages offer these capabilities, others do not, demonstrating yet another place where systems integration offers the best solution. In the future these capabilties will be contained in GIS, or even in the models themselves. The editors hope that this again demostrates the value of integrating software rather than duplicating systems. Finally, the editors in Chapter 12 record the last of the lessons of GIS/EM4, with a synopsis of the valuable workgroup discussions held in Banff, with the synthesis statement of the conference, and with some thoughts on future directions. The year 2001 finds the GIS/EM integrated science approach very much on the rise. We hope that this volume finds a role as a promoter of sound modeling, real data, and excellent science. The synergy of the participants at GIS/EM has been energizing. It is also greatly encouraging. With these tools and approaches, how many of our earth's problems can be solved, and how many in time to leave future inhabitants today's quality of life? The road ahead is clear, and the editors encourage the young and bold to step into the metaphorical puddle and start the journey.

## References

Clarke, K. C., Parks, B. O. and Crane, M. P. (2000a) "Integrating Geographic Information Systems (GIS) and Environmental Models. Preface: A Perspective on GIS-
Environmental Model Integration (GIS/EM)," *Journal of Environmental Management*, vol. 59, pp. 229–233.

Clarke, K. C., Parks, B. O. and Crane, M. P. (2000b) "Selected papers from the Fourth
International Conference on Integrating GIS and Environmental Modeling (GIS
EM4)," *Transactions in GIS*, vol. 4, no. 3, pp. 177–180.

Parks, B. O., Crane, M.P., and Clarke, K.C., eds. (2001). Integrating GIS and Environmental Modeling: Problems, Prospects, and Research Needs. *Proceedings of the
4th International Conference (GIS/EM4)*. September 2-8, 2000, Banff, Alberta,
Canada. CD-ROM. Conference Secretariat, University of Colorado, Boulder, Colorado.

Chapter authors present in Banff, September 2000.

# 1

# *Modeling the Environment with GIS: A Historical Perspective from Geography*

## 1.0 INTRODUCTION

This chapter provides a historical overview of both environmental modeling and the development of geographic information systems (GISs). The ongoing evolution of GIS is also described. Special attention is paid to the four GIS and environmental modeling conferences that were held in 1991, 1993, 1996, and 2000. There is a brief discussion of the tight, loose, and multiple coupling debates that are concerned with the extent to which GIS and environmental modeling are integrated. The chapter concludes with a comparison of the present state of the art in modeling the environment using GIS followed by a speculative look at the future of the subdiscipline of GIS and environmental modeling. As the various topics are introduced, the reader is directed to other chapters in the book that deal with these issues in more depth and detail. We aim to show here, and in the remaining chapters, that linking GIS and environmental modeling is a demanding task. Our environment varies and changes continuously in space and time. It involves the action and interaction of many variables representing both living (including human) and non-living systems. These are the characteristics that make environmental modeling within a GIS framework so demanding. The following discussion shows that GIS are particularly effective at handling spatial data and that, over the years, they have gradually become more adept at handling spatial analysis. GISs are not well designed for handling either time or the interactions of continuously changing variables. By contrast, environmental models have been effective for many years at handling time and variable interactions—especially systems-based models—but such models rarely include spatial information either for the purpose of analysis or even for the display of their results. The ongoing challenge has been to integrate these two vastly different approaches that seek to provide an understanding of our environment.

Both GIS and environmental modeling are relatively new areas of research that have made extraordinary progress in the last few decades. Both are likely to make a significant impact on how we can manage our scarce and dwindling resources. Using these tools, using this science (see Section 1.3.1 for an introduction to GIScience), and integrating these approaches with the analysis of socioeconomic systems, we should be able to stem the tide of environmental degradation and pollution and protect endangered habitats. We should be able to provide an environmental future that successive generations will be able to enjoy in perpetuity. Using, developing, and extending the tools and science of GIS and environmental modeling are challenging but ultimately altruistic tasks. Read the chapter, read the book, and find out how you can be part of the solution and not part of the problem in protecting the environment that is ours to enjoy.

## 1.1 GIS AND ENVIRONMENTAL MODELING: HISTORICAL LEGACY

Isaac Newton is reputed to have acknowledged his debt to his peers and forbears with the admission that his success in driving back the frontiers of knowledge was due to fact that he stood on the "shoulders of giants" (Newton, 1675/6). Despite Chorley and Kates's (1969, p. 1) comment that all scholars tend to think of themselves as living in times of intellectual revolution and renewal, in reality no academic or scientific venture is achieved in isolation and in most instances the success stories of today can trace their origins back many decades to the innovations of the past. So it is with GIS and environment modeling. Discussions of the development of GIS can be found in the standard textbooks of the field (see, for example, Clarke, 1999, pp. 7–10), in various articles (Tomlinson, 1984; Waters, 1998a), and in volumes that are devoted completely to documenting the history of GIS (Foresman, 1998). Here no attempt will be made to delve into the earliest and speculative origins of the key developments in GIS, but the interested reader should consult Unit 23, *History of GIS*, in the National Center for Geographic Information and Analysis (NCGIA) online introduction to GIS (this unit is in the original 1990 version of the NCGIA's core curriculum and may be found at http://www.geog.ubc.ca/courses/klink/gis.notes/ncgia/ toc.html).

Four developments were of particular importance. First, there were the conceptual innovations in overlay modeling popularized by Ian McHarg (1969, especially pp. 31–41) and his associates. Clarke (1999, p. 7) describes the original contributions of Jacqueline Tyrwhitt to this methodology. Steinitz et al. (1976) provide a complete history of the map overlay approach and suggest that Tyrwhitt's 1950 essay in the *Town and Country Planning Textbook*, published by the Architectural Press, London, was the first explicit discussion of the overlay technique despite McHarg's implied statement that his first use of the method had been novel (McHarg, 1996, p. 338). Tomlinson (1999, p. 17) explains that it was his company, Spartan Air Services of Ottawa, that in 1962 first proposed computerizing the overlay methodology.

Second was the development of the idea that modeling as opposed to "mere description" was an informative approach that would provide useful insights. The third development involved the introduction of analytical and empirical approaches as an aid in understanding geographical and environmental data. The fourth development was the rise of systems thinking. These four areas were developed largely independently of each other and there was, in addition, a significant divide separating research contributions and communications in Europe and North America. Each of these four contributions is now considered in turn.

### 1.1.1 The Idea of the Overlay: The Beginnings of Integration

In 1969 Ian McHarg published his seminal text, *Design with Nature* (McHarg, 1969). This book popularized the widespread application of overlay or "layer cake" methods (McHarg, 1996, p. 258) for determining, among other things, the location and routing of transportation corridors and other linear infrastructure. McHarg considered the various physical, environmental, and other variables, such as the degree of slope, the drainage characteristics, and the presence of historical buildings, that influenced costs associated with the construction of a road. For each of these variables, he then produced transparencies of those areas with high, medium, and low costs. High costs received the darkest shade and low costs the lightest. The resulting transparencies were then laid one on top of the other to visually determine a corridor-like route that snaked its way through the lightest areas of the combined maps. McHarg developed a complex process for choosing variables that involved citizen and expert opinion or preference and for determining high, medium, and low decision thresholds (McHarg, 1996, p. 338–339) but the actual overlay process was essentially the mechanical forbear of the overlay combinations common in today's GISs. In more recent years, Tomlin played an important role in popularizing Map Algebra methods, which permit mathematical operations to be performed on combinations of maps ("layers" or "coverages") in a GIS (Tomlin, 1990).

### 1.1.2 Modeling in the Geographical Sciences Before the Computer

Within geography, the intellectual foundations of modeling were developed first in Haggett's text *Locational Analysis in Human Geography* (1965). Haggett not only provided the original conceptualization of the human and physical environment into movements, networks, nodes, hierarchies, and surfaces, a presage of the cartographic primitives of points, lines, and areas, but he also gave an early discussion of the importance of models in geography (in this text the modeling of human systems is described in detail by Grove in Chapter 7, and the modeling of the physical environment is covered in Chapter 8 by Mitasova). This discussion later became the focus of Chorley and Haggett's (1967) pioneering and iconoclastic text, *Models in Geography*. As these authors noted, many other disciplines in both the natural and social sciences had already developed their own, somewhat discipline-specific, model-based approaches to their subject matter (for example, in geology by Krumbein and Graybill, 1965; and in sociology by Meadows, 1957). Conversely, other disciplines, such as archaeology and anthropology, were quick to follow geography in embracing a modeling approach to their data (see Clarke, 1968, 1972, and 1977 for the former and Dalton, 1981, for the latter).

      Chorley and Haggett, editors of *Models in Geography*, noted that modeling emphasizes a rather than an idiographic view of the world. The former, concerned with the development of generalizations, laws, and theories, is traditionally seen as the domain of the sciences, and the latter, which emphasizes individual cases and the unique, is the approach usually associated with the humanities (for more detailed definitions and discussion see Johnston et al., 2000, pp. 555 and 370–371, respectively). Haggett and Chorley (1967, p. 21) adopted Skilling's (1964) broad view that a model can be "a theory or a law or an hypothesis or a structured idea." It is also an idealization of the world, a template that allows the researcher to emphasize those aspects of a database considered to be important and to ignore the supposedly unimportant aspects, the noise. Even a traditional map or a coverage in a GIS may be considered to be a model (Board, 1967; Helmer, 1988). The model is therefore a subjective abstraction of the world and in recent years it is also a view of the world that can be operationalized, now usually as a computer-based implementation (see Giudici, Chapter 5, in this volume for a detailed discussion of the complete model abstraction, definition, development, calibration, and validation process).

      Originally models were physical representations of the world about us. In some instances they still are. It still makes sense to use "real" models in simulations of car accidents (crash-test dummies). Chorley (1967), in his discussion of models in physical geography, described physical representations of the environment such as flumes and wind tunnels, and Morgan (1967) described a series of physical analogues that may be used to model the environment.

      The ideas introduced by Chorley and Haggett in the 1960s were revisited and reviewed at a conference held twenty years later in Oxford, England. Many of the papers from that conference were subsequently published in the volume *Remodelling Geography* (Macmillan, 1989a). For some this may seem a lengthy hiatus, but the theoretical concerns of modeling in geography were always front and center on the research agenda of those geographers adopting a scientific approach to the discipline. In the intervening years notable texts on modeling in geography and its subdisciplines were published, including the volume by Thomas and Huggett (1980) and that by Woldenberg (1985) in geomorphology. Two years after the publication of *Models in Geography*, the *Progress in Geography* volumes were produced. Between 1969 and 1976 nine volumes were published, and this subsequently led to the launch of two quarterly journals *Progress in Human Geography* and *Progress in Physical Geography* (Haggett and Chorley, 1989, p. xix). In 1977 a long awaited second edition of *Locational Analysis in Human Geography* (Haggett et al., 1977) was published. In the preface to the second edition the authors contrasted the differences between the editions in terms of a military metaphor. The first edition was a report from an active battlefront, and in hindsight these "dispatches home" now appeared to resemble those of a euphoric expeditionary force. Haggett and his coauthors described the material from the second edition as more akin to grinding trench warfare.

      Unanticipated problems with a model-based geography were now becoming clearly identified, but if the difficulties and limitations of a scientifically oriented geography were now more obvious than before, new tools were being developed to address these difficulties, and the emerging popularity of GIS was one of the more prominent of these approaches.

### 1.1.3 Analytical, Empirical, Analog, and Other Types of Modeling

Geography and the spatial sciences have had a long history of emphasizing both the analytical and empirical approaches in modeling. Analytical modeling emphasizes abstraction from the real world and usually the building of mathematical and computer-based models, while empirical modeling emphasizes experience and observation of the real world. To some extent these two approaches may be equated with the deductive (the process of reasoning from agreed-upon, general principles or premises in order to arrive at inferred or deduced consequences) and inductive (the process of reasoning from particular instances to general principles) approaches to science, respectively. Although some researchers (such as Openshaw, 1989) have argued for a return to induction and data driven models, as Chorley (1964; and also in Berry and Marble, 1968) and Harvey (1969) have both commented, rarely are these methods used in isolation.

Many of the best-known models in the discipline of geography began with empirical observation of the real world and then only later were developed into models that could be formulated algebraically. Thus Richardson's (1961, p. 169) observations on the measured lengths of coastlines led to an empirically deduced law that the length of a geomorphic feature is inversely proportional to the length or sensitivity of the measuring device (see Rapoport et al.'s, 1961, p. 139 comments on Richardson's work). Mandlebrot (1982), whose career in fractal analysis was strongly influenced by Richardson's work (Lauwerier, 1991, p. 29), showed that this could be due to the property of self-similarity allowing a variety of geomorphic features to be modeled as fractals (Klinkenberg and Goodchild, 1992). Consequently, fractal analysis has now become a standard operation in some GIS packages (see, for example, Idrisi32: http://www.clarklabs.org/). Von Thunen's Isolated State is another example (see Abler et al.'s [1971] discussion and the NCGIA's Lab Exercises Report 96-12 GIS Laboratory Exercises: Introduction to GIS described on the Internet at http://www.ncgia.ucsb.edu/education/ed.html).

Other widely used models and mathematical representations of the environment began with a deductive derivation that was then tested and examined in the real world. Scheidegger's (1991, pp. 94-103) discussion of the reduction of rocks in the process of slope formation is based on known chemical processes and by equations from physics describing the surficial movement of particles. These equations are then empirically tested in the field. Christaller's central place theory (CPT) is an example in which a geometrical model of the arrangement of settlements is deduced mathematically and then is examined for validity in Southern Germany; see Christaller, 1966, for a discussion.

Richardson's and Von Thunen's inductive models have stimulated the deduction of mathematical models of fractal analysis and of locational rent, respectively (see Klinkenberg and Goodchild, 1992 for the former and Haggett et al., 1977, p. 204 for the latter; also refer to Mackey, Chapter 10, for a discussion of contemporary approaches to land use and land cover modeling). Similarly, both Scheidegger's slope models and CPT have inspired scores of empirical studies of slope development and of settlement patterns, respectively (see Scheidegger, 1991, pp. 94–156, and Berry, 1967, for respective reviews). The iterative movement from an inductive to a deductive approach and back again is well emphasized in Chorley's (1964) original discussion. The researchers' starting point is perhaps what should determine how their approach is characterized, but while their formal starting point may be deductive, as in the case of Christaller, who is to say (perhaps not even the researcher) that it was not informed by empirical observation of some specific instance in the real world? George (1967, p. 46) states that "a science would be ideally made up of a set of inductive generalizations which form the basis of a deductive argument." The subtle distinction between these two forms of reasoning is perhaps what led Arthur Conan-Doyle to assume, erroneously, that his famous detective was using only a deductive approach when instead Holmes always argued from observed facts and observations of the real world, an inductive approach, before then applying deductive logic to these observations.

Massam (1975) provides a more extensive typology of modeling procedures, which includes mechanical, geometrical, heuristic, numeric-analytical, simulation, and intuition. Some of these approaches are used only rarely in today's environmental models. Despite the work of Schneider (1971), which argued for the superiority of intuitive over computer-based solutions, today's environmental models are too complex to be amenable to intuitive solutions. The superiority of intuitive solutions was not true when Schneider was writing, and it is even less true today. Nevertheless, intuitive models may be of interest for comparing optimized solutions to real-world, human behavior. Intuitive models may be useful in the more demanding analysis of integrated economic and environmental systems (Maheddrarajah, et al., 1999; and see Section 1.6 of this chapter). The difficulty when using intuitive methods is in determining the importance of the influential, independent variables. The so-called revealed and stated preference methods popular in transportation science may be deployed (Ortuzar and Willumsen, 1994, pp. 252–415) but they are not without their limitations.

Many of the classical models of geography, such as CPT and Von Thunen's isolated state, have strong geometrical components and can be analyzed in these terms. Mathematical models such as linear programming models can often be spatialized, and the geometry of optimal solutions can lend insight into the nature of the models (Wirasinghe and Waters, 1983). GIS itself is founded on algorithms rooted in computational geometry, but these purely geometrical models have tended to receive less attention in recent years than in the early days of modeling. Exceptions are the work of Skupin and Buttenfield (1996, 1997), who have attempted to represent the relationships between qualitative variables in spatial terms and, more important for environmental modeling, the work of Dieckmann et al. (1999) on the geometry of ecological interactions.

Mechanical models and analogs are of less significance than they were in the 1960s when Morgan (1967) was writing. Recently, Turkle (1995, pp. 63–66) has documented the move away from physical methods toward computer simulations in the teaching of chemistry and physics to students at the Massachusetts Institute of Technology, Cambridge, Massachusetts. Massam's remaining categories of heuristic, numeric-analytical, and simulation have become increasingly popular among environmental modelers. Heuristic approaches usually involve an algorithm or procedure that specifies a series of steps that allows a researcher to approach or converge on an optimal solution when the procedure is iterated repeatedly (Church, 1999). Numeric-analytical models are often defined by a series of equations that optimize a predefined objective function, as in the standard programming methods such as linear programming and integer programming. Mathematical models have proven popular in predator-prey dynamics and in metapopulation models in ecology (see, for example, Batschelet, 1979, pp. 369–73 and Hanski, 1997, respectively;

and in this volume see Mackey, Chapter 10, for a discussion of ecological system modeling).

Simulation models are becoming increasingly popular as they can be used to replicate experiments in the field and can then be used to extend these experiments in time perhaps over 100 or more generations of a species. The replication of field results allows the scientist confidence in the results of the simulation, while the development of a simulation over many generations of a species permits the extension of the model so as to determine the long-term effects of environmental changes (see the discussion in Steinberg and Kareiva, 1997, pp. 319–325). The most recent developments in simulation with respect to GIS are discussed next.

### 1.1.4 The Systems Approach to Modeling

Another intellectual forbear of today's use of GIS-based models for understanding environmental processes was the introduction of systems analyses in the 1960s (see Emery, 1969; von Bertalanffy, 1962; Chorley's 1962 paper on General Systems Theory, Scheidegger 1991, p. 248 and Warntz, 1973 on the same topic; and for a later retrospect and overview see Coffey, 1981). Although the foundations of systems thinking had been developed by Wiener in the late 1940s (Wiener, 1948) it was not until Jay Forrester created the mainframe-based computer language Dynamo, which implemented his systems dynamics methodology, that systems thinking became a practical proposition for applied science. Forrester quickly produced a series of books that described the application of the Dynamo language to simulation in the study of industrial, urban, and finally world dynamics (Forrester, 1961, 1969, 1971, respectively). The simulations of the various world systems gained universal attention in *The Limits to Growth* study commissioned by the Club of Rome organization (Meadows et al., 1972).

The publication of the *Limits to Growth* was greeted by a storm of academic criticism and unfavorable analogies, including the observation that the research represented little more than Malthus with a computer (Cole et al., 1973; it should be noted that not all criticism was negative see—for example, Perlman, 1976, for a more positive view). This negative criticism and the difficulty of implementing the mainframe-based Dynamo language meant that the popularity of the systems dynamics approach grew only slowly until the Dynamo language was eventually implemented in the Stella program on the personal computer (first on the Macintosh and then on the Windows platform during the 1980s; see Meadows et al., 1992, p. 252, for details). The migration from mainframe to desktop computer and the resulting popularization of the primary tool for implementing systems-based modeling paralleled the evolution of GIS during the same time period. In 1980 Huggett published a general introduction to systems analysis for geographers (Hugget, 1980). More recently systems thinking has been enriched by the introduction of first catastrophe theory (Wilson, 1981a) and then chaos theory (see Gleick, 1987, for an introduction and Long, 1999, p. 378, for an environmental modeling application). These developments together with the renewed popularity of systems dynamics modeling (Meadows, et al., 1992; Mayer, 1990) have led to a resurgence in systems-based approaches. Any systems-modeling will benefit from animation and visualization procedures in order to provide insight and understanding of the numerical results. The topic of visualization is reviewed by Kraak (1999) and his collaborators in a special issue of the *International Journal of Geographical Information Science*, and in this volume by Copsey, Chapter 11.

Chorley and Kennedy (1971) presented a conceptual overview of how the systems approach might be used as a basis for modeling in physical geography. They identified eleven types of systems of increasing complexity in the physical and human environments (Chorley and Kennedy, 1971, pp. 3–4). Only the first four levels of complexity—morphological, cascading, process-response, and control systems—were considered to be of primary interest to the physical geographer. Bennett and Chorley (1978) broadened and extended this research to environmental systems. They also greatly developed the mathematical sophistication of the tools that they brought to bear on the problem of systems analysis. Until that time many geographers and scientists had emphasized statistical analysis at the expense of mathematical tools, and yet ironically the modeling community had emphasized mathematics at the expense of statistical analysis. Sumner (1978) provided a primer on mathematical tools for physical geographers and Wilson and Kirkby (1975, 1980) for both physical and human geographers. More recent primers and tutorials showing how these conceptual developments could, in their simplest incarnations, be turned into computerized models may be found in the work of Kirkby et al. (1987) and of Hardisty et al. (1993). More complex and sophisticated approaches are found in the work of Zannetti (1993) and Lane et al. (1998).

## 1.2 THE EVOLUTION OF GIS

The history of GIS has been described in a number of publications. Tomlinson (1984) provides an early account, and for more recent treatments see Coppock and Rhind (1991) and Waters (1998a). Foresman's (1998) comprehensive edited volume on the history of GIS is introduced with a foreword by Ian McHarg.

Even during the 1960s, in the early days of the world's first GIS, the Canada Geographic Information System, which had a strong environmental thrust (Tomlinson, 1984, 1998), there was a concern with the analysis of the spatial data sets stored within the GIS. Nevertheless, until the late 1980s, many private sector organizations and public sector institutions either did not explore the full potential of GIS for analysis and modeling or alternatively refused to purchase a system in the first place because they believed GIS lacked these all important capabilities (Lane and Hartgen, 1989, provide a detailed discussion of the GIS adoption process).

In the following discussion only trends in the evolution of GIS that have tended to support the use of GIS for environmental modeling are described.

### 1.2.1 The Rise of GIScience

In recent years there has been a resurgence of the view that geographic information systems research should be based on geographic information *science* (thus the *International Journal of Geographic Information Systems* [IJGIS] and the journal *Cartography and Geographic Information Systems* both changed the word *Systems* in their titles to *Science*). Goodchild promoted this idea with a seminal article published in the *IJGIS* (Goodchild,

1992). Discussion of the concept on the GIS-L list server produced heated debate and is itself analyzed in the article by Wright et al. (1997).

That GIScience is now entering the mainstream and becoming an increasingly widely accepted idea is shown by the debut of the first in a new series of GIScience Conferences. The NCGIA-sponsored GIScience 2000 Conference was the First International Conference on Geographic Information Science, held in Savannah, Georgia. The expectation is that these conferences will complement and support those in the GIS/EM series (see Section 1.3) and like them will become an outlet for the best research in the field (see http://www.giscience.org/ for details).

The concept of GIScience has been addressed and promoted in recent articles by Goodchild et al. (1999) and Mark (2000). Goodchild et al. (1999b) identify three approaches that have been used to define GIScience. The first was Goodchild's seminal definition of the subject (Goodchild, 1992), where he identified those disciplines that might contribute to the development of GIScience. These included disciplines that traditionally have focused on geographic information (GI): cartography, surveying, remote sensing, geodesy, and photogrammetry. In addition, there are those disciplines that have not been as important in the past but that are now regarded as making a substantial contribution: statistics (spatial statistics), economics (information economics), cognitive science (spatial cognition), psychology (environmental, developmental, and social), and mathematics (geometry). Goodchild (1992) also added geography and computer science.

The second attempt to define the nascent GIScience discipline mentioned by Goodchild et al. (1999b) is that of the U.S.-based University Consortium for Geographic Information Science (UCGIS). In 1996 ten topics were defined: (1) spatial data acquisition and integration; (2) distributed computing; (3) extensions to geographic representation; (4) cognition of geographic information; (5) interoperability of geographic information (see Section 1.5); (6) scale; (7) spatial analysis in a GIS environment; (8) the future of the spatial information infrastructure; (9) uncertainty in spatial data and GIS-based analysis; and (10) GIS and society. In 1998 each of these topics were refined into a series of "white papers," available at http://www.ucgis.org/research98.html and now under reconsideration.

Some of these topics (4 and 10, for example) are likely to have only limited appeal to environmental scientists and modelers, and many natural scientists would perhaps have liked to have seen more emphasis on physical and environmental science. In 1999 the UCGIS, at its summer assembly, focused its research efforts on how GIScience might relate to a number of research application areas. A series of reports are available at the UCGIS Web site (http://www.ucgis.org/apps99.html) and have also appeared in print (see the special UCGIS issue of the *Urban and Regional Information Systems Association Journal*, 2000, vol. 12, no. 2), but only Wilson et al.'s (2000) work on water resources relates to the natural environment.

If this is to some extent true of the work of the UCGIS, with its heavy focus on the information and human sciences, it is perhaps an even more valid statement with respect to the third major attempt to define GIScience, the Varenius project. The goal of advancing GIScience was the primary focus of the NCGIA's three year (1997–2000) Varenius project (Goodchild, et al., 1999b; http://www.ncgia.ucsb.edu/varenius/varenius.html). The Varenius project is named after the seventeenth century Dutch geographer Bernardus Varenius, whose magnum opus, *Geographia Generalis*, with its emphasis on the principles of scientific observation and reasoning, has been credited with inspiring Newton's scientific work (see Warntz, 1989; http://www.ncgia.ucsb.edu/varenius/Bernhard.html).

Varenius was one of Newton's "giants." In Varenius's work geography was divided into two parts: *general geography,* which was concerned with earth measurement and the study of environmental processes, and *special geography,* which dealt with the unique characteristics of places. For Goodchild and his colleagues GIScience can also be split into the general and special categories, avoiding the sterile division of the subject into human and physical geography, a concern that had been a constant challenge for Chorley and Haggett and that they too had hoped to address through methodology (Haggett and Chorley, 1969). Today it remains a challenge for environmental modeling if such research is to be politically and socially relevant. The Varenius project, with its emphasis on individuals (cognitive science, environmental psychology, and linguistics), systems (computer and information science), and geographies of the information society, may have only limited appeal to natural scientists.

As GIScience matures it will become defined by its hardware technologies (e.g., the nature of its computing, remote sensing, and global positioning devices: Abler, 1987; O'Callaghan 1999), by its software (e.g., the available geographic information systems, spatially aware statistical analysis programs [see Section 1.3.2], and modeling languages, among others), by its conceptualizations of the world (namely its theories, models, and algorithms, such as dynamic systems behavior, temporal 3-D representations of reality, and adaptive-genetic algorithms) and, finally, by its paradigms (these are problem exemplars that are considered to be research imperatives: Kuhn, 1971; Waters, 1993). All four aspects of GIScience are likely to show both continuous evolution as well as dramatic shifts from time to time. These defining characteristics of GIScience will have a constraining influence on what aspects of the environment are modeled, for to a person with a hammer, everything appears to be a nail. The fierce desire for a true GIScience would appear to be a necessary condition to support the new emphasis on a scientific approach to modeling within GIS research. Without one the other can hardly exist.

## 1.2.2 The Emergence of Spatially Aware Statistical Analysis (SASA) and Other Forms of Spatial Analysis

Statistical methods are invariably seen as a hallmark of science and the scientific method (Waters, 1994). Such methods guarantee the ability to replicate and verify previous research. Holland et al. (1986) note that "training in statistics has a demonstrable effect on the way people reason about a vast range of events in everyday life." For an in-depth discussion of the scientific method, see Chalmers (1982), and for a review of various modeling frameworks, paradigms, and approaches see Couclelis, Chapter 2.

That traditional statistical methods are less than appropriate for spatial data has been recognized for a long time. Gould (1970) noted that rarely are standard statistical methods useful because spatial data is so commonly spatially autocorrelated. As Goodchild (1986) has stated in his monograph on spatial autocorrelation, the existence of spatial dependence within a data set can be seen as an opportunity for providing evidence concerning the underlying generative processes but it can also be an insurmountable problem for formal statistical testing, in part because of the loss of an unknown number of degrees of freedom (Anselin, 1998; Goodchild, 1986; Griffith, 1999).

In recent years stand-alone statistical packages such as SpaceStat (http://www.spacestat.com/) have become available for carrying out SASA procedures, and some mainstream statistical packages such as S-Plus (http://www.splus.mathsoft.com/) have incorporated SASA methods. The S-Plus spatial statistics module even has a link to Arc-View, a GIS software package (http://www.esri.com). Since the use of such software is a complex task that usually requires new data structures and the use of new statistics that are difficult to interpret, the NCGIA, together with its associated Center for Spatially Integrated Social Science (CSISS), has sponsored a series of summer workshops to provide professionals and academics working in this area the opportunity to upgrade their skills (http://www.ncgia.ucsb.edu/CSISS/workshops). Moreover, professional graduate degrees in GIS (see Waters, 1999a, for a discussion of a number of these graduate programs), which recently have become somewhat of a growth industry now routinely offer courses in spatial, statistical analysis as part of the requirements for students taking these programs. Some GIS software packages that are oriented toward environmental modeling, such as Idrisi32, have also increased the scope of spatial analysis operations in their latest releases (http://www.clarklabs.org/) but still lack the ability to carry out such procedures as spatially aware regressions (see Lark, 2000, for an application in GIS and environmental modeling) and thus they still include qualifying remarks about the use of traditional, nonspatially aware, statistical analysis when used with spatial data. Finally, it may be observed that standard texts for senior undergraduates and graduates working in the area of spatial analysis are now available (Fotheringham et al., 2000) as are complex and exhaustive treatments (Arlinghaus, 1996; Griffith, 1999).

Besides spatial dependence, environmental GIS data often present other difficulties. In much environmental modeling a digital elevation model (DEM) is used. Elevation is useful as an independent variable in its own right for such environmental analyses as habitat modeling, but it can also be used to calculate other independent variables such as slope and aspect. Slope, like variables that represent percentages, is a constrained distribution (values must fall within the range 0 to 90 degrees), and aspect is an example of a variable that forms a circular distribution (values go from 0 to 360 and then back to 0 degrees). Few statistics texts address the problem of calculating statistics for such variables, although Burt and Barber (1996) provide one exception while Zar (1999, pp. 278–280 and 592–660) provides another in a thorough and comprehensive review of descriptive and inferential statistics and associated significance tests for this type of data.

Finally, it should be noted that interest in analyzing local forms of spatial dependence has burgeoned in recent years. Anselin (1998b) provides a detailed review and Openshaw (1998) describes his geographical analysis machine which, may now be downloaded from his Web site at http://www.geog.leeds.ac.uk/staff/s.openshaw/s.openshaw.html).

## 1.3 THE GIS AND ENVIRONMENTAL MODELING (GIS/EM) CONFERENCES: NUMBERS 1, 2, 3, AND 4

Conferences whose primary concern has been the use of GIS for management of environmental information date back to such early work as by Roger Tomlinson, who in 1972 edited the Proceedings of the Second International Geographical Union Symposium on GIS. The first volume of the two-volume proceedings was titled *Environment Information Systems* (Tomlinson, 1972). The GIS and environmental modeling conferences, which have sought to improve the complementarity of spatial analyses offered by GIS and the anticipatory capability offered by environmental models later, made a significant departure from prior environmental approaches. Since 1991 there have been four of these GIS and environmental modeling conferences. In September 2000, the last of these conferences, which have routinely attracted the best researchers from around the world, was held in Banff, Alberta, Canada. The first two conferences resulted in the publication of two influential and enormously successful books. The Proceedings of the Third Conference were published on CD-ROM, as were those for the Banff Conference.

### 1.3.1 The GIS/EM 1 Conference

The first international conference on Integrating GIS and Environmental Modeling was held in Boulder, Colorado, September 15–18, 1991. This conference was organized as an independant, cross-disciplinary, and cross-agency undertaking by scientists working with the National Center for Geographic Information and Analysis (NCGIA), U.S. Environmental Protection Agency, U.S. Geological Survey, Department of Energy, and National Oceanic and Atmospheric Administration. Over 600 individuals from government, universities, and private enterprize attended this conference. Many of the presenters were then invited to submit their contributions for publication in the highly regarded *Environmental Modeling with GIS* (GIS/EM1, in retrospect) text (Goodchild et al., 1993).

In their preface, the editors noted that the book focuses on "contemporary modeling in natural science as related to global change research, land and water resource management, and environmental risk assessment (this focus on natural and physical sciences would later be allowed to broaden)." The editors sought to integrate GIS and environmental simulation models. Specifically they wanted to improve interdisciplinary understanding of GIS in relation to simulation models that were then in use in the natural sciences, to identify needs and opportunities for integration, and to generate the interest and enthusiasm in these issues that would drive the research frontier forward (a review of integrative environmental modeling is provided by Frysinger in Chapter 9; see also Abel et al., 1998).

The GIS/EM1 text began with a series of six introductory essays. The first by Eddy (1993) described six imperatives for environmental research: (1) continuous monitoring the Earth's vital signs; (2) need to recovering the past history of climate and other environmental changes; (3) need for an earth system science; (4) need for realistic earth system models; (5) need for a global data and information system for environmental data; and (6) need to train new scientists in the GIS and environmental modeling fields.

In his opening chapter Goodchild (1993a) reviewed the state of GIS for environmental modeling. He drew an important distinction between two primary conceptualizations of models in GIS: field models and object models. If a variable could be considered to be distributed as a continuous function such as atmospheric pressure or a soil class, then the database objects were simply the creations of the data modeling process. A set of objects could then be used to model the variation in the variable, and this was termed a layer or coverage. The resulting models were then known as layer, coverage, or field models. By contrast, an object could be defined a priori, ahead of the modeling process. For example, the object Lake Minnewanka has its own identity that is entirely independent of any attempt to define a discrete land/water variable in that part of Alberta, Canada. Sometimes the lake may be larger/higher or smaller/lower or indeed it may be frozen, but it is always Lake Minnewanka. A key difference between the field and object representations was that in the object view of the world it was entirely possible for two objects to occupy the same space, whereas this could not occur in the field representation. Thus the object Lake Minnewanka could have a jetty jutting out into it. This would not be possible if we had a water layer and a land layer and a jetty layer, as would be the case with a field representation.

Goodchild (1993a) described six common field models in use in GIS: (1) irregular point sampling; (2) regular point sampling; (3) contours; (4) polygons; (5) cell grid; and (6) triangular net (TIN models). In GIS, models 2 and 5 are traditionally referred to as raster models, whereas models 1, 3, 4, and 6 are known as vector models (Copsey, Chapter 11, discusses gridding methods and TIN models). That fields continue to occupy the interest of the GIS community was demonstrated by the recent specialist meeting on The Ontology of Fields, which was held by the NCGIA as part of the Varenius project in 1998 (a copy of the report by Donna Peuquet, Barry Smith, and Berit Brogaard may be found at http://www.ncgia.ucsb.edu/~vanzuyle/varenius/ontologyoffields_rpt.html).

Object models are characterized as points, lines, and areas, but many GIS databases will make no distinction between these two different views of the world (see Varma, Chapter 4). Thus a line could represent a field view of the world as, for example, in the case of a contour or an object view as in the case of a road. As Goodchild argued, the object view is commonly used to model artifacts. Field and object models are usually characterizations of two-dimensional data. A third type of model, network models could be used to describe one-dimensional linear features such as road systems, and they have traditionally been important in transportation studies (Miller, 1999). One of the main challenges for GIS and environmental modeling in the future will be to incorporate fully three-dimensional models of the environment (see O'Callaghan, 1999, p. 17).

In a second paper Goodchild (1993b) introduced the topic of models of error and spatial data quality within GIS analysis. Since that time this work has been extended by Journel (1996), who has used geostatistical models to simulate uncertainty, and by Heuvelink (1998), who provides a detailed review of current practices and software availability in this area (in this volume see Chapter 9 by Frysinger).

GIS/EM 1 also began the debate over loose versus tight coupling (see Section 1.5) and many of the individual contributions addressed these and related issues (Avissar, 1993; Gao et al., 1993; Maidment, 1993, especially p.156; Schimel and Burke, 1993; Skelly et al., 1993). That the early work of Chorley and Haggett had not completely resolved the debate over modeling by the advent of GIS/EM 1 was demonstrated by

Moore et al.'s (1993, pp. 197–198) detailed discussion of the characteristics of mathematical models, which they suggested should include parsimony, modesty, accuracy, and testability. GIS/EM 1 also introduced researchers to a variety of environmental models and to issues surrounding the development of scientific databases (for example, Kineman, 1993) and presented state-of-the-art discussions of the research frontier in spatial statistics by many of the world's leading experts (for example, Anselin, 1993; Cressie, 1993; Englund, 1993) and a discussion of kriging by ver Hoef (1993). In the present volume Copsey (Chapter 11) discusses kriging and other geostatistical methods.

### 1.3.2 The GIS/EM 2 Conference

By the end of the first conference, participants developed strong expectations resulting in the second GIS and Environmental Modeling Conference, held in Breckenridge, Colorado, in 1993. This produced another highly regarded text: *GIS and Environmental Modeling: Progress and Research Issues* (Goodchild et al., 1996). This book was organized into three sections: Environmental Databases and Mapping; Environmental Modeling Linked to GIS; and Building Environmental Models with GIS. These overarching topics were introduced by two overview chapters that were then followed by individual papers emphasizing applications and case studies. In the short time separating the two conferences little had changed. Main problems acknowledged by researchers at the conference involved how to integrate the static, atemporal, two-dimensional GIS software with the dynamic, three-dimensional worlds of the environmental modelers (the current state of the art in dynamic systems modeling of three-dimensional GIS, which also includes the fourth-dimension of time, is discussed by Rogowski in Chapter 6 of this book).

The first section of the book sought to answer the question how GIS could be used as a tool to produce spatial data for environmental models? It opened with a paper by Bretherton (1996) advocating a limits to growth world-view in which human-environmental interactions were the primary subject of the analysis and social-natural science models were the main method of understanding the world. Many of the papers in this first section of the book were concerned either with data infrastructure issues (Goodchild, 1996a; Steyaert, 1996) or with data quality issues (Aspinall and Pearson, 1996; Burrough et al., 1996).

Papers in the second section of the book were mainly concerned with GIS applications that were used in environmental modeling to manage spatial data (Battaglin et al., 1996), integrate diverse data types (French and Read, 1996; Saghafian, 1996; Vieux et al., 1996), conduct spatial analyses (Krummel et al., 1996; Mackey et al., 1996), or visualize results (Hay et al., 1996; see Copsey, Chapter 11, for current research on visualizing environmental data).

The third and final section of the Proceedings of the Breckenridge conference dealt with many advanced issues, including the embedding and integration of environmental models within GIS (Bromberg et al., 1996; Cowan et al., 1996; Fedra, 1996; Kessell, 1996; Lam and Pupp, 1996; Maidment, 1996; Raper and Livingstone, 1996a). New research themes and innovative methods were also addressed in this section, including the use of cellular automata (Clarke and Olsen, 1996; and see Couclelis, Chapter 2), new approaches for handling space and time dynamically or as continuously varying variables (Kemp, 1996; Mitasova et al., 1996; Yeh and de Cambray, 1996), object-oriented approaches (Bennett, et al., 1996; Crosbie, 1996), parallel processing of data (LaPoton and McKim, 1996), use of bootstrapping techniques (Reitsma, 1996), and several papers discussing theory and related aspects of hierarchies (Csillag, 1996; Johnson, 1996; Stoms et al., 1996). There was a renewed emphasis on GIScience and scientific approaches to modeling (Smith et al., 1996).

This second volume from what had now become a GIS and environmental modeling conference series left little doubt that, while the first meeting had set forth the research agenda facing environmental modelers who wished to use GIS, the attendees at the second conference had already begun to address many of the more challenging problems previously identified. Much work remained to begin to address GISs challenging underlying problems when used for modeling, and these problems were further refined and explored in the proceedings of the second meeting.

### 1.3.3 The GIS/EM 3 Conference

In response to contiuning demand, a third conference was held in Santa Fe, New Mexico in 1996. To encourage and better contribute to increased representation of scientific work and data in digital, rather than paper form, proceedings of the conference were not published as a book but instead on CD-ROM. Details on how to acquire the CD-ROM or about the conference may be found at the following Web site: http://www.ncgia.ucsb.edu/conf/SANTA_FE_CD-ROM/santa_fe.html. The conference began with a series of technical workshops followed by three full days of paper sessions with about twelve sessions per day and about four papers per session.

By 1996 there had evolved many environment models that could be linked, to a greater or lesser degree, within a GIS framework. For instance, in one of the plenary presentations at the start of the conference, Wilson (1996) discussed the characteristics of six such landsurface/subsurface models (USLE, ANSWERS, AGNPS, CMLS, LEACHM, and TOPMODEL; see Wilson, 1996, for details). But despite significant evolution among environmental models generally, Wilson was able to note many remaining unresolved data integration issues, including widely different data structures between models and different methods of organizing spatially distributed model inputs.

Paper sessions at the Santa Fe Conference included discussions of the following topics (references are examples only since many topics were addressed by numerous authors at the conference; the reader is urged to explore the CD-ROM for this conference):

1. Climate and weather interpolation, including smart interpolators that attempt to incorporate information about the generating process (see Willmott, 1996). Spatial interpolation has received considerable attention recently in the literature (see the electronic journal *Journal of Geographic Information and Decision Analysis* at http://publish.uwo.ca/~jmalczew/gida.htm for a complete issue devoted to this topic [volume 2, number 2, December 1998]; see also Mitas and Mitasova, 1999; and see Copsey, Chapter 11, in this volume for a discussion of interpolation methods).

2.  Spatial analysis and data exploration using custom designed programs such as XGobi (Majure et al., 1996) being dynamically linked to standard industry software such as ArcView (in this volume exploratory data analysis is covered in Chapter 11 by Copsey).

3.  The linkage of qualitative data within a GIS for decision-making purposes (see Fujisaka et al., 1996).

4.  Data access issues, including the importance of the Internet (McCauley et al., 1996; and many others) and the difficulty of implementing and maintaining user-pay policies for data (Barringer et al., 1996).

5.  Sampling concerns, including the use of genetic algorithms to select a representative sample (Jarvis et al., 1996).

6.  The use of artificial intelligence (AI) approaches such as decision tree analysis (DTA), neural networks (NNs), and expert systems for modeling environmental processes such as land degradation by soil erosion and for controlling GIS operations (Ellis, 1996; Wachter-Harms and Wendholt, 1996; among others).

7.  Distributed computing and parallel processing (El Haddi et al., 1996).

8.  Concern over biodiversity and the role that GIS and environmental modeling might play in maintaining such diversity (see, for example, Church et al., 1996).

9.  Spatially explicit analysis (van Horssen, 1996).

10.  Integrated systems of coupled GIS and environmental models (see, for example, Bian et al., 1996).

11.  Spatial-temporal modeling within a GIS (see Raper and Livingstone, 1996b; Yuan, 1996; among others).

12.  Cellular-based models were also discussed at the Santa Fe conference. Although Clarke and Olsen (1996) had introduced cellular-based models at GIS/EM 2, there was now a complete session devoted to this topic. This session included a paper by Clarke et al. (1996) on calibrating cellular automaton models of urban growth as well as models that incorporated systems dynamics models using the Stella software (Westervelt and Hopkins, 1996; see Section 1.1.4 for a discussion of systems dynamics and the Stella software; see also Couclelis, Chapter 2). Theoretical developments from earlier research were now being incorporated into sophisticated models of environmental processes within a GIS framework.

13.  Agent-based modeling using such computer environments as the Swarm system (Kohler et al., 1996; a workshop on this topic was also presented at the conference by Chris Langton of the Santa Fe Institute; see also Couclelis, Chapter 2).

14.  Collaborative approaches and computing environments (Carver et al., 1996; Sandhu and Treleaven, 1996).

It is easy to see that many of these topics are recurring themes from earlier meetings, so presumably, they still presented researchers with ongoing challenges. Some topics, such as agent-based modeling, were completely new approaches to many attendees at the conference which its designers hoped would provide new options to modeling environments in which the behaviors of many individual entities interact in complex ways. Finally, some old approaches to modeling processes, such as systems dynamics, were resurrected and recast within a GIS framework (Westervelt and Hopkins, 1996).

The last day of the third GIS/EM conference began with a plenary session discussing new directions in GIS that would be likely to have an impact on the world of environmental modeling (Goodchild, 1996b). These included the use of GIS for data collection by scientists while they are in the field; the "mainstreaming" of GIS into applications such as Microsoft Excel's (http://www.microsoft.com/catalog/), introduction of mapping functions and the popularization of car navigation systems; the increasing importance of digital, geographic technologies in communication; the distribution of data using client-server technologies, thus allowing spatial data to be maintained at a high level of quality; and use of object-oriented technology to encapsulate processes with data.

Essentially, Goodchild was arguing that geographic information technologies would play an important role in every step of the life cycle of environmental data from data capture, through its maintenance and analysis, and finally for its dissemination to end users. Goodchild also argued that the future should include interoperability among data models, software packages, and computing platforms so that GIS software would become easier to use regardless of the computational environment in which end users were working (interoperability is discussed in Section 1.3.5).

### 1.3.4 The GIS/EM 4 Conference

The fourth conference on integrating GIS and environmental modeling was held in Banff, Alberta, in Canada's environmentally stressed Banff National Park (Waters, 2000). The pre-conference Web site may be found at http://www.Colorado.EDU/research/cires/banff/. This conference again included a remarkable diversity of papers on an extensive range of environmental topics. Perhaps six different trends can be detected in the papers presented.

First there was the continuing popularity of many new methods for solving environmental problems within a GIS/modeling framework. These included the use of genetic algorithms, tree regression analysis, cell-based modeling, three-dimensional and temporal modeling, object-based analysis, expert and knowledge-based systems, visualization methodologies, and intelligent agents. Second, some completely new techniques and approaches were introduced, including the use of Internet-based solutions, generalized regression procedures, inexact-fuzzy multiobjective programming, multivariate adaptive regression splines, the role of socioeconomic drivers, virtual reality modeling and virtual landscapes and, finally, the use of Web-based, distance learning systems.

Third, some techniques and methodologies were introduced that were relatively new to integrated GIS and environmental modeling but had been used outside this research area for a number of years. These included cumulative effects modeling. Fourth, many of the current environmental concerns of society, such as modeling the presence of fecal coliform in watersheds, flood inundations, the impact of rising sea levels on coastal environments, and the health impacts of airborne pollutants, were featured frequently in the presentations. Gradually, a move toward the integration of the modeling of socioeconomic and natural systems may be seen to be occurring. This is addressed in Section 1.3.6.

Fifth, there was a great geographical diversity in the topics addressed, demonstrating unequivocally that the integration of GIS and environmental modeling was now a global concern being actively pursued in universities and research centers around the world. Finally, there was a strong interest in component-based modeling and the issue of interoperability. These new concerns are discussed in Section 1.5.

### 1.3.5 Related Conferences: Related Publications

The GIS/EM conferences have been distinguished by their strong emphasis on a rigorous, state-of-the-art scientific approach and the integration of environmental models within a GIS framework. Other conferences have also considered the role of GIS and the environment. Among the more important of those familiar to this author have been the International Symposia on Environmental Concerns in Rights-of-Way Management. The first of these conferences was held in 1976 (Tillman, 1976). Since then they have been held every three years, the most recent being the seventh symposium, which was held in Calgary, Alberta, Canada, in September 2000 (http://www.rights-of-way-env.com). The proceedings of this conference will be published by Elsevier and the proceedings of the sixth conference were also published by Elsevier (Williams et al., 1997).

Also important is the series of International Conferences on Wildlife Ecology and Transportation (ICOWET I, II, and III) that began in 1997 (see the conference Web site at http://www.dot.state.fl.us/emo). These conferences are now held annually. The most recent (Evink et al., 1999) had numerous papers concerned with the integration of GIS and environmental modeling (see, for example, Alexander and Waters, 1999; Kautz et al., 1999; and Klein, 1991). Similarly, with a strong emphasis on environmental modeling and somewhat less on GIS are the Roads, Rails, and Environment annual conference series, which by 1999 had held four conferences (http://cmiae.hypermart.net.conferences.htm).

There have also been a number of one-time-event conferences that have had a strong emphasis on environmental modeling using GIS. One of the more significant of these was the Decision Support—2001 Conference held in 1994 (Power et al., 1995). This conference featured papers on object-oriented modeling, autonomous and intelligent agent-based models, the Swarm modeling language, knowledge-based systems, intelligent visualization systems, neural networks, expert systems, and case-based modeling. Some of these papers were also published in the journals *Mathematical and Computer Modelling* and *AI Applications in Natural Resource Management.* Many of the annual conferences that have GIS as their main theme and focus have published papers on GIS and environmental modeling. Of particular note have been the Vancouver GIS conferences (though occasionally they have been held in Toronto), which have a strong resource management emphasis. Each of these conferences has produced an annual, print-based proceedings, although in 2000, when the fifteenth conference was held, the proceedings were published on CD-ROM, a format to which most conferences are now moving (see the Events link at http://www.geoplace.com for information on these conferences). The Environmental Systems Research Institute (ESRI) holds an annual user's conference at which many papers discuss GIS and environmental modeling issues (see, for example, Jennifer Benaman's presentation at http://www.crwr.utexas.edu/gis/gisenv98/watqual/benaman/enviro/sld001.htm).

Various texts and edited collections of papers have contributed to the development and use of GIS for environmental modeling. In the former category is Burrough's (1986) landmark volume, *Principles of Geographical Information Systems for Land Resources Assessment,* although by the time the second edition had been published the last part of the title had been dropped (Burrough and McDonnell, 1998). As representatives of latter category Ripple (1987), Lane et al. (1998), and Thill (1999) have all made key contributions.

## 1.4 THE TIGHT, LOOSE, AND MULTIPLE COUPLING DEBATES

Some of the earliest discussions of the difficulty of coupling models together across varying temporal and spatial scales, across different environmental and comuputing systems, and together with standard GIS packages are provided by Steyaert (1993), Parks (1993), and Dangermond (1993). Steyaert (1993, p. 21) stated unequivocally that "the dominant theme with the highest research priority is the development of integrated systems models, also referred to as coupled systems models." Steyaert discussed the linkage of "cross-disciplinary models" such as atmosphere and biosphere models (see also McGuffie and Hendersen-Sellers, 1997; and in this volume Mackey, Chapter 10, provides a review of global climatic models) and discussed ways in which various temporal and spatial scales may be incorporated into such models. Parks (1993, p. 33) argued that modeling and GIS technology both benefit by their linkage and co-evolution. Hartkamp et al. (1999) suggest that the term linking be used to refer to the simple passing of input and output between GIS and model, that the term combining be used to refer to the automatic exchange of data and GIS functionality between GIS and model, and that the term integration should be used to refer to the complete embedding of a model within a GIS or vice versa.

Dangermond (1993), of ESRI (manufacturers of the ARC/INFO GIS software) discussed the difficulties of linking GIS and environmental models from the GIS developers' perspective. In 1993 these problems included the static nature of most GIS software, which had great difficulty in handling temporally changing data; the problem of accessing and acquiring data; the great variability among hardware, software, communications, and other standards; the virtual ignoring of ecological, social, cultural, visual, and other relationships in environmental decision support models for such problems as facility location and site assessment (see Copsey, Chapter 11, for examples of the latter); and finally the deterministic representation of stochastic variables. In the twenty-first century these difficulties are now being addressed, but there is still much work to do (see Corbett, Chapter 3 for a discussion of spatial environmental decision support systems).

One bright hope for the future of strongly integrated problem solving environments is suggested by current concern over interoperability. Within the computer science community this concern has been paramount for years. In 1991 *Byte* magazine published a series of articles detailing concerns over interoperability. Donovan (1991, p. 185) stated that it was "ironic that the real effect of computing is too often to prevent the sharing of data." He noted that the "problems contributing to incompatibility are rooted in incompatible architectures, data formats, and communications protocols." Nance (1991) explained that there is a hierarchy of difficulty in linking systems. Beginning with the simplest and ending with the most complex, these were identified as physical, data, network, transport, session, presentation, and application links. If GIS and environmental models are to be coupled seamlessly, then the linking must include all these levels, including the very highest and the complex, application level. A similar hierarchy for GIS interoperability was proposed by Bishr et al. (1999, p. 57). *Byte* magazine, now an electronic journal (http://www.byte.com/), has returned to the theme of interoperability on a number of occasions, including a cover article on Windows 98 and Windows NT 5 that included a discussion of Microsoft's Distributed Component Object Model system (http://www.byte.com/art/9801/sec5/art2.htm). Remarkably, ESRI's online glossary for GIS (http://www.esri.com/library/glossary/glossary.html) does not include an entry on interoperability, although one may be found in the Association for Geographic Information's online dictionary (http://www.geo.ed.ac.uk/agidict/welcome.html). ESRI's glossary does include an entry on inter-application communication, the highest level of the interoperability hierarchy. That GIS scientists are taking interoperability ever more seriously is shown by two recent and influential publications on this topic: Goodchild et al. (1999) and Vckovski (1998). In addition, the most recent version of ARC/INFO GIS software, version 8, uses a component-object model software architecture and the Microsoft Component Object Model standard (Maguire, 1998/1999; also http://www.esri.com/software/arcinfo/customization.html).

As new partnerships are formed in the computer and GIS industries (Waters, 1998b), and as the leading companies come to dominate their sectors of the industry, it might be expected that the need for interoperability would be met. That this is not yet the case is partly due to new software developments, such as Extensible Markup Language (XML; and see Waters, 1999b), partly due to new hardware developments such as wireless modes of operation (see the wireless consortium at the InterOperability Lab: http://www.iol.unh.edu/consortiums/index.html; also Waters, 1998c), and partly due to new modes of operation such as the widespread use of the Internet and distributed users of GIS (Clayton and Waters, 1999).

In 1993 links between environmental models and GIS were tenuous at best. Since then functionality of GIS has improved and the ability to import data from and export data to environmental models has increased significantly (see Varma, Chapter 4, for a discussion of data sources with respect to environmental modeling). How tightly coupled the models should be is still a point of discussion. Tight coupling of an environmental model within a specific GIS package implies a lack of flexibility that may hinder unfettered research experimentation. This debate was continued at the second and third GIS/EM conferences and was an ongoing concern at the fourth conference. It is likely to remain an issue of contention well into the first decade of the new century.

## 1.5 SIGNPOSTS TO THE FUTURE

A more detailed discussion of future research needs in GIS and environmental modeling is contained in the final chapter of this book (Parks, Crane, and Clarke, Chapter 12). Here it may simply be noted that there is a need to continue pushing the research frontier forward in areas that include, continuous space-time modeling (see, for example, http://kabir.cbl.umces.edu/SME3/Smod.html) and parallel processing, where recent developments will speed both the compute and input-output operations within a GIS (Healey et al., 1998). A more significant, demanding, and amorphous challenge lies in the integration of the modeling of environmental and economic systems as a single entity. Chorley and Kennedy (1971, p. 4) defined the remaining seven levels of complexity in systems as self-maintaining systems, plants, animals, ecosystems, humans, social systems, and human ecosystems, respectively. The human ecosystem, which includes the natural environment, is the most complex to analyze, model, and control.

Chorley (1969) made one of the first attempts to integrate the analysis of social and natural systems. His book *Water, Earth and Man* was subtitled *A Synthesis of Hydrology, Geomorphology and Socio-Economic Geography,* but there was no use of GIS. This theme was continued almost a decade later in Bennett and Chorley's (1978, p. xii) text *Environmental Systems*, where one of the explicit aims was to develop "an integrated theory relating social and economic theory. . . to physical and biological theory." But still there was no use of the power of GIS. At a similar time others too argued, if somewhat more circuitously, for an integration of ecology and economics within a systems framework (Chapman, 1977, p. 409). A few years later Wilson (1981b) was able to bring a variety of formidable mathematical tools to his analysis of environmental systems, including dynamical approaches using microsimulation, difference and differential equations, and catastrophe theory. Environmental systems were defined as natural, socioeconomic, and integrated, and although his moorland and water resource systems showed a remarkable degree of integration, there was no use of GIS.

Lincoln (1993) discusses the future of environmental modeling under three primary headings, each of which consists of three aspects. The first heading is Capability which includes: computing performance; the changing and increasingly sophisticated science that can be embedded in environmental models; and a series of "Grand Challenges," which, according to Lincoln, includes the tight coupling of models and new methods of spatial analysis, including the finite element techniques of multigridding and adaptive gridding, which can speed up algorithms in a manner similar to the way in which quadtrees reduce storage in a GIS (http://hpcc.arc.nasa.gov/insights/vol5/overflow.htm). Lincoln's second heading for considering the future of environmental modeling is Uniformity, and here he considers databases (including those in GIS), nomenclature, and interfaces. Uniformity is perhaps another way of saying interoperability and indeed it also influences the tight coupling of GIS and environmental models. Lincoln's final heading is Accessibility, where reduced costs, increased user friendliness, and more effective networking are the main issues to be addressed in future environmental modeling scenarios.

Where GIS and environmental modeling approaches are applied in the developing world, issues of "appropriate technology" are likely to loom large (Smillie, 1991, pp. 89–93). In both the developed and developing worlds, as natural science and social science models of the environment become more fully integrated, ethics, ideologies, and philosophies will play an increasingly important role in model construction and interpretation (Fairweather et al., 1999; Pratt et al., 2000; Samson and Pitt, 1999).

Kurzweil (1999) has argued that the pace of intellectual and scientific development will accelerate exponentially in the coming decades. In the sphere of environmental modeling and GIS, both disciplines are likely to benefit from significant developments in the fields of hardware and software engineering and in data and model integration. As the first decade of the new millennium develops, many of the participants of the GIS/EM 4 conference appear optimistic. It may well be possible to avoid the "grinding trench warfare" described by Haggett et al. (1977, p. ix) in the preface to the second edition of *Locational Analysis in Human Geography* and to make some remarkable new progress. Indeed Macmillan (1989b, p. 310) argued that if a third British, models-in-geography conference were to be held another twenty years later, in 2007 (the previous conferences had been held in 1967 and 1987), it might well be dominated by powerful GIS with sophisticated and highly integrated mathematical and statistical modeling tool chests. Such new tools are hard to imagine today, but the likelihood is that if they are to occur within this timeframe they will emerge from research initiatives similar to the NCGIA's Varenius project, that seeks to broaden GIS to include use of geo-technology in human decision-making.

## 1.6 ACKNOWLEDGMENTS

I acknowledge the help and support of my graduate students, Shelley Alexander and Amanda Fischl, whose research focuses on the issues of environmental modeling. Thanks also Bob Maher and three anonymous referees, who kindly read over my manuscript and offered comments, ideas, and suggestions.

## 1.7 REFERENCES

Abel, D. J., Ooi, B. C., Tan, K-L., and Tan, S. H. 1998. Towards Integrated Geographical Information Processing. *International Journal of Geographical Information Science*, 12, 353–372.

Abler, R. F. 1987. The National Science Foundation National center for Geographic Information and Analysis. *International Journal of Geographical Information Systems*, 1, 303–326.

Alexander, S., and Waters, N. 1999. Decision Support Applications for Evaluating Placement Requisites and Effectiveness of Wildlife Crossing Structures. Pp. 237–246 in Evink et al., 1999, op. cit.

Anselin, L. 1993. Discrete Space Autoregressive Models. Pp. 454–469 in Goodchild, Parks, and Steyaert, 1993, op. cit.

Anselin, L. 1998. *Spatial Econometrics: Methods and Models*. Dortrecht, The Netherlands: Kluwer Academic.

Anselin, L. 1998. Exploratory Spatial Data Analysis in a Geocomputational Environment. Pp. 77–94 in Longley, P. A., Brooks, S. M., McDonnell, R. and Macmillan, B., eds., *Geocomputation: A Primer*, New York: Wiley.

Arlinghaus, S. L. (ed) 1996. *Practical Handbook of Spatial Statistics*. Boca Raton, Florida: CRC Press.

Aspinall, R. J., and Pearson, D. M. 1996. Data Quality and Spatial Analysis: Analytical Use of GIS for Ecological Modeling. Pp. 35–38 in Goodchild, Steyaert, and Parks, et al., 1996, op. cit.

Avissar, R. 1993. An Approach to Bridge the Gap Between Micro-Scale Land-Surface Processes and Synoptic-Scale Meteorological Conditions Using Atmospheric Models and GIS: Potential for Applications in Agriculture. Pp. 123–34 in Goodchild, Parks, and Steyaert, 1993, op. cit.

Barringer, J., et al. 1996. Environmental Data Access in New Zealand 1985–1995: User-Pay to Open Access and the WWW. Santa Fe CD-ROM, available from National Center for Geographic Information and Analysis, University of California, Santa Barbara, CA.

Batschelet, E. 1979. *Introduction to Mathematics for the Life Scientists* (3rd ed.). New York: Springer-Verlag.

Battaglin, W. A., Kuhn, G., and Parker, R. 1996. Using GIS to Link Spatial Data and the Precipitation Runoff Modeling System: Gunnison River Basin, Colorado. Pp. 159–164 in Goodchild, Steyaert, and Parks, et al., 1996, op. cit.

Bennett, D. A., Armstrong, M. P., and Weirich, F. 1996. An Object-Oriented Model Base Management System for Environmental Simulation. Pp. 439–444 in Goodchild, Steyaert, and Parks, et al., 1996, op. cit.

Bennett, R. J., and Chorley, R. J. 1978. *Environmental Systems: Philosophy, Analysis and Control.* London: Methuen.

Berry, B. J. L. 1967. *Geography of Market Centers and Retail Distribution.* Englewood Cliffs, NJ: Prentice-Hall.

Berry, B. J. L. and Marble, D. F. 1968. *Spatial Analysis: A Reader in Statistical Geography.* Englewood Cliffs, NJ: Prentice-Hall.

Bian, L., Sun, H., Blodgett, C., Egbert, S., Li, W-P, Ran, L-M., and Koussis, A. 1996. An Integrated Interface System to Couple the SWAT Model and ArcInfo. Santa Fe CD-ROM, available from National Center for Geographic Information and Analysis, University of California, Santa Barbara, CA.

Bishr, Y. A., Pundt, H., Kuhn, W., and Radwan, M. 1999. Probing the Concept of Information Communitiesq—A First Step Toward Semantic Interoperability. Pp. 55–70 in Goodchild, Egenhofer, Fegeas, and Kottman, 1999, op. cit.

Board, C. 1967. Maps as Models. Pp. 671–725 in Chorley and Haggett, eds., 1967, op. cit.

Bretherton, F. 1996 Why Bother? Pp. 3–6 in Goodchild, et al., 1996, op. cit.

Bromberg, J. G., McKeown, R., Knapp, L., Kittel, T. G. F., Ojima, D. S. and Schimel, D. S. 1996. Integrating GIS and the CENTURY Model to Manage and Analyze Data. Pp. 429–432 in Goodchild, Steyaert, and Parks, et al., 1996, op. cit.

Burrough, P. A. 1986. *Principles of Geographical Information Systems for Land Resources Assessment.* Oxford, England: Oxford University Press.

Burrough, P. A., and McDonnell, R. A. 1998. *Principles of Geographical Information Systems* (Second Edition). Oxford, England: Oxford University Press.

Burrough, P. A., Van Rijn, R. and Rikken, M. 1996. Spatial Data Quality and Error Analysis Issues: GIS Functions and Environmental Modeling. Pp. 29–34 in Goodchild, Steyaert, and Parks, et al., 1996, op. cit.

Burt, J. E., and Barber, G. M. 1996. *Elementary Statistics for Geographers* (2nd ed.). New York: Guilford Press.

Carver, S., Frysinger, S., Madison, J., and Reitsma, R. 1996. Environmental Modeling and Collaborative Spatial Decision-Making: Some Thoughts and Experiences Arising from the I-17 Meeting. Santa Fe CD-ROM, available from National Center for Geographic Information and Analysis, University of California, Santa Barbara, CA.

Chapman, G. P. 1977. *Human and Environmental Systems: A Geographer's Appraisal.* London: Academic Press.

Chorley, R. J. 1962. *Geomorphology and General Systems Theory.* U.S. Geological Survey, Professional Paper 500-B.

Chorley, R. J. 1964. Geography and Analog Theory. *Annals, Association of American Geographers,* 54, 127–137.

Chorley, R. J. 1967. *Models in Geomorphology.* In R. J. Chorley and P. Haggett (*eds.*), Models in Geography, London: Methuen.

Chorley, R. J. (ed.) 1969. *Water, Earth and Man: A Synthesis of Hydrology, Geomorphology and Socio-Economic Geography.* London: Methuen.

Chorley, R. J., and Haggett, P.(eds.) 1967. *Models in Geography.* London: Methuen.

Chorley, R. J., and Kates, R. W. 1969. Introduction. Pp. 1-7 in Chorley, ed., 1969, op. cit.

Chorley, R. J., and Kennedy, B. A. 1971. *Physical Geography: A Systems Approach.* London: Prentice Hall.

Christaller, W. 1966. *Central Places in Southern Germany* (translated by Baskin, C. W.). Englewood Cliffs, NJ: Prentice-Hall.

Church, R., Stoms, D., Davis, F., and Okin, B. J. 1996. Planning management Activities to Protect Biodiversity with a GIS and an Integrated Optimization Model. Santa Fe CD-ROM, available from National Center for Geographic Information and Analysis, University of California, Santa Barbara, CA.

Church, R. L. 1999. Location Modeling and GIS. Pp. 293–303 in Longley, Goodchild, Maguire, and Rhind, eds., 1999, op cit.

Clarke, D. L. 1968. *Analytical Archaeology.* London: Methuen.

Clarke, D. L. 1972. *Models in Archaeology.* London: Methuen.

Clarke, D. L. 1977. *Spatial Archaeology.* London: Academic Press.

Clarke, K. C. 1999. *Getting Started with Geographic Information Systems* (2nd ed.). Upper Saddle River, NJ: Prentice Hall.

Clarke, K. C., and Olsen, G. 1996. Refining a Cellular Automaton Model of Wildfire Propagation and Extinction. Pp. 333–38 in Goodchild, et al., 1996, op. cit.

Clarke, K. C., Hoppen, S. and Gaydos, L. 1996. Methods and Techniques for Rigorous Calibration of a Cellular Automaton Model of Urban Growth. Santa Fe CD-ROM, available from National Center for Geographic Information and Analysis, University of California, Santa Barbara, CA.

Clayton, D., and Waters, N. Distributed Knowledge, Distributed Processing, Distributed Users: Integrating Case Based Reasoning and GIS for Multicriteria Decision Making. Pp. 275–308 in Thill, 1999, op. cit.

Coffey, W. 1981. *Geography: Towards a General Spatial Systems Approach.* London: Methuen.

Cole, H. S. D., Freman, C., Jahoda, M. and Pavitt, K. I. R. (eds.)1973. *Thinking About the Future: A Crtique of the Limits to Growth.* London: Chatto and Windus.

Coppock, J. T., and Rhind, D. W. 1991. The History of GIS. Pp. 21–43 in Maguire, D. J., Goodchild, M. F., and Rhind, D. W. (eds.), *Geographical Information Systems: Principles and Applications,* New York: Wiley.

Cowan, D. D., Grove, T. R., Mayfield, C. I., Newkirk, R. T., and Swayne, D. A. 1996. An Integrative Information Framework for Environmental Management and Research. Pp. 423–428 in Goodchild, Steyaert, and Parks, et al., 1996, op. cit.

Cressie, N. 1993. Geostatistics: A Tool for Environmental Modelers. Pp. 414–421 in Goodchild, Parks and Steyaert, 1993, op. cit.

Crosbie, P. 1996. Object-Oriented Design of GIS: A New Approach to Environmental Modeling. Pp. 383–386 in Goodchild, Steyaert, and Parks, et al., 1996, op. cit.

Csillag, F. 1996. Variations on Hierarchies: Toward Linking and Integrating Structures. Pp. 433–438 in Goodchild, Steyaert, and Parks, et al., 1996, op. cit.

Dalton, G. 1981. Anthropological Models in Archaeological Perspective. Pp. 17-48 in I. Hodder, G. Isaac, and N. Hammond (eds.), *Pattern of the Past: Studies in Honour of David Clarke*, Cambridge, England: Cambridge University Press.

Dangermond, J. 1993. The Role of Software Vendors in Integrating GIS and Environmental Modeling. Pp. 51–56 in Goodchild, Parks, and Steyaert, 1993, op. cit.

Dieckmann, U., Law, R., and Metz, J. A. J. (eds.), 1999. *The Geometry of Ecological Interactions: Simplifying Spatial Complexity.* Cambridge, England: Cambridge University Press.

Donovan, J. W. 1991. Interoperability: the Unfulfilled Promise. *Byte*, 16, #12, 185.

Eddy, J. A. 1993. Environmental Research: What We Must Do. Pp. 3-7 in Goodchild, Parks, and Steyaert, 1993, op. cit.

El Haddi, A. A., Shekhar, S., Hills, R., and McManamon, A. 1996. Mirror-Image Round Robin Spatial Data Partitioning: A Case Study with Parallel SEUS. Santa Fe CD-ROM, available from National Center for Geographic Information and Analysis, University of California, Santa Barbara, CA.

Ellis, F. 1996. The Application of Machine Learning Techniques to Erosion Modeling. Santa Fe CD-ROM, available from National Center for Geographic Information and Analysis, University of California, Santa Barbara, CA.

Emery, F. E. (ed.) 1969. *Systems Thinking.* Penguin, Harmondsworth, Middlesex, England.

Englund, E. J. 1993. Spatial Simulation: Environmental Applications. Pp. 432–437 in Goodchild, Parks and Steyaert, 1993, op. cit.

Evink, G. L., Garrett, P., and Zeigler, D. (eds.) 1999. *Proceedings of the Third International Conference on Wildlife Ecology and Transportation.* FL-ER-73-99. Tallahassee, FL: Florida, Department of Transportation.

Fairweather, N. B., Elworthy, S., Stroh, M., and Stephens, P. H. G., (eds.) 1999. *Environmental Futures.* New York: St. Martin's Press.

Fedra, K. 1996. Distributed Models and Embedded GIS: Integration Strategies and Case Studies. Pp. 413–418 in Goodchild, Steyaert, and Parks, et al., 1996, op. cit.

Foresman, T. W. 1998. *The History of Geographic Information Systems: Perspectives from the Pioneers.* Upper Saddle River, NJ: Prentice Hall.

Forrester, J. W. 1961. *Industrial Dynamics.* Cambridge, MA: MIT Press.

Forrester, J. W. 1969. *Urban Dynamics.* Cambridge, MA: MIT Press.

Forrester, J. W. 1971. *World Dynamics.* Cambridge, MA: Wright-Allen Press.

Fotheringham, A. S., Brunsdon, C., and Charlton, M. 2000. *Quantitative Geography: Perspectives on Spatial Data Analysis.* London: Sage.

French, D. P., and Reed, M. 1996. Integrated Environmental Impact Model and GIS for Oil and Chemical Spills. Pp. 197–198 in Goodchild, et al., 1996, op. cit.

Fujisaka, S, Thomas, N., and Crawford, E. 1996. Deforestation in Two Brazilian Amazon Colonies: Analysis Combining Farmer Interview and GIS. Santa Fe CD-ROM, available from National Center for Geographic Information and Analysis, University of California, Santa Barbara, CA.

Gao, X., Sorooshian, S., and Goodrich, D. C. 1993. Linkage of a GIS to a Distributed Rainfall-Runoff Model. Pp. 182-87 in Goodchild, Parks and Steyaert, 1993, op. cit.

George, F. H. 1967. The Use of Models in Science. Pp. 43–56 in Chorley and Haggett, (eds.), 1967, op. cit.

GIS World. 1995. Ian McHarg Reflects on the Past, Present and Future of GIS. *GIS World,* 8 (10): 46-49.

Gleick, J. 1987. Chaos: Making a New Science. New York: Viking Penguin.

Goodchild, M. F. 1986. *Spatial Autocorrelation.* Concepts and Techniques in Modern Geography, Number 47, Norwich, England: GeoBooks.

Goodchild, M. F. 1992. Geographical Information Science. *International Journal of Geographical Information Systems*, 6, 31–45.

Goodchild, M. F. 1993a. The State of GIS for Environmental Problem-Solving. Pp. 8-15 in Goodchild, Parks, and Steyaert, 1993, op. cit.

Goodchild, M. F. 1993b. Data Models and Data Quality: Problems and Prospects. Pp. 94-103 in Goodchild, Parks, and Steyaert, 1993, op. cit.

Goodchild, M. F. 1996a. The Spatial Data Infrastructure of Environmental Modeling. Pp. 11-16 in Goodchild, Steyaert, and Parks, et al., 1996, op. cit.

Goodchild, M. F. 1996b. Directions in GIS. Santa Fe CD-ROM, available from National Center for Geographic Information and Analysis, University of California, Santa Barbara, CA.

Goodchild, M. F., Parks, B. O., and Steyaert, L. T. (eds.) 1993 *Environmental Modeling with GIS.* New York: Oxford University Press.

Goodchild, M. F., Steyaert, L. T., Parks, B. O., Johnston, C., Maidment, D., Crane, M., and Glendinning, S. (eds.) 1996 *GIS and Environmental Modeling: Progress and Research Issues.* New York: Wiley.

Goodchild, M. F., Egenhofer, M. J., Kemp, K. K., Mark, D. M., and Sheppard, E. 1999. Introduction to the Varenius Project. *International Journal of Geographical Information Science*, 13, 731–745.

Goodchild, M. F., Egenhofer, M., Fegeas, R., and Kottman, C. (eds.) 1999. *Interoperating Geographic Information Systems.* Boston: Kluwer Academic Publishers.

Gould, P. 1970. Is Statistix Inferens the Geographical Name for a Wild Goose? *Economic Geography,* 46, 439–448.

Griffith, D. A. 1987. *Spatial Autocorrelation: A Primer.* Association of American Geographers, Washington, D.C.: Resource Publication.

Griffith, D. A. 1999. *A Casebook for Spatial Statistical Data Analysis.* Oxford, England: Oxford University Press.

Haggett, P. 1965. *Locational Analysis in Human Geography.* London: Arnold.

Haggett, P., and Chorley, R. J. 1967. Models Paradigms and the New Geography. In R. J. Chorley and P. Haggett (eds.) *Models in Geography,* London: Methuen.

Haggett, P., and Chorley, R. J. 1969. *Network Analysis in Geography.* London: Arnold.

Haggett, P., Cliff, A. D., and Frey, A. 1977. *Locational Analysis in Human Geography* (Second Edition). London: Arnold.

Hanski, I. 1997. Predictive and Practical Metapopulation Models: The Incidence Function Approach. Pp. 21-45 in D. Tilman and P. Kareiva, *Spatial Ecology: The Role of Space in Population Dynamics and Interspecific Interactions,* Princeton, NJ: Princeton University Press.

Hardisty, J., Taylor, D. M., and Metcalfe, S. E. 1993. *Computerised Environmental Modelling: A Practical Introduction Using Excel.* New York: Wiley.

Hartkamp, A. D., White, J. W., and Hoogenboom, G. 1999. Interfacing Geographic Information Systems with Agronomic Modeling: A Review. *Agronomy Journal,* 91, 761–72.

Harvey, D. 1969. *Explanation in Geography.* London: Edward Arnold.

Hay, L., Knapp, L. and Bromberg, J. 1996. Integrating GIS, Scientific Visualization Systems, Statistics, and an Orographic Precipitation Model for a Hydroclimatic Study of the Gunnison River Basin. Pp. 235–238 in Goodchild, Steyaert, and Parks, et al., 1996, op. cit.

Healey, R., Dowers, S., Gittings, B., and Mineter, M. (eds.) 1998. *Parallel Processing Algorithms for GIS.* London: Taylor and Francis.

Helmer, O. 1988. Using Expert Judgment. Pp. 87-119 in Miser, H. J., and Quade, E. S. *Handbook of Systems Analysis.* New York: North-Holland.

Heuvelink, G. B. M. 1998. *Error Propagation in Environmental Modelling with GIS.* Bristol, PA: Taylor and Francis.

Holland, J. H., Holyoak, K. J., Nisbett, R. E., and Thagard, P. R. 1986. *Induction: Processes of Inference, Learning and Discovery.* Cambridge, MA: MIT Press.

Huggett, R. 1980. *Systems Analysis in Geography.* Oxford, England: Oxford University Press.

Jarvis, C. H., Stuart, N., Kelsey, J., and Baker, R. H. A. 1996. Towards a Methodology for Selecting a "Characteristic" Sample from an Existing Database: An Evolutionary Approach. Santa Fe CD-ROM, available from National Center for Geographic Information and Analysis, University of California, Santa Barbara, CA.

Johnson, A. R. 1996. Spatiotemporal Hierarchies in Ecological Theory and Modeling. Pp. 451–456 in Goodchild, Steyaert, and Parks, et al., 1996, op. cit.

Johnston, R. J., Gregory, D., Pratt, G., and Watts, M. (eds.) 2000. *The Dictionary of Human Geography* (4th ed.). Oxford, England: Blackwell.

Journel, A. G. 1996. Modelling Uncertainty and Spatial Dependence: Stochastic Imaging. *International Journal of Geographical Information Systems,* 10, 517–522.

Kautz, R., Gilbert, T., and Stys, B. 1999. A GIS Plan to Protect Fish and Wildlife Resources in the Big Bend Area of Florida. Pp. 193–208 in Evink et al., 1999, op. cit.

Kemp. K. K. 1996. Managing Spatial Continuity for Integrating Environmental Models with GIS. Pp. 339–344 in Goodchild, Steyaert, and Parks, et al., 1996, op. cit.

Kessell, S. R. 1996. The Integration of Empirical Modeling, Dynamic Process Modeling, Visualization, and GIS for Bushfire Decision Support in Australia. Pp. 367–372 in Goodchild, Steyaert, and Parks, et al., 1996, op. cit.

Kineman, J. J. 1993. What is a Scientific Database? Design Considerations for Global Characterization in the NOAA-EPA Global Ecosystems Database Project. Pp. 372–378 in Goodchild, Parks and Steyaert, 1993, op. cit.

Kirkby, M. J., Naden, P. S., Burt, T. P., and Butcher, D. P. 1987. *Computer Simulation in Physical Geography.* New York: Wiley.

Klein, L. 1999. Usage of GIS in Wildlife Passage Planning in Estonia. Pp. 179–184 in Evink et al., 1999, op. cit.

Klinkenberg, B. and Goodchild, M. F. 1992. The Fractal Properties of Topography—A Comparison of Methods. *Earth Surface Processes and Landforms,* 17, 217–234.

Kohler, T. A., Van West, C. R., Carr, E. P., and Langton, C. G. 1996. Agent-Based Modeling of Prehistoric Settlement Systems in the North American South West. Santa Fe CD-ROM, available from National Center for Geographic Information and Analysis, University of California, Santa Barbara, CA.

Kraak, M-J 1999. Visualization for Exploration of Spatial Data. *International Journal of Geographic Information Science,* 13, 285–288.

Krumbein, W. C., and Graybill, F. A. 1965. *An Introduction to Statistical Models in Geology.* New York: McGraw-Hill.

Krummel, J. R., Dunn, C. P., Eckert, T. C. and Ayers, A. J. (1996). A Technology to Analyze Spatiotemporal Landscape Dynamics: Application to Cadiz Township (Wisconsin). Pp. 169-174 in Goodchild, Steyaert, and Parks, et al., 1996, op. cit.

Kuhn, T. S. 1971 *The Structure of Scientific Revolutions.* Chicago: University of Chicago Press.

Kurzweil, R. 1999. *The Age of Spiritual Machines: When Computers Exceed Human Intelligence.* New York: Viking Penguin.

Lam, D., and Pupp, C. 1996. Integration of GIS, Expert Systems, and Modeling for State-of-Environment Reporting. Pp. 419–422 in Goodchild, Steyaert, and Parks, et al., 1996, op. cit.

Lane, J. S., and Hartgen, D. T. 1989. *Factors Affecting the Adoption of Information Systems in State DOTs.* Transportation Publication Report No. 18, Transportation Academy, Department of Geography and Earth Sciences, University of North Carolina at Charlotte, Charlotte, NC.

Lane, S., Richards, K., and Chandler, S. (eds.) 1998. *Landform Monitoring, Modelling and Analysis.* New York: Wiley.

LaPoton, P. J., and McKim, H. L. 1996. Optimal Classification Techniques for Feedforward Parallel Distributed Processors. Pp. 405–412 in Goodchild, Steyaert, and Parks, et al., 1996, op. cit.

Lark, R. M. 2000. Regression Analysis with Spatially Autocorrelated Error: Simulation Studies and Application to Mapping of Soil Organic Matter. *International Journal of Geographical Information Science,* 14, 247–264.

Lauwerier, H. 1991. *Fractals: Images of Chaos.* London: Penguin.

Lincoln, N. 1993. The Future of Environmental Modeling. In Zannetti, P., ed., *Environmental Modeling, Vol. 1, Computer methods and Software for Simulating Environmental Pollution and Its Adverse Effects,* Pp. 513–532. New York: Elsevier Applied Science.

Long, N. V. 1999. Modelling Global Common Fisheries Exploitation and Regulation. Pp. 369–386 in Mahedrarajah, Jakeman, and McAleer, 1999, op. cit.

Longley, P. A., Goodchild, M. F., Maguire, D. J., and Rhind, D. W. (eds.) 1999. *Geographical Information Systems* (2nd ed. ). New York: Wiley.

Mackey, B. G., Sims, R. A., Baldwin, K. A., and Moore, I. D. 1996. Spatial Analysis of Boreal Forest Ecosystems: Results from the Rinker Lake Case Study. Pp. 187–190 in Goodchild, Steyaert, and Parks, et al., 1996, op. cit.

Macmillan, B. (ed.) 1989a. *Remodelling Geography.* Oxford: Blackwell.

Macmillan, B. 1989b. Modelling Through: An Afterword to Remodelling Geography. Pp. 291-313 in Macmillan, 1989a, op. cit.

Mahedrarajah, S., Jakeman, A. J., and McAleer, M. (eds.), 1999. *Modelling Change in Integrated Economic and Environmental Systems.* New York: Wiley.

Maidment, D. R. 1993. GIS and Hydrological Modeling. In Goodchild, Parks and Steyaert, Pp. 147–167, op. cit.

Maidment, D. R. 1996. Environmental Modeling within GIS. In Goodchild, Steyaert, and Parks, et al., Pp. 315–324, op. cit.

Maguire, D. J. 1998/1999. ARC/INFO Version 8: Object-Component GIS. *ARC News,* 20, no. 4, pp. 1, 2, and 5.

Majure, J. J., Cressie, N., Cook, D., and Symanzik, J. 1996. GIS, Spatial Statistical Graphics and Forest Health. Santa Fe CD-ROM, available from National Center for Geographic Information and Analysis, University of California, Santa Barbara, CA.

Mandelbrot, B. B. 1982 *The Fractal Geometry of Nature.* New York: Freeman.

Mark, D. (ed.), 2000. Geographic Information Science: Critical Issues in an Emerging Cross-Disciplinary Research Domain. *URISA Journal,* 12, 45-54.

Massam, B. 1975. *Location and Space in Social Administration.* London: Arnold.

Mayer, L. 1990. *Introduction to Quantitative Geomorphology.* Englewood Cliffs, NJ: Prentice Hall.

McCauley, J. D., Navulur, K. C. S., Engel, B. A., and Srinivasam, R. 1996. Serving GIS Data Through the World Wide Web. Santa Fe CD-ROM, available from NCGIA, University of California, Santa Barbara, CA.

McGuffie, K., and Hendersen-Sellers, A. 1997. *A Climate Modelling Primer.* New York: Wiley.

McHarg, I. L. 1969. *Design with Nature.* New York: Doubleday.

McHarg, I. L. 1996. *A Quest for Life: An Autobiography.* New York: Wiley.

Meadows, D. H., Meadows, D. L., Randers, J., and Behrens III, W. W. 1972. *The Limits to Growth.* New York: Universe Books.

Meadows, D. H., Meadows, D. L., and Randers, J. 1992. *Beyond the Limits: Confronting Global Collapse, Envisioning a Sustainable Future.* Toronto: McClelland & Stewart.

Meadows, P. 1957. Models, System and Science. *American Sociological Review,* 22, 3–9.

Miller, H. 1999. Potential Contributions of Spatial Analysis for Geographic Information Systems for Transportation (GIS-T). *Geographical Analysis,* 31, 373–399.

Mitas, L. and Mitasova, H. 1999. Spatial Interpolation. Pp. 481–492 in Longley, Goodchild, Maguire and Rhind, op.cit.

Mitasova, H., Mitas, L., Brown, W. M., Gerdes, D. P., Kosinovsky, I. And Baker, T. 1996. Modeling Spatial and Temporal Distributed Phenomena: New Methods and Tools for Open GIS. Pp. 345–352 in Goodchild, Steyaert, and Parks, et al., op. cit.

Moore, I. D., Turner, A. K., Wilson, J. P. Jenson, S. K., and Band, L. E. 1993. GIS and Land-Surface-Subsurface Modeling. Pp. 196–230 in Goodchild, Parks and Steyaert, 1993, op. cit.

Morgan, M. A. 1967. Hardware Models in Geography. In R. J. Chorley, and P. Haggett (eds.) *Models in Geography,* London: Methuen.

Nance, B. 1991. Interoperability Today. *Byte,* 16, no. 12, 187–196.

Newton, I. 1675/6. Letter to Robert Hooke. (Cited in *The Concise Oxford Dictionary of Quotations,* 1981, Oxford University Press, Oxford; this dictionary notes that the idea of standing on the shoulders of giants is originally attributed to Bernard of Chartres, c. 1130; so even in this Newton was borrowing).

O'Callaghan, J. 1999. Technology Changes Everything. P. 17 in Longley, Goodchild, Maguire, and Rhind, eds., 1999, op cit.

Openshaw, S. 1989a. Computer Modelling in Human Geography. Pp. 70–88 in Macmillan, ed., 1989, op. cit.

Openshaw, S. 1998b. Building Automated Geographical Analysis and Explanation Machines. Pp. 95–115 in Longley, P. A., Brooks, S. M., McDonnell, R. and Macmillan, B., eds., *Geocomputation: A Primer,* New York: Wiley.\

Ortuzar, J. de D., and Willumsen, L. G. 1994. *Modelling Transport* (2nd ed.). New York: Wiley.

Parks, B. O. 1993. The Need for Integration. Pp. 31–34 in Goodchild, Parks and Steyaert, 1993, op. cit.

Perlman, L. J. 1976. *The Global Mind: Beyond the Limits to Growth.* New York: Mason Charter.

Power, J. M., Strome, M., and Daniel, T. C. 1995. *Proceedings of Decision Support – 2001.* Bethesda, MD: American Society for Photogrammetry and Remote Sensing.

Pratt, V., Howarth, J., and Brady, E. 2000. *Environment and Philosophy.* New York: Routledge.

Raper, J., and Livingstone, D. 1996a. High-Level Coupling of GIS and Environmental Process Modeling. Pp. 387–390 in Goodchild, Steyaert, and Parks, et al., 1996, op. cit.

Raper, J., and Livingstone, D. 1996b. Spatio-Temporal Interpolation in Four Dimensional Coastal Process Models. Santa Fe CD-ROM, available from National Center for Geographic Information and Analysis, University of California, Santa Barbara, CA.

Rapoport, A., von Bertalanfly, L., and Meier, R. L. 1961. Editorial Comments on The Problem of Contiguity. *General Systems Yearbook,* 6, 139–140.

Reitsma, R. 1996. Bootstrapping River Basin Models with Object Orientation and GIS Technology. Pp. 457–462 in Goodchild, Steyaert, and Parks, et al., 1996, op. cit.

Richardson, L. F. 1961. The Problem of Contiguity: An Appendix to Statistics of Deadly Quarrels. *General Systems Yearbook,* 6, 140–187.

Ripple, W. J. 1987. *Geographic Information Systems for Resource Management: A Compendium.* American Society for Photogrammetry and Remote Sensing and American Congress on Surveying and Mapping, Falls Church, VA.

Saghafian, B. 1996. Implementation of a Distributed Hydrologic Model within GRASS. Pp. 205–208 in Goodchild, Steyaert, and Parks, et al., 1996, op. cit.

Samson, P. R., and Pitt, D. 1999. *The Biosphere and Noosphere Reader.* New York: Routledge.

Sandhu, R., and Treleaven, P. 1996. Client-Server Approaches to Model Integration Within GIS. Santa Fe CD-ROM, available from National Center for Geographic Information and Analysis, University of California, Santa Barbara, CA.

Scheidegger, A. E. 1991. *Theoretical Geomorphology* (3rd ed.). New York: Springer-Verlag.

Schimel, D. S., and Burke, I. C. 1993. Spatial Interactive Models of Atmosphere-Ecosystem Coupling. Pp. 284–89 in Goodchild, Parks and Steyaert, 1993, op. cit.

Skelly, W. C., Hendersen-Sellers, A., and Pitman, A. J. 1993. Land Surface Data: Global Climate Modeling Requirements. Pp. 135–41 in Goodchild, Parks and Steyaert, 1993, op. cit.

Skilling, H. 1964. An Operational View. *American Scientist,* 52, 388A–396A.

Skupin, A., and Buttenfield, B. P. 1996. Spatial Metaphors for Visualizing Very Large Data Archives. *Proceedings of GIS/LIS '96,* 607–617, Association of American Geographers, Washington, D.C.

Skupin, A., and Buttenfield, B. P. 1997. Spatial Metaphors for Visualizing Information Spaces. *AutoCarto,* 13, 116–125.

Smillie, I. 1991. *Mastering the Machine. Poverty, Aid and Technology.* Boulder, CO: Westview Press.

Smith, T. R., Su, J., Saran, A., and Sastri, A. M. 1996. Computational Modeling Systems to Support the Development of Scientific Models. Pp. 373–382 in Goodchild, Steyaert, and Parks, et al., 1996, op. cit.

Steinberg, E. K., and Kareiva, P. 1997. Challenges and Opportunities for Empirical Evaluation of "Spatial Theory." Pp. 318–332 in D. Tilman and P. Kareiva (eds.), *Spatial Ecology: The Role of Space in Population Dynamics and Interspecific Interactions*, Princeton, NJ: Princeton University Press.

Steinitz, C., Parker, P., and Jordan, L. 1976. Hand-Drawn Overlays: Their History and Prospective Uses. *Landscape Architecture,* 66, 444–455.

Steyaert, L. T. 1993. A Perspective on the State of Environmental Simulation Modeling. Pp. 16–30 in Goodchild, Parks and Steyaert, 1993, op. cit.

Steyaert, L. T. 1996. Status of Land Data for Environmental Modeling and Challenges for Geographic Information systems in Land Characterization. Pp. 17–28 in Goodchild, Steyaert, and Parks, et al., 1996, op. cit.

Stoms, D. M., Davis, F. W., and Hollander, A. D. 1996. Hierarchical representation of Species Distribution for Biological Survey and Monitoring. In in Goodchild, Steyaert, and Parks, et al., 1996, op. cit.

Sumner, G. N. 1978. *Mathematics for Physical Geographers.* London: Arnold.

Thill, J-C. 1999. *Spatial Multicriteria Decision Making and Analysis: A Geographic Information Science Approach.* Aldershot, England.

Thomas, R. W., and Huggett, R. J. 1980. *Modelling in Geography: A Mathematical Approach.* Totowa, NJ: Barnes and Noble.

Tillman, R. (ed.) 1976. *Proceedings of the First National Symposium on Environmental Concerns in Rights-of-Way Management.* Mississippi State University, MS.

Tomlin, C. D. 1990. *Geographic Information Systems and Cartographic Modeling.* Englewood Cliffs, NJ: Prentice-Hall.

Tomlinson, R. F. (ed.) 1972. *Proceedings of the International Geographical Union on Geographical Data Sensing and Processing: Volume 1, Environment Information Systems.* Second Symposium on Geographical Information Systems, Ottawa, August, 1972, National Technical Information Service, Springfield, VA.

Tomlinson, R. F. 1984. Geographic Information Systems-A New Frontier. *The Operational Geographer*, 5, 31–36.

Tomlinson, R. F. 1998. The Canada Geographic Information System. Chapter 2, pp. 21–32, in Foresman, 1998, op. cit.

Tomlinson, R. F. 1999. How It All Began and the Importance of Bright People. Pp. 17-18 in Longley, Goodchild, Maguire and Rhind, 1999, op.cit.

Turkle, S. 1995. *Life on the Screen: Identity in the Age of the Internet.* New York: Simon and Shuster.

van Horssen, P. 1996. Ecological Modelling in GIS. Santa Fe CD-ROM, available from NCGIA, University of California, Santa Barbara, CA.

Vckovski, A. 1998. Special Issue: Interoperability in GIS. *International Journal of Geographical Information Science,* 12, 297–298.

ver Hoef, J. M. 1993. Universal Kriging for Ecological Data. Pp. 447–453 in Goodchild, Parks and Steyaert, 1993, op. cit.

Vieux, B. E., Farajalla, N. S. and Gaur, N. 1996. Integrated GIS and Distributed Storm Water Runoff Modeling. Pp. 199–204 in Goodchild, et al., 1996, op. cit.

von Bertalanffy, L. 1962. General System Theory: A Critical Review. *General Systems*, 7, 1–20.

Wachter-Harms, T. and Wendholt, B. 1996. Controlling a GIS by the Expert System EXCEPT. Santa Fe CD-ROM, available from National Center for Geographic Information and Analysis, University of California, Santa Barbara, CA.

Warntz, W. 1973. New Geography as General Spatial Systems Theory–Old Social Physics Writ Large? Chapter 5, pp. 89-126, in Chorley, R. J., ed., *Directions in Geography,* London: Methuen.

Warntz, W. 1989. Newton, the Newtonians, and the Geographia Generalis Varenii. *Annals of the Association of American Geographers*, 79, 165-91.

Waters, N. M. 1993. GIS: Paradigms Lost. *GIS World,* 6(8): 64.

Waters, N. M.1994. Statistics: How Much Should a GIS Analyst Know? *GIS World,* 7(3): 62.

Waters, N.M. 1998a. Geographic Information Systems. *Encyclopedia of Library and Information Science*, 63, 98–125.

Waters, N. M. 1998b. GIS Alliances: Wave of the Future or Antitrust End Run?,*GIS World,* 11, no.5, 40–41.

Waters, N. M. 1998c. GIS Enters the Wireless World. *GIS World,* 11, no.7, 40–42.

Waters, N. M. 1999a. Creating a New Master's Degree in GIS Proves Rewarding. *GEOWorld*, 12, no. 9, 30–31.

Waters, N. M. 1999b. Is XML the Answer for Internet-Based GIS? *GEOWorld* 12, no.7, 32–33.

Waters, N. M. 2000. Healy Park: Can GIS save Banff from "the invigorating scent of dollar bills?" *GEOWorld,* 13, no. 3.

Westervelt, J. D., and Hopkins, L. D. 1996. Facilitating Mobile Objects within the Context of Simulated Landscape Processes. Santa Fe CD-ROM, available from NCGIA, University of California, Santa Barbara, CA.

Wiener, N. 1948. *Cybernetics.* New York: Wiley.

Williams, J. R., Goodrich-Mahoney, J. W., and Wisniewski, J. R. (eds.) 1997. *The Sixth International Symposium on Environmental Concerns in Rights-of-Way Management.* New York: Elsevier.

Willmott, C. 1996. Smart Interpolation of Climate Variables. Santa Fe CD-ROM, available from National Center for Geographic Information and Analysis, University of California, Santa Barbara, CA.

Wilson, A. G. 1981a. *Catastrophe Theory and Bifurcation: Applications to Urban and Regional Systems.* London: Croom Helm.

Wilson, A. G. 1981b. *Geography and the Environment: System Analytical Methods.* New York: Wiley.

Wilson, A. G., and Kirkby, M. J. 1975, 1980. *Mathematics for Geographers and Planners.* Oxford, England: Oxford University Press.

Wilson J. P. 1996. GIS-Based Land Surface/Subsurface Modeling: New Potential for New Models. Santa Fe CD-ROM, available from National Center for Geographic Information and Analysis, University of California, Santa Barbara, CA.

Wilson, J. P., Mitasova, H., and Wright, D. J. 2000. Water Resource Applications of Geographic Information Systems. *URISA Journal,* 12 (2): 61–79.

Wirasinghe, S.C., and Waters, N.M. 1983. An Approximate Procedure for Determining, the Number, Capacities and Locations of Solid Waste Transfer Stations in an Urban Region. *European Journal of Operations Research*, 12, 105–111.

Woldenberg, M. J. (ed.) 1985. *Models in Geomorphology.* Allen and Unwin, Boston.

Wright, D. J., Goodchild, M. F., and Proctor, J. D. 1997. Demystifying the persistent ambiguity of GIS as tool versus science. *Annals of the Association of American Geographers*, 87, 346–362.

Yeh, T-S, and de Cambray, B. 1996. Time as a Geometric Dimension for Modeling the Evolution of Entities: A Three-Dimensional Approach. Pp. 397-404 in Goodchild, Steyaert, and Parks, et al., 1996, op. cit.

Yuan, M. 1996. Temporal GIS and Spatio-Temporal Modeling. Santa Fe CD-ROM, available from National Center for Geographic Information and Analysis, University of California, Santa Barbara, CA.

Zannetti, P. 1993. *Environmental Modeling–Volume 1: Computer Methods and Software for Simulating Environmental Pollution and Its Adverse Effects.* New York: Elsevier.

Zar, J. H. 1999. *Biostatistical Analysis* (4th ed.). NJ: Prentice Hall.

## 1.8 ABOUT THE CHAPTER AUTHOR

 Nigel Waters is Professor in the Department of Geography at the University of Calgary. He is the Director of the Masters in Geographic Information Systems Program and Director of the Transportation Theme School in the Faculty of Social Sciences. Since 1989 he has been a contributing editor to *GEOWorld* (formerly *GIS World*), for which he writes the *Edge Nodes* column. He has written numerous articles on geographic information systems, transportation, and the impact of transportation on the environment.

# 2

# *Modeling Frameworks, Paradigms, and Approaches*

## 2.0 INTRODUCTION

This chapter provides a general conceptual background to the more specific discussions of GIS-based environmental modeling presented in this book. It addresses three main themes central to environmental modeling: spatial models, complex systems, and geocomputation. The discussion covers static and dynamic, discrete and continuous, and social and natural system models, and helps put in perspective the diverse modeling approaches and issues presented in the following chapters. This chapter highlights the systems perspective with an emphasis on spatiotemporal environmental systems representations. General classes of models rather than any specific ones are discussed, in particular, systems dynamics models under the top-down paradigm, and cellular automata and agent-based models under the bottom-up paradigm.

### 2.0.1 Environmental Models in Research and Policy

A model is an abstract and partial representation of some aspect or aspects of the world "that can be manipulated to analyze the past, define the present, and to consider possibilities of the future" (Smyth, 1998, p.191). According to other definitions, models are devices for producing missing data about the past or the present, and for anticipating data about possible futures; or, as one great twentieth-century physicist put it, models simply are frameworks for organizing knowledge. Note that these latter definitions do not presuppose that models must resemble the real world in any form or fashion: indeed, for some model theorists, the "real world" is nothing more than the universe of potentially acquirable data (Zeigler et al., 2000). This noncommittal view allows us to skirt the difficult philosophical question of what the world is "really" like, helps explain why very different models of the same phenomenon may be equally true, and focuses our attention on the performance of models as predictive or explanatory devices.

More specifically, what do we mean by environmental models? Any aspect of the earth's environment may be the focus of environmental modeling: A definition can hardly get broader than this. Atmospheric and hydrological processes, land-surface–subsurface processes, biological and ecological systems, natural hazards, and ecosystems management issues are all popular themes. Environmental models cover a full range of geographic scales, from the local to the regional to the global. Moreover, they cover a wide range of input domains: natural and human, biotic and abiotic, atmospheric, oceanic, terrestrial and socioeconomic. Because the environment is a synthesis of all these domains, environmental models often combine several aspects from one or more of these areas. Thus we have models of the effects of climate change on biota, of fire and forest regeneration, of the interdependence of hydrology and ecosystems, of atmospheric circulation and industrial pollution, or of fisheries under the impact of different fishing policies.

There are many reasons why interest in environmental models has greatly increased in the past decade or so. Mounting environmental consciousness in the industrialized regions of the world and an increasing interest among public and private funding agencies to support modeling work related to environmental issues have attracted large numbers of capable researchers. At the same time appropriate techniques, computational resources, and especially data of suitable quality and quantity have become widely available, so that the gap between the desirable and the feasible in environmental modeling keeps decreasing. Geographic Information Systems (GIS), with their power to integrate diverse databases, undoubtedly played a key role in this development and have become the core technology of environmental modeling research.

While environmental models integrating natural processes have already achieved a certain maturity, those seeking to combine human and natural processes are still in their infancy. It is fair to say that today's frontier in environmental modeling lies at the natural-human interface: This is where some of the most important problems and some of the more interesting research issues are to be found. Researchers and funding agencies alike seem increasingly willing to invest in such cross-disciplinary research, despite the still strong institutional and intellectual barriers separating academic fields. Major examples of integrated human-natural environmental modeling research recently funded by the U.S. National Science Foundation include the two Urban LTER (Long Term Ecological Research) projects at the University of Maryland and Arizona State University (see http://baltimore.umbc.edu/lter, http://caplter.asu.edu, or http://www.ecostudies.org/bes), and some of the projects from the 1998 Urban Change competition (see, for example, Alberti, 1999, and the Urban Change Integrated Modeling Environment project at http://www.geog.ucsb.edu/~kclarke/ucime/). Similar efforts are underway in the Netherlands (White and Engelen, 1999) and other European countries.

### 2.0.2 What Makes a "Good" Environmental Model?

Environmental models are developed for research or policy purposes. While the line between the two is a fine one in applied fields, there are certain criteria that make a model more suitable for one or the other purpose. All models must be built on good science, they all need to be based on good data, and they all must deal with good problems. Research models are expected to exhibit a higher degree of scientific rigor and to contribute some original theoretical insights or technical innovation. In policy models originality is less of an issue (often, the more tried and true a model is, the better!), but transparency, manipulability, and the inclusion of key policy variables are especially important.

Clearly, research models can have significant policy implications (as is the case with the global climate models developed in the past decade) just as policy models can make original contributions to the science of environmental modeling. The next two sections will help clarify these points. Section 2.1 discusses the characteristics and aspects of environmental models, while Section 2.2 focuses on the systems and subsystems that are the objects of modeling. Starting with the key role of GIS, Section 2.4 extends the discussion to the contribution of *geocomputation* to both research and policy-oriented environmental modeling. Finally, Section 2.5 concludes with the brief review of the accomplishments and future challenges of the field.

## 2.1 THE NATURE OF ENVIRONMENTAL MODELING

### 2.1.1 Characteristics of Environmental Models

Environmental models make up a distinct family different from other classes of models in either the natural or the social sciences. They tend to be data-intensive, cross-disciplinary, dynamic, and complex. They often integrate subsystems from several different domains without the support of widely accepted theoretical frameworks to lend credibility to the attempted synthesis. Their need for explicit spatial as well as temporal dimensions increases their complexity. They depend on data of very variable quality gleaned from a wide variety of sources. As a result, they face issues of uncertainty and fuzziness to a greater extent than either traditional natural science models, which deal with more homogeneous, usually "cleaner" data sets, or than social science models, which often aim at producing qualitative rather than strictly quantitative results. Moreover, because of their direct or indirect policy implications, environmental models don't have the right to be wrong!

### 2.1.2 Facets of Environmental Modeling

Clearly environmental modeling presents special challenges. Distinguishing the wheat from the chaff is not always easy. This is why it is all the more important for model builders to be aware of the different paradigms and approaches that underlie the wide variety of environmental models competing for attention and funding these days. This section outlines a general framework to help put such models in perspective and to help recognize the strengths and weaknesses of each. It is based on Smyth's (1998) framework for general geographic modeling adapted to the environmental domain.

According to Smyth (1998, p. 192), it is convenient to think of the modeled world as a *microworld* defined by an *ontology* consisting of contents, spatial structure, temporal structure, "physics" (rules of behavior), and rules of inference or logic. The notion of an artificial world or microworld, borrowed from artificial intelligence, is useful for reminding us that models are not the real thing, and that they need to be internally consistent. A closed microworld is autonomous in that the behaviors within it are completely specified through its initial definition without further reference to the external world. A traditional cellular automaton model (see Section 2.3.2) is a good example: Once the initial conditions, neighborhood template, and transition rules have been specified, the model will enfold all its possible behavior independently of anything external.

Other closed microworlds well known to geographers are the classic models of location and land use proposed by Christaller, von Thuenen, and Weber. Similarly, ecologists are familiar with simple predator-prey models that work out the evolution of interdependent populations within predetermined environments. Most environmental models, however, correspond to open microworlds, admitting elements, relationships, causation, and spatiotemporal and logical structure from outside of their ontological specification (Smyth, 1998, p. 193). More commonly these are referred to as open systems, though this terminology is less clear about the fact that it is the model, rather than the part of the world modeled, that is open. Once the ontology of a model has been clarified, it must then be expressed in a formal system that will eventually be translated into a computational model and a concrete implementation.

The first things defining the ontology of an environmental model are the *entities* within it. For example, in an integrated urban-environmental model, the entities may be roads, built-up areas, different categories of land uses, streams, slopes, bodies of water, forests, wildlife populations, and the like. These will most conveniently be modeled as objects, though other representations are possible. Entities primarily have an identity on the basis of which their other aspects can be defined. Relevant aspects of entities include versions (alternative descriptions that may indicate uncertainty as to some properties of an entity: Which western boundary of this urban area is the correct one for 1990?); class membership (about which there may be confusion: Is that patch pine or spruce forest?); alternatives regarding the attributes and spatial and temporal descriptions of these entities; and the structures and configurations (e.g., hierarchies of different kinds) that may relate the different entities in an ontology. Anderson classes of land use/land cover are a well-known such hierarchy whereby classes are subdivided into more and more detailed categories ("aggregation hierarchy"). Another kind of hierarchy binds together elements that compose a whole. For example, an ecosystem can be decomposed into its constituent entities (forest, grassland, water bodies, soils, animal and plant species, etc.).

What entities will be included in an environmental model will depend to some extent on its intended principal use (i.e., research or policy support). A policy-oriented model must involve entities and attributes of entities that can be manipulated by policy makers (policy variables). For example, a slope cannot be manipulated but a road network can; the age of a tree stand cannot be manipulated but its acreage can; and so on. A good policy model will include policy variables that have a significant effect on the behavior of the model. Observing how the microworld is affected by manipulating these variables can give decision makers insights into how to act in the real world. Such requirements for practically useful variables do not apply to models developed primarily for research purposes, where description, explanation, or prediction are the main goals.

The *spatial and temporal structure* of geographic models in general has been the subject of considerable research (Egenhofer and Golledge, 1998). Several kinds of conceptualizations and frameworks have been proposed for both space and time. Of special interest to the present discussion is the fact that environmental models typically consist of several different modules or subsystems, each based on its own spatial and temporal framework. Problems may arise when these frameworks are very different and perhaps inconsistent with one another in terms of scale, metricity, topological structure, reference frame, and other such properties.

The *physics* of a microworld are the rules of evolution of entities and of interaction between or among entities-that determine what can happen within it: They govern the possible behaviors of a model. These rules may be expressed mathematically (e.g., in the form of differential or difference equations), or computationally (e.g., in the form of if-then statements and other such specifications). In a physical model the physics are literally based on actual physical theory (e.g., fluid dynamics, mechanics, electromagnetism, or the theory of chemical interactions), but in most other cases (and models) the term is to be understood metaphorically.

Environmental models often include modules or submodels backed by the rigor of physical theory, but being strongly cross-disciplinary and synthetic, environmental science (and modeling) is by and large a theory-poor domain. This means that a typical environmental model will mix together several kinds of "physics", some based on causal hypotheses (A appears to cause B), some on statistical regularities (A is statistically associated with B), others on empirical rules of thumb (when A, usually B), and others still on arbitrary rules of behavior specified by the modeler (if A is the case, then do B). Combining such a variety of partial "physics" into a whole free of internal contradictions is a delicate task for which few guidelines exist, and which becomes more challenging as the aspects to be brought together are drawn from domains more remote from one another. Thus it is one thing to integrate a submodel of rainfall with one of runoff to determine flooding potential in an area, and quite another to combine a model of job growth and one of species extinction into a framework for exploring the environmental effects of urban development.

The *logic* of a microworld completes Smyth's (1998) pentad of a model's ontology. This is what allows new facts to be deduced about a microworld from a given configuration. In the case of mathematical models, the logic is indistinguishable from the physics as they are both implicit in the formalisms used. In the case of simulation models, however, the two are distinct. For example, the logic may include default rules to help decide what will happen in situations where behavior is underdetermined (e.g., in case of a tie), or to help determine which aspects of a microworld must be changed in order to reach a configuration with specific properties. The latter kind of question is particularly pertinent in policy-oriented models, where the interest is not just in possible futures but also in desirable ones and in the means necessary to reach these.

Several formal logics have been developed that in principle can provide rich enough inference mechanisms for environmental simulation models. Most of these are quite complex, however, and not yet widely used (Worboys, 1995). There will be an increasing need to study the logics of environmental modeling microworlds as more models are developed integrating drastically different kinds of processes (e.g., socioeconomic and physical). Formal modeling theories such those by Zeigler et al. (2000), and Casti (1997) can greatly assist the development of logically coherent environmental microworlds, as can the closely related perspective of systems theory, briefly discussed in Section 2.2.1.

Smyth's model specification sequence involves, in addition to the choice of an ontology, the following steps: expression of the ontology in a formal system; development of a computational model of the formal system; and realization of the computational model in a concrete implementation (Smyth, 1998, p. 204). These topics are explicated in detail in other chapters of this book and will not be further discussed here. As we will see in Section 2.4, recent computational advances have largely blurred the distinctions among the steps of this sequence. Logically, however, all these aspects remain a necessary part of modeling even if they do not constitute explicit, separate steps.

## 2.2 COMPLEX ENVIRONMENTAL SYSTEMS

### 2.2.1 Systems Theory: Philosophy and Key Concepts

In this section the emphasis shifts from the model to the system modeled. This is a subtle change of focus since the system is itself a kind of model: We model an environmental system that is a representation of some part of the real world. The word system goes back to the ancient Greeks and literally means something that hangs (or stands) together. In its modern scientific usage a system is something made up of a set of elements and relations among these elements. This may sound vague, but in fact many systems can be defined precisely by families of differential or difference equations where the variables are the elements and the relations are defined by the mathematical operations on the variables.

More generally, the elements of a system are the entities in a microworld's ontology, and these are connected to one another through relations of causation, influence, or dependence that together determine the physics of that microworld. Positive and negative feedback, stimulus and response, activation, inhibition, and so on are more specific terms for different kinds of relationships in a system, usually quantifiable. These, along with characterizations of behavior such as self-organization, steady state, instability, oscillation, emergence, and chaos, have become an integral part of the system modeler's vocabulary.

What is less well known is that these technical terms have strong philosophical underpinnings in the work of Ludwig von Bertalanffy (1968) dating from the 1930s. Being a theoretical biologist, von Bertalanffy was aware of the limitations of the reigning reductionist paradigm (i.e., the attempt to reduce all scientific explanation to physical laws) in helping describe and explain living things. His general system theory, composed of system science, system technology, system philosophy, and system epistemology, was proposed as a new way of thinking about, or paradigm for, the kinds of complex phenomena that biologists and other modern scientists in both the natural and the social sciences were studying. Von Bertalanffy's key insight was that there are laws of systems *qua* systems, in the abstract, regardless of the domain (economics, engineering, ecology, biology, physics, sociology, and so on) from which particular applications may be drawn. This was a highly ambitious program aiming at unifying all science under the systems paradigm. That integrated environmental modeling is possible at all is evidence of the basic correctness of general system theory's key premise—that systems behave like systems no matter what they are made of. A number of broad methodological concepts, such as systems thinking, the systems approach, the systems perspective, etc., have been derived from principles of general system theory. Their proponents make fine distinctions between them that need not concern us here.

Wilson (1981) distinguishes three basic classes of systems. *Simple* systems have few components and few variables (e.g., the solar system) and are studied with the traditional methods of classical physics. Systems of *disorganized complexity* have very large numbers of components and variables, but the couplings between them are either weak or random. Gases are typical examples of such systems that are studied primarily with statistical methods: hence the field of statistical mechanics and theoretical principles such as entropy maximization. The third class is that of systems of *organized complexity*. These systems too have large numbers of components, but there are multiple couplings and interactions among them that need to be considered explicitly. In these systems, the whole is more than the sum of the parts, meaning that their properties cannot be deduced from an understanding of their components. These systems of organized complexity are the ones the systems paradigm is specifically concerned with. Research into *complex systems* of all sorts has flourished in the second half of the twentieth century and has stimulated the interest in the study of environmental systems, which provide some of the most challenging examples of complex systems known.

There are two complementary methodological and theoretical approaches to the study of complex systems. The first is the top-down approach, exemplified by *systems analysis* and *systems dynamics*. The second is the bottom-up approach, emphasized in *complexity theory*. Both are directly relevant to environmental modeling and are briefly discussed next.

## 2.3 ENVIRONMENTAL PHENOMENA AND COMPLEX SYSTEMS

### 2.3.1 The Top-Down Approach: Decompose and Conquer

More familiar than general system theory is systems analysis, one of the many fields under von Bertalanffy's original paradigm. Systems analysis gained prominence after World War II primarily through the spectacular successes of engineering disciplines in building and controlling very complicated systems, from guided missiles to computers. At the same time systems analysis and the closely related field of operations research were being applied to "soft" fields such as planning, management, and the modeling of human decision making. The possibility to apply the same basic concepts and methods to natural, engineered, and social systems was thus practically established. An excellent though somewhat dated introduction to systems analysis for environmental scientists, addressing both natural and social systems, can be found in Wilson (1981). More abstract but still readable expositions of systems theory and systems analysis are found in Casti (1989) and a variety of other sources.

The basic principle behind systems analysis is that complex systems or problems can be decomposed into simpler subsystems (or subproblems), which themselves may be subdivided into even simpler subsubsystems, until a level is reached where the component parts may be treated as elementary. This approach differs substantially from reductionism in that the focus is not on the elementary components but on the relationships among components and assemblies of components at and between the levels of the (de)composition hierarchy. For example, in an ecological model, a spatial-scale decomposition may consider the interactions of individual organisms within patches at one level, and of populations in an ecosystem at another level, the ecosystem being much more than a sum of ecological patches. By contrast, in a reductionist ontology, the focus would be on aggregation hierarchies that allow properties of entities at one level to be generalized into properties of groups of these entities. Systems analysis may be static, clarifying the internal structure of a complex system, or dynamic, seeking to derive forecasts regarding future system behavior. This latter case is embodied in the paradigm of systems dynamics.

The history of modeling large-scale systems dynamics began in the 1960s with the publication of Forrester's series of three increasingly ambitious simulation models: industrial dynamics, urban dynamics, and world dynamics (Forrester, 1975). The latter is one of the earliest integrated environmental models developed, bringing together global population growth, agricultural food production, industrial production, pollution, etc. (however, *world dynamics* was not a spatial model). Each of these models consisted of several major coupled subsystems, which in turn included large numbers of linked components. Forrester's models were heavily criticized at the time but made some lasting contributions to complex systems research. Of particular relevance to environmental modeling was the demonstration that complex dynamic models of integrated systems could be built with the same basic techniques and vocabulary used in modeling a variety of physical systems, thus opening the way for the integration of social and natural science modeling. Coupled system models in the tradition of Forrester's are still routinely used. The widely popular modeling platform STELLA® is a direct implementation of Forrester's approach, and several texts are available to teach related methods (see Odum and Odum, 2000)

Most systems dynamics models represent phenomena in analogy with complicated hydrological cascading systems, where *flows* (of matter, energy, animals, people, money, etc.) move between *storage compartments* within which these quantities are created or transformed. Storage compartments are connected to one another through the flows that are resolved into outputs (quantities flowing out of one compartment) and inputs (quantities flowing into another). Most such models assume an external compartment representing the environment of the model (i.e., the rest of the world within which the system studied is embedded and where some of the flows originate and end [open systems]). In physical systems the transfer and storage of mass and energy is governed by two groups of physical laws: laws of conservation and laws of transport (also called process and flow laws). The logic is the same for social and ecological systems, although there are no corresponding rigorous laws of process or flow for these. Discrepancies in the degree of reliability of physical, biological, and social-science model submodels of environmental system models constitute one of the challenges of integrated systems modeling.

### 2.3.2 The Bottom-Up Approach: Complexity Theory

Mathematically, complex systems are characterized by multiple nonlinearities and feed-backs. Physically, these formal properties are associated with active exchanges of matter, energy, and information between systems and their environment. These characteristics of complex systems often lead to phenomena such as self-organization, chaos, adaptation, emergence, lock-in, bifurcating trajectories, and other types of surprising and unexpected behaviors unknown to classical science.

In the late twentieth century a new science of complexity developed around the study of these interesting types of dynamics. The new science of complexity and its hallmark phenomenon, chaos, produced a long series of articles and books that captured the imagination of scientists and laypeople alike (Waldrop, 1992). The Santa Fe Institute was founded in 1984 by an interdisciplinary group of high-flying scientists to foster theoretical research into complex systems. The complexity theory paradigm pioneered by the Santa Fe Institute has led to some very elegant mathematical work, to a large volume of interesting publications, and to the definition of a new field (artificial life), but has also been criticized as a rather hollow intellectual fad (Horgan, 1995). Just how relevant is complexity theory to environmental modeling?

There is little doubt that the environmental sciences deal with systems that are very complex by any definition. Unpredictable outcomes, major consequences of relatively small disturbances, unanticipated side effects, the fragility of large environmental systems or, conversely, the robustness of others that appear very delicate, all provide empirical evidence of complex systems behavior. Examples of self-organization abound in the natural world: grains of sand forming crescent-shaped dunes; birds flying in triangular formations; surface runoff converging to a few discrete channels. Phillips (1999) distinguishes eleven distinct kinds of self-organization that are relevant to landscapes, which may be further aggregated into two broad categories: those that tend to create order and regularity in the landscape, and those that result in greater diversity and differentiation. On the other hand, outside physics and the laboratory, there are not that many empirical examples of some of the 'sexier' aspects of complex systems behavior (in particular chaos) in environmental systems. For example, the wide fluctuations of animal populations in landscapes were earlier seen as a striking case of chaos in action. But as Zimmer (1999) reports, citing a well-known ecologist, "there is no unequivocal evidence for the existence of chaotic dynamics in any natural population." Indeed natural systems are often found to totter on the brink on chaos without quite plunging into it—a sign of nature's resiliency, if not wisdom, in the face of constant perturbations and assaults.

Very often the interesting dynamics of complex systems are the macroscale outcomes of simple interactions among microlevel system components. Neural networks are well-known examples of complex structures capable of highly organized behavior resulting from the parallel operation of large numbers of interconnected neurons. When the microlevel interactions are restricted to neighboring elements, the resulting system is intrinsically spatial and under appropriate conditions it will produce spatial organization at the macrolevel: molecules in a solution yielding regular patterns on the surface of the liquid, pigment activators and inhibitors creating stripes or dots on animal skins, segregated neighborhoods unexpectedly arising from simple preferences of individuals for particular levels of racial mix among their immediate neighbors. Cellular automata (CA) are a well-known class of complex systems models that embody that principle, generating macroscale spatial patterns in a gridded space through the parallel operation of microscale rules involving local neighbors (Wolfram 1984, 1986). CA models have become increasingly popular with environmental modelers because of their direct compatibility with raster GIS, their ability to make use of detailed spatial data, and their conceptual simplicity, which contrasts with the extreme diversity and complexity of the patterns they are capable of generating. CA models have proven equally suitable for the simulation of physical and social spatial processes and are thus particularly well suited for integrated environmental modeling (Clarke and Gaydos, 1998; White and Engelen, 1997).

Agent-based simulation is a more recent development in complexity theory that also involves generating complex macroscale behavior through modeling microscale interactions. Agents are interacting entities that may be sentient (people, animals, organizations, etc.) or nonsentient (any kind of physical or computational objects) and their interactions may or may not be in space. As in CA models (and unlike neural net models), interactions among agents can be based on arbitrarily complex rules. Unlike CA, where the interacting elements are localized cells, interacting agents may be mobile in space, thus providing a straightforward, intuitive way of modeling actually moving organisms or other objects. Applications are already appearing in the environmental modeling domain, such as the simulation of the impacts of visitors on wildland settings by Gimblett et al. (1999).

Critics of complexity theory point out that the wide variety of surprising behavior exhibited by mathematical and computational complex system models is rarely found in the empirical world. The criticism is primarily directed at the bottom up paradigm, which tends to produce especially skittish models, very sensitive to initial conditions and to small variations in the interaction rules. Real-world populations rarely crash and then explode chaotically, local interaction rules are rarely as simple as these models would have it, landscapes don't easily jump from one state of organization to the next, neither natural nor social systems tend to get into runaway positive feedback loops, and emergence seems to be just what the current model cannot explain. In other words, there is much more stability in the real world than complexity theory would have us think. Some authors contrast model complexity (or model chaos) with actual system complexity (Malanson, 1999), and Goldenfeld and Kadanoff (1999, p. 88) warn researchers "not to model bulldozers with quarks." The issue here is that there is a right level or levels of description for every phenomenon that must be judiciously chosen for the model to work. This is one of the many lessons that modelers are learning from complexity theory. There are other lessons: that nature can produce complex structures even in simple situations and can obey simple principles even in complex situations; that each complex system is different so that similar-looking systems may develop in very different ways (Goldenfeld and Kadanoff, 1999). Even though we may never enjoy the intellectual comfort of scientific laws when modeling complex environmental systems, complexity theory has made us more aware of the subtlety, diversity, and interconnectedness of the phenomena we study and has contributed many powerful concepts, modeling approaches, and techniques.

## 2.4 ENVIRONMENTAL MODELING AND GEOCOMPUTATION: MODELING WITH THE COMPUTER

Modeling without the computer is inconceivable in the environmental domain these days, except perhaps when developing abstract conceptual models. In many ways the computer has made environmental modeling possible, and not just because of computing in its original, number-crunching sense. In this data-dominated age, GIS has given modelers the possibility to handle arbitrarily detailed data at any spatial scale and is unquestionably a major driver behind the current blossoming of environmental analysis and modeling. Statistical, mathematical, graphics, and visualization software has complemented the increasingly sophisticated capabilities of commercial GIS, providing additional power to modelers and literally new ways of looking at data. But modeling with the computer means something more than that.

In Section 2.1 computation was mentioned as a separate stage that is reached late in the development of a model, after the real modeling work has been completed. This view is less and less tenable as the computer increasingly becomes an integral component of the modeling of complex systems and processes instead of a tool for handling data and solving equations. More and more, computer simulation replaces analytic model development as the systems modeled become larger, more integrated, and more complex. Step 2 of Smyth's (1998) sequence in particular, the formalization of the model's ontology, tends to be merged with step 3, the computational expression of the model. Formal languages such as Haskell and Gofer have been developed that are at the same time algebras and computer languages. Map algebra (Tomlin, 1990) and image algebra (Ritter, Wilson, and Davidson, 1990), which are formalisms operating on spatial elements as the variables, have been extended into dynamic spatial modeling languages such as PCRaster (Burrough, 1998). In addition, graphical, icon-based model construction environments are attracting increasing attention (Maxwell and Costanza, 1997) and in simple forms even begin to be available bundled with commercial GIS software (e.g., Environmental Systems Research Institute (ESRI)'s ModelBuilder available with ArcView Spatial Analyst 2.0). More generally, computer visualization has substituted the advanced pattern-recognizing capabilities of the human eye-brain system for much of the deductive work that characterizes traditional scientific analysis. All these developments aim at making environmental modeling easier, more widely available, and more of a collaborative enterprise than it could ever be under the traditional approach.

Just as the logical steps of model building are being compacted and merged, the distinction made in traditional model ontologies among entities, relationships, physics, and logic is becoming less and less sharp with the growing popularity of object-oriented languages and their recent extensions. In this approach objects are defined in terms of the possible and allowable behaviors of the corresponding real-world entities. Object-orientation, with its emphasis on representing quasiautonomous units within models and submodels, facilitates a bottom-up approach to complex system modeling and especially favors agentbased simulation. Indeed, object orientation has been loosely defined as "the software modeling and development discipline that make it easy to construct complex systems out of individual components" (Khoshafian, 1993, p. 6). Further developments currently underway in Internet-oriented computer languages are likely to affect modeling practices even further. XML (Extended Markup Language) and its derivatives, for example, are considered especially well suited for sharing and handling geographic data over the Internet, opening up the possibility for environmental models that are not just dynamic but are themselves dynamically evolving with the contributions of modelers "anywhere, any time" (Lowe, 2000).

Computation is thus taking a life of its own in environmental modeling, increasingly driving rather than just supporting modeling efforts. Some may view this as a regrettable development that threatens to distance environmental modeling from proper scientific practice—except that, of course, similar developments are taking place to a greater or lesser extent in practically all areas of science. Researchers in the spatial sciences are converging around the notion of *geocomputation*, a novel concept whose definition keeps evolving just as its subject matter does (Couclelis, 1998a; Longley et al., 1998). Originally no more than a diverse grab-bag of computational techniques, geocomputation eventually came to be identified with GIS practice for some people but now seems to have acquired an identity of its own as the convergence of computer science, geography, information science, mathematics, and statistics. Once its theoretical potential is fully unfolded, geocomputation may become synonymous with the computational theory of complex spatiotemporal processes (Couclelis, 1998b). If that's the case, environmental modeling will surely become geocomputation's proudest application field and may benefit immeasurably from a formal convergence of all the relevant technical fields.

## 2.5 CONCLUSION

These are heady days for environmental modeling. The talent and the money are there, we're drowning in environmental data, the growing wonders of geocomputation keep us on our toes, big problems await our wisdom for their solution, and the policy makers may for once be listening! There are things we already do very well, and there are things we will need to do even better. Summing up such a vast, multifaceted, fast-moving field is too much of a challenge for any single person to undertake, but here are some points to take home from the discussion in this chapter.

First, environmental modeling is primarily an applied field addressing problems that are directly or indirectly of considerable societal importance. Environmental models need to be policy relevant. This does not mean that we are expected to build "answer machines" but rather that our models must be good enough to be taken seriously in the policy process. Speaking from experience, King and Kraemer (1993, p. 356) list three roles a model must play in a policy context: A model should clarify the issues in a debate; it must be able to enforce a discipline of analysis and discourse among stakeholders; and it must provide an interesting form of advice, primarily in the form of what *not* to do—since no politician in his or her right mind will ever simply do what is suggested by a model. The properties a good policy model needs to have been known since the time of Lee (1973) and his 'requiem for large-scale models': transparency, robustness, reasonable data needs, appropriate spatiotemporal resolution, and the inclusion of enough key policy variables to allow for some of the really interesting policy questions to be explored.

Second, the fact that environmental modeling is primarily an applied field does not exonerate it from the need to be theoretically well grounded. It is all too easy to write and calibrate simulation models that do neat things on a computer screen but have an underlying ontology less plausible than that of a computer game. Since it is extremely unlikely that we will ever have a "theory of everything" in the environmental domain, we must get better at piecing together the wide array of partial theories (some highly respected and reliable, others very controversial and unreliable) contributed by the physical, biological, technological, and social subfields of the environmental sciences. Integrated environmental modeling is more than a matter of opting for integrated rather than coupled models: rather, it has to do with making sure that the patchwork of concepts, ontologies, approaches, laws, rules of thumb, degrees of confidence, and spatiotemporal structures that may come together within a single framework respects the strengths and weaknesses of each part and yields a whole that logically hangs together (Couclelis and Liu, 2000). Formal theories of modeling and someday soon perhaps a more mature science of geocomputation should provide the foundation for scientifically rigorous environmental models. These will help put back together the world we tried to understand by pulling it apart into smaller and smaller subdisciplinary pieces, thus fulfilling von Bertalanffy's vision of science unified though system theory.

Citing King and Kramer (1993, p. 353) once again (who paraphrase Dickens), "the Spirit of Modeling produces the shadows of what Might be, only. No one knows what Will be." How true. Illuminating the shadows of what might be is what environmental modeling is all about.

## 2.6 REFERENCES

Alberti, M. 1999. Modeling the Urban Ecosystem: A Conceptual Framework. *Environment and Planning B* 26, no. 4, 605–630.

von Bertalanffy, Ludwig 1968. *General System Theory*. Harmondsworth, Middlesex, England: Penguin Books Ltd.

Burrough, Peter A. 1998. Dynamic Modelling and Geocomputation. In *Geocomputation: A Primer*, P. Longley, S. Brooks, B. Macmillan, and R. McDonnell (eds.), pp. 165-191, New York: Wiley.

Casti, John L. 1989. *Alternative Realities Mathematical Models of Nature and Man*. New York: Wiley.

Casti, John L. 1997. *Would-Be Worlds: How Simulation Is Changing The Frontiers of Science*. New York: Wiley.

Clarke, Keith C., and Gaydos, Leonard J. 1998. Loose-Coupling a Cellular Automaton Model and GIS: Long-term Urban Growth Prediction for San Francisco and Washington/Baltimore. *International Journal of Geographical Information Science,* 12, no. 7, 699–714.

Couclelis, Helen. 1998a. Geocomputation and Space. *Environment and Planning B: Planning and Design*, 25th Anniversary Issue, 41–47.

Couclelis, Helen. 1998b. Geocomputation in context. In *GeoComputation: A Primer*, Paul Longley, Susan Brooks, Rachael McDonnell, and Bill Macmillan (eds.), pp. 17–29. New York: Wiley.

Couclelis, Helen, and Liu, Xiaohang 2000. The Geography of Time and Ignorance:Dynamics and Uncertainty in Integrated Urban-Environmental Process Models. In *Proceedings, 4th International Conference on Integrating GIS and Environmental Modeling* (GIS/EM4): Problems, Prospects and Research Needs. Banff, Canada, September 2–8, 2000 (CD-ROM).

Egenhofer, Max J., and Golledge, Reginald G. 1998. *Spatial and Temporal Reasoning in Geographic Information Systems*. New York: Oxford University Press.

Forrester, Jay W. 1975. *Collected Papers of Jay W. Forrester*. Cambridge, MA: Wright-Allen Press.

Gimblett, H. Randy, Richards, Merton T., and Itami, Robert. 1999. A Complex Systems Approach to Simulating Human Behaviour Using Synthetic Landscapes. *Complexity International* 6, January.

Goldenfeld, Nigel, and Kadanoff, Leo P. 1999. Simple Lessons from Complexity. *Science,* 284, no. 2, 87–89.

Horgan, John. 1995. From Complexity to Perplexity: Can Science Achieve a Unified Theory of Complex Systems? *Scientific American*, June, 104–109.

Khoshafian, S. 1993. *Object-Oriented Databases*. New York: Wiley.

King, John Leslie, and Kraemer, Kenneth L. 1993. Models, Facts, and the Policy Process: the Political Ecology of Estimated Truth. In *Environmental Modeling with GIS,* Michael F. Goodchild, Bradley O. Parks, and Louis T. Steyaert (eds.), pp. 353–360. New York: Oxford University Press.

Lee, Douglass B., Jr. 1973. Requiem for Large-Scale Models. *AIP Journal*, May, 163–177.

Longley, Paul A., Brooks, Sue M., McDonnell, Rachael, and Macmillan, Bill. 1998. *Geocomputation: A Primer*. Chichester, England: Wiley.

Lowe, Jonathan L. 2000. What the XML are Scalable Vector Graphics? *Geo Info Systems*, May, 46–50.

Malanson, George P. 1999. Considering complexity. *Annals of the Association of American Geographers* 89, no. 4: 746–753.

Maxwell, T., and Costanza, R. 1997. A Language for Modular Spatio-Temporal Simulation. *Ecological Modeling*, 103, no. 2-3, 105–113.

Odum, Howard T., and Odum, Elisabeth C. 2000. *Modeling for All Scales: An Introduction to System Simulation*. New York: Academic Press.

Phillips, Jonathan D. 1999. Divergence, convergence, and self-organization in landscapes. *Annals of the Association of American Geographers* 89, no. 3: 466–488.

Ritter, G. X., Wilson, J. N., and Davison, J. L. 1990. Image Algebra: An Overview. *Computer Vision, Graphics and Image Processing,* 49, 297–331.

Smyth, C. Stephen. 1998. A Representational Framework for Geographic Modeling. In *Spatial and Temporal Reasoning in Geographic Information Systems,* Max J. Egenhofer and Reginald G. Golledge (eds.), pp. 191–213. New York: Oxford University Press.

Tomlin, C. D. 1990. *Geographic Information Systems and Cartographic Modeling*. Englewood Cliffs, NJ: Prentice Hall.

Waldrop, Mitchell M. 1992. *Complexity: The Emerging Science at the Edge of Order and Chaos.* New York: Simon & Schuster.

White, R., and Engelen, G. 1997. Cellular Automata as the Basis of Integrated Dynamic Regional Modelling. *Environment and Planning B: Planning and Design*, 24, no. 2, 235–246.

White, Roger, and Engelen, Guy 1999. High Resolution Integrated Modelling of the Spatial Dynamics of Urban and Regional Systems. Paper presented at *GeoComputation 99*, Mary Washington College, VA.

Wilson, A. G. 1981. *Geography and the Environment Systems Analytical Methods.* Chichester, England: Wiley.

Wolfram, Stephen. 1984. Cellular Automata as Models of Complexity. *Nature,* 311, no. 4, 419–424.

Wolfram, Stephen. 1986. *Theory and Application of Cellular Automata.* Singapore: World Scientific Publishing Co.

Worboys, Michael F. 1995. *GIS: A Computing Perspective.* London: Taylor & Francis.

Zeigler, Bernard P., Praehofer, Herbert, and Kim, Tag Gon. 2000. *Theory of Modeling and Simulation: Integrating Discrete Event and Continuous Complex Dynamic Systems* (2nd ed.). New York: Academic Press.

Zimmer, Carl. 1999. Life after chaos. *Science,* 284 (5411), no. 2, 83–86.

## 2.7 ABOUT THE CHAPTER AUTHOR

Helen Couclelis is Professor in the Department of Geography at the University of California, Santa Barbara. She has published numerous articles on urban and regional modeling and planning, geographic information science, spatial cognition, and the methodology and philosophy of the geographic sciences. Prior to her appointment at UC Santa Barbara she spent several years as planning and policy consultant to the Greek Government. She has been associated with the National Center for Geographic Information and Analysis (NCGIA) since its inception, formerly as Associate Director, and currently as a member of the Executuve Committee of NCGIA's Center for Spatially Integrated Social Science (CSISS). She is co-editor of the journal *Environment and Planning B: Planning and Design.*

# 3

# Spatial Decision Support Systems and Environmental Modeling: An Application Approach

## 3.0 SUMMARY

A spatial decision support system (SDSS) is a computer-based system designed to assist decision making. Typically, such a system will include spatial data relevant to the decisions, analytic tools to process the data in ways meaningful for decision makers, and output or display functions. Thus, an SDSS has considerable overlap with the functionality of a geographic information system (GIS). According to the National Center of Geographic Information and Analysis (NCGIA) Core Curriculum in Geographical Information Systems, Unit 127 (Malczewski, 1997), an SDSS is an "interactive, computer-based system designed to *support* a user or group of users in achieving a higher *effectiveness* of decision making while solving a *semi-structured spatial decision problem*" (emphasis as per the original document).

There are many examples of SDSSs for specific decisions in the environmental domain, particularly in the areas of crop, livestock, flood, and forest management. However, many decisions in environmental management frequently require consensus building among diverse stakeholders. Furthermore, the management options being addressed often are difficult to quantify both because of their complexity and because input data have high

51

levels of uncertainty. This situation suggests a need for SDSSs that are easily accessed by diverse users who require flexible tools that can address a wide range of issues. This objective, SDSS for a wide range of users, guides this chapter's organization.

Our vision of the necessary SDSS characteristics is significantly influenced by our effort to deliver an agricultural and natural resource–oriented SDSS for wide public deployment in Africa. The design audience is typically nontechnical and has limited means to interact with the developers.

An object-oriented software development approach, including extensive use of an object model, was used because of its flexibility and robustness. Not only does this enable use of third-party tools (controls and objects), but the objects, controls, and object model are available for third parties to independently customize. We call our multiuse object-oriented SDSS the Almanac Characterization Tool or ACT (Corbett et al., 1999).

## 3.1 INTRODUCTION

Decision makers in environmental fields face the difficult challenge of anticipating the potential biophysical and socioeconomic impacts of management and policy interventions over regions that may vary dramatically in terms of climate, soils, topography, land use, and other factors. In the past, potential impacts were often presented through static maps of variables such as disease risk, crop suitability, or erosion susceptibility. The maps were produced using costly and time-consuming procedures that hid many assumptions from the decision makers and, just as important, from people affected by the decisions, including farmers, foresters, conservation specialists, and the general public.

Leung (1997) addressed a host of conceptual, theoretical, systems development, and applications perspectives on intelligent spatial decision support systems. This monograph covers a range of issues, from symbolic approaches to spatial knowledge to fuzzy logic, management of uncertainty, neural networks, spatial inference, and data models and data structures. From these more theoretical constructs evolve tools to minimize and report error propagated in an SDSS, handle integration of data from various scales, and construct more optimal query and visualization tools.

Applications of these theoretical constructs are enabled by advances in GIS and environmental modeling. These tools offer a flexible capability to test scenarios and understanding of how environmental factors such as precipitation, topography, and land use interact. Power (1997) defined a decision support system (DSS) as "an interactive computer-based system intended to help managers make decisions." Thus, a GIS can be considered a decision support system. However, a more functional view is to recognize an SDSS as an information system that combines selected functions of a GIS with software tools specifically designed to support decision making. Daniel (1992) noted that many of the distinctions between GIS and SDSS are derived from the differences in their target audiences. While a GIS is typically operated by technicians focusing on details of database and tool development, the user of an SDSS seeks only a handful of intuitive operations that yield quick and effective answers.

Many decision support tools focus on a narrow range of tasks and are limited to the data required for the relevant decisions. This approach works well if decision rules are well established and uncertainty in data is low—and the users are well informed. For many decisions relating to the natural environment, however, quantitative rules are lacking or include many approximations, and data are inexact due largely to difficulty of measurement over large regions and to uncertainties introduced by temporal variation in climate. Furthermore, many decisions are negotiated among stakeholders with diverse, if not opposing, interests. Thus, some people prefer the term "discussion support system" to emphasize that the tools should be used to develop a consensus through full participation of stakeholders (Hammer et al., 2000).

These considerations suggest a need for SDSSs that are readily accessed by users with varying objectives and expertise. Our approach in this chapter is to present an SDSS for agriculture, designed specifically for resource managers and researchers working—and living—in Africa, as one example of an SDSS (the ACT, as implemented in the United States Agency for International Development [USAID]-supported Africa Country Almanac Series; see Corbett et al., 1999). The ACT is distributed with a large set of foundation data on climate, topography, infrastructure, land use, and other variables. Many of these data sets were produced through environmental modeling, and a common theme is the conversion of point, vector, or polygon data to continuous surfaces that are more easily used in decision making. The core tools of this object-oriented SDSS include basic spatial searching and querying tools, as well as more complex functions such as the site similarity tool, which can be viewed as a simple but flexible environmental model. Furthermore, the ACT design exploits a software "shell" that permits incorporation of more complex models.

This chapter outlines how flexible and user-friendly tools for spatial analysis can be linked with diverse sets of data to provide decision makers with SDSSs that greatly facilitate a dynamic approach to decision making for agriculture and related environmental issues. Environmental modeling (from creation of climate surfaces to simple models of the growing season, to sophisticated simulation models) becomes an integral facet of an SDSS as a source both of data and of tools for the decision support process. Recent technological advances in software development methodologies are discussed because they offer an avenue for greatly facilitating objective-driven targeting and classification of the spatial environment.

## 3.2 A FLEXIBLE, MULTIUSE SPATIAL DECISION SUPPORT SYSTEM

Robert Fri, writing in the foreword to *Tools to Aid in Environmental Decision Making* (Dale and English, 1999), described one of the target audiences of that volume as the people who provide the "link between analytic and the ultimate decision maker." He emphasized that the tools will not get used to improve the process unless these people use them. Developers of SDSS should seek to create systems that seamlessly link data and tools to support decision making. Who will actually run the SDSS will set many design criteria, as there are tremendous design implication differences between SDSS for the GIS technicians themselves versus SDSS for wide, public dispersal. The SDSS we envision encapsulates the dynamism that necessarily evolved in building an SDSS for wide public deployment.

In one context, an SDSS for environmental research is a tool for data synthesis oriented toward spatial data on the environment. In *Data Smog: Surviving the Information Glut*, Shenk (1997) described the too-common situation of data and stimulation overflow and resulting nondecisions due to overload: Information is not necessarily power. For effective decision support, data on a given location must be summarized and delivered to stakeholders in a timely fashion.

Spatial data and derived information exist in vast quantities, and software tools offer immense power for utilizing these resources. However, we have only just begun to assemble these elements into sufficiently flexible SDSSs. Many application specific SDSSs exist (e.g., Arentze et al., 1996; Carver, 1991; Lal et al., 1991; Luitjen et al., 2000; Stuth and Lyons, 1993;). Although effective within their discipline, these systems address a limited suite of decisions. Building even a fraction of possible capabilities into a flexible SDSS for environmental issues requires careful design, planning, and cooperation in data and tool sets, as well as presentation to decision makers.

Our experiences in creating and delivering an agricultural and natural resource management SDSS for countries in Africa allow us to emphasize an underrepresented aspect of SDSS implementation: If the goal is adoption and use of an SDSS and the target is, as Daniel (1992) emphasizes, users who want quick and effective tools to deliver "answers," then a focus on simplifying complex environmental assessments is paramount. The design of the SDSS user interface is crucial. The science of environmental modeling, assessment, and characterization requires tools to organize, synthesize, and manage. There is no limit to possible forms and functions, but effective SDSS for nontechnical users requires a streamlined, industry standard "client" for the target users. The client serves as the interface among the data, models, and tools and the user, or, as Malczewski (1997) describes in his DDM paradigm (dialog, data, and modeling) in the principles of SDSS section of Unit 127 of the NCGIA core curriculum, our "client" is equivalent to the "dialog generation and management system."

If the SDSS is designed as a powerful extension to a GIS to be operated by technical "insiders," then less emphasis should be placed on the client. If, however, the goal of the SDSS is adoption and use by a broad, more public and less technical audience, then the value of an open architecture increases dramatically and there must be specific emphasis on client design. This chapter focuses on the latter case.

### 3.2.1 Enabling Technologies

To deliver object-oriented, open architecture SDSS to a wide, nontechnical target audience, we exploited recent technological innovations to link the application architecture intimately to the data structure via an object model (Figure 3.1). This enhances the theoretical construct presented by Malczewski (1997) in that the three-tiered structure (data base management system, model base management system, dialog generation and management system) is expanded through the use of objects and object models to create a more flexible and less hierarchical architecture. An object model provides systematic mechanisms to deliver an open, flexible design to SDSS users. Individual components are created as controls, each of which has properties and methods. Through these properties and methods and with access to the object model, development of new capabilities within an SDSS can take place completely independent of the original programming team. For example, the object that takes position (e.g., "latitude and longitude") and provides access

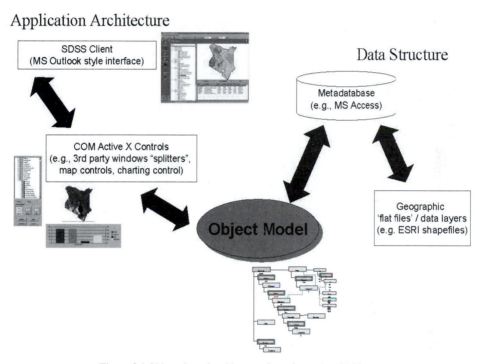

**Figure 3.1** Object-riented architecture for software in a SDSS.

to any spatial database in the metadatabase is accessible to the user of the SDSS, who can add a new data layer and a new environmental model and connect the spatial data to the model through the properties and methods of the existing object. In a similar vein, 'exposing' the object model and the properties and methods of specific controls (objects) opens the possibility to build and improve an SDSS continually without the requirement of direct collaboration with the original developers. Intellectual property is protected because a third party can access the properties and methods but cannot change them. Furthermore, because the application structure exploits controls as objects, third parties could simply use a selection of those objects and create wholly new applications, again building and improving upon previous work.

This object orientated programming (OOP) approach is used for the ACT, including

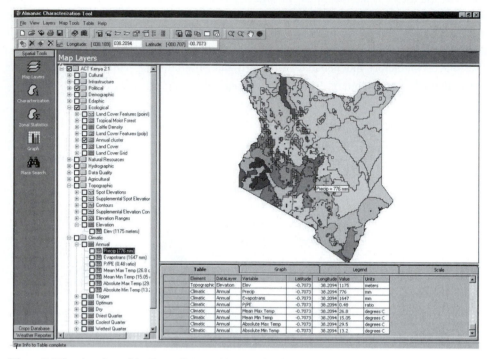

**Figure 3.2** User interface of the Kenya Country Almanac showing tool bars, task bars and division into split windows. Mapped data are mean annual precipitation for each climatic cluster (Wards minimum variance). Data in table reflect information from selected site.

an object model. This improves software requirements engineering (Graham, 1998) and is implemented with component object model (COM) technology. OOP also simplifies use of multiple programming languages and interconnection of modules (Williams and Kindel, 1994). Interconnectability permits third-party software developers to introduce specific tools, including models, without having to release their source code. Tools can communicate with one another so long as they use the same object model.

For the client, and within the broad concept of user friendliness, one can identify components such as consistency, forgiveness, simplicity, feedback, and robustness (Microsoft Corporations, 1995). To provide consistency and simplicity, the design of the ACT user interface follows the tool bar and split-window layout found in many World Wide Web sites and used in software packages such as Microsoft Outlook (Figure 3.2). Icons displayed on tool bars follow generic standards as much as possible. Feedback is provided through the status bar, which displays progress in loading files or processing queries. This example of the ACT client in Figure 3.2 serves to demonstrate a key design goal of this particular SDSS: flexibility. The design and target tools for ACT met certain goals. However, any particular user in a specific geographical area could have quite specific needs both in data and tools. The open architecture permits groups to add separate modules linked via the object model. On the left side of Figure 3.2, there are three modules in this particular instance of the client: Spatial Tools, Crops Database, and the Weather

Reporter. Each of these modules is completely independent; they access different databases and provide different tools, but they share a common interface and are linked both spatially (for example, a meteorological station in the weather reporter can be selected in the Spatial Tools by geographic location) and aspatially (a variety in the crop database can selected for yield and maturity date—a theme).

In a complex software system such as the ACT, robustness is a major concern. Besides following recommended practices for iterative design, coding, and testing of software (e.g., McDonald, 1996), ACT development relies heavily on commercial software components for routines such as map display, graphing, and control of printing. This allows in-house programming to focus on simplification of complex environmental interactions (the environmental modeling) while tapping into the robust programming capabilities of specialty companies.

Rapid distribution of data is required for many applications of SDSS used for environmental issues. The Internet offers an obvious vehicle for rapid diffusion of data sets smaller than one or two megabytes. However, for users in remote locations or requiring larger volumes of spatial data, CD-ROM remains the best option for distribution of data and tools. A third option emerges through the COM compliance of each control: Web interfaces to the object model can be written either as COM compliant controls served up through custom interfaces or written in HTML (hypertext markup language) or Java.

### 3.2.2 Foundation Database

Earlier in this chapter we made a distinction between GIS and SDSS primarily on the basis of the user's profile. For an SDSS such as the ACT, users cannot be expected to assemble the baseline data necessary to characterize agroecosystems (these are the activities of the GIS technicians). Moreover, up-to-date and reliable information about our environment is crucial for enabling stakeholders to reach a consensus and take appropriate actions. Scholes et al. (1994) argued that climate is a first-order determinant of ecosystem character, with edaphic factors second, followed by human intervention and other natural disturbances. This hierarchy of ecosystem characteristics is captured in the concept of foundation data used in the ACT, which includes climate, topography, soils, land use, as well as census and other population and socioeconomic variables (Table 3.1). One goal of the ACT is to create significant efficiencies in the decision-making process by paying the transaction costs of assembling generally accepted databases only once by providing these data preloaded in the SDSS.

The foundation data of an SDSS can be thought of as spatial data about which there is minimal controversy—they provide an accepted baseline and should move stakeholders closer toward a goal of equity in decision making (Corbett and Dyke, 1999). Resources freed from data compilation can be better invested in improving data quality or obtaining additional data.

Much of the foundation data is produced with environmental models (Table 3.1). Methods such as splining (Hutchinson, 1995) and geostatistics are used to create climate

**Table 3.1**. Selected data included in the Kenya Country Almanac (Corbett et al., 1999) as an example of foundation data for a spatial decision support system. (Complete descriptions of data sets, including sources, are given as metadata in the Almanac).

| Variables | Description | Use of Environmental Modeling |
|---|---|---|
| Monthly surfaces | Precipitation, potential evapotranspirations, and maximum and minimum temperature on a 5-km grid | Interpolation from climate normals |
| Growing season surfaces | Various mathematical relationships between precipitation and potential evapotranspiration | Identifying onset and length of growing season from monthly surfaces |
| Elevation | Mean elevation on a 1-km grid | Modeled from point data and locations of drainages and watershed boundaries |
| Human population density | Density on a 1-km grid for 1960, 1970, 1980, and 1990 | Modeled from census data that were normalized to decades and distributed spatially with a model accounting for effects of access |
| Infrastructure | Locations of roads, airports, railroads, utility lines | Calculation of indices of spatial access based on location and quality of transportation networks |
| Soils | FAO soil taxonomy | Calculation of typical water holding capacity from texture information. |
| Land cover | USGS Anderson level II | Modeled from AVHRR data |

surfaces by interpolating monthly values of normals from individual stations. The monthly data are modeled further to describe growing seasons based on water availability or temperature constraints (e.g., Corbett and O'Brien, 1997). Other examples of modeled data include human population density estimated from census data (Deichmann, 1994) and land cover derived from satellite imagery (USGS EROS Data Center, 2000). A common objective in many modeling efforts is to convert point-, vector-, or polygon-based data into surfaces of continuous data that are more readily used in decision support tools.

### 3.2.3 Primary Analytical Functions—The Tool Set

A GIS is often defined as having four major capabilities: data entry, data management, analysis, and presentation of outputs. A multiuse SDSS differs from a GIS in that data entry is deemphasized, analytic tools are simplified but given powerful goal-specific capabilities, and presentation of outputs is restricted to the basic needs of decision makers. This view assumes that users of an SDSS will be able to access full GIS capabilities as needed.

Given the specification that a multiuse SDSS is packaged with foundation data, data entry capabilities are an intermediate priority. Users should be able to import new sets of spatial data in standard GIS formats, including associated metadata. Since many datasets are linked with established points or polygons (e.g., monitoring sites or municipal boundaries), an SDSS should also permit updating or appending attribute data of features already in the SDSS. Furthermore, since GPS permits researchers and managers to collect high-quality spatial data at point locations, tools for importing data from GPS units and similar sources are also justified.

Analytic tools form the core of any DSS. What distinguishes a multiuse SDSS for environmental issues (as opposed to a highly tailored SDSS designed for technicians to run to assist with the management of, say, a particular watershed) is the emphasis on identifying a suite of analytic tools that are relevant and powerful, yet easily used. For example, the core tools of the ACT permit.

1. Identifying all features in the database related to a given site.
2. Identifying all areas "similar" to a given site based on user-defined criteria of similarity
    (e.g., all areas that have a mean growing season temperature within plus or minus 10 percent of the temperature of the reference site).
3. Defining zones based on user defined criteria (e.g., all areas with greater than 300 mm of cool season precipitation and average district maize yields greater than 3000 kg/ha).
4. Calculating summary statistics for a zone based on user-selected attributes
5. Producing maps, graphs, and charts of attributes of selected features (both temporal and spatial).

Beyond these basic functions, there is scope for adding a wide range of more specialized tools, many of which are forms of environmental modeling. More sophisticated generic procedures for synthesizing data might include principal component analysis and clustering. Various approaches for allowing users to interpolate data from point locations to surfaces could be provided.

In agriculture and natural resource management, many groups are working on systems that link process-based models to GIS (Collis and Corbett, 1999; Hartkamp et al., 1999). Such models find applications ranging from guiding management decisions at the field level to providing input for development strategies at the country level. Recognizing the diversity of applications and the need for detailed inputs, the International Consortium for Agricultural Systems Applications (ICASA) promotes standards for data interchange and software modules for crop modeling (Bouma and Jones, 2000; Jones, 1999). ICASA emphasizes a "tool-kit" concept whereby users can selectively install models and associated tools for handling model inputs and outputs.

### 3.2.4 Auxiliary Databases, Documentation, and Tools

There often are large sets of auxiliary data and information that are too loosely linked to locations to be incorporated into a GIS but that still hold interest for decision makers using an SDSS. Examples include descriptions of ongoing conservation or development projects, sets of daily weather data, and detailed listings of results of agricultural trials.

An object-oriented design enables access to a wide range of databases and documents in electronic format. For example, a database of maize varietal trials for all of Africa is provided with each county database accessed by the ACT. These data, coupled with a climatic characterization of each site where a trial was conducted, allow decision makers in agricultural research and extension and in disaster mitigation to identify varieties that might be adapted to target regions (e.g., if drought is in the seasonal forecast, what varieties might be better adapted to a specific production area?). Building on this type of database, the International Crop Information System (ICIS) is being developed using the ACT software shell and has the objective of providing data ranging from genetics to farmer evaluations (ICIS Project, 1999).

The open structure of the ACT shell also allows incorporation of decision support or other tools that are not directly linked to spatial data. For example, the ACT's Weather Reporter extracts data from World Meteorological Organization records and provides graphical views of the data and summary analyses using statistics calculated according to user specifications. CIMMYT (International Maize and Wheat Improvement Center) has built a farming systems research database. These modules as stand-alone products have an inherent value. Embedding the modules in an SDSS enables a whole suite of potentially enlightening analytical steps from identifying a spatial pattern relating farm resources to markets, climate, soils, and seasonal yield prediction models to identification of spatial patterns in temporal variation in the meteorological database. Finally, the Impact Assessment Group at Texas A&M University recently tested incorporation of the Development and Health Survey (Macro International, 2000) in the ACT SDSS for Kenya. We hypothesize that there will be tremendous insight gained when human health surveys with detailed child nutritional measures are linked with agricultural technology and ecological characterization.

# 3.3 APPLICATIONS OF SDSS USING ENVIRONMENTAL MODELS

### 3.3.1 Sample Stratification and the KARI/CIMMYT Maize Database Project

Agroecological classification schemes such as Sombroek et al. (1982) and Jaetzold and Schmidt (1983) are widely used by Kenyan agricultural institutions. The systems identify broad zones of similarity for crop adaptation, which are used in many decisions in research and extension. Experience shows, however, that these schemes perform poorly for targeting cultivars of crops. The zones are too general, and it is difficult to rework the base data for specific objectives.

Hassan (1998) noted the potential for intensifying maize production in the well-watered highlands of Kenya. This raised the question of whether the Kenya Agricultural Research Institute (KARI) was producing the appropriate cultivars for farmers and whether existing materials were being targeted to priority environments. To address these questions, the Kenya Maize Data Base Project (MDBP) assembled data on climate, topography, soils, infrastructure, population density (from the census of Kenya), and maize distribution (interpreted from aerial photos). Geo-referenced farmer- and village- level surveys were conducted to characterize farmers' practices and resources. The sample frame was defined using climate surfaces to identify approximate zones matching the categories of maize germplasm: lowland, midaltitude, highland, and dry areas. These data were then combined with an air-photo interpreted maize density database and a human population density database.

KARI had released improved varieties for each of the four zones. However, the survey identified a divergence between farmer practices and existing zone-based targeting of varieties. In the midaltitude zone, farmers often planted late-maturing maize hybrids that were developed for the highlands. A flexible SDSS allowed for iterative improvements in the delimitation of zones through consultation with maize breeders. The result was identification of a transitional zone that fell between the midaltitude and highland zones, and KARI subsequently established a breeding program at Kakamega to represent this zone.

The outputs of the Kenya MDBP confirmed the usefulness of an SDSS and of an agroclimatological approach to planning and evaluation of agricultural research. The experience highlighted the utility of a flexible system allowing open discussion with key stakeholders, in this case KARI maize breeders and the farmers. Environmental modeling was used to produce the initial climate surfaces and the derived seasonal climates.

### 3.3.2 Technology Targeting and Extrapolation Domains

Many decisions on collection and use of crop genetic resources are influenced by the assumption that the origin of an accession is significant because the material has evolved adaptations to that locality. Thus, given the latitudes and longitudes where accessions were collected along with spatial databases on climate, soils, topography, or other factors of interest, it should be possible to characterize accessions in terms of expected adaptation.

To better evaluate strategies for increasing wheat production in Ethiopia, we examined relations between wheat germplasm collection sites and climatic conditions as indicated by interpolated climate surfaces. For a set of 180 accessions, it appeared that most wheat materials in Ethiopia were adapted to conditions where, during the wettest quarter, total precipitation is between 300 and 900 mm and the mean minimum temperature is between 6 and 14°C. Two accessions whose reported collection sites correspond to temperatures over 17°C were thought to be mislocated, confirming the utility of these exercises for identifying problem data.

These ranges were further examined in consultation with wheat researchers using the Ethiopian Country Almanac, an implementation of the ACT. Although drought is a major constraint to wheat production, the results suggested that high nighttime temperature is a dominant factor that limits the potential area for wheat production. Thus, depending on whether researchers seek to expand area or increase productivity per se, different plant breeding strategies should be pursued.

In the consultation process, the question arose as to whether the set of collections provided a representative sample of Ethiopian wheat environments. Taking advantage of the Almanac foundation data, it was shown easily that many collections originated near major road networks. This highlighted a series of questions concerning strategies for germplasm collecting, which would affect decisions on subsequent germplasm collection and maintenance activities. Major efficiencies can be realized in agricultural and natural resource institutions through incorporation of a flexible SDSS. Resources can be targeted much more efficiently (spatial sample frames), and project results stored in a comprehensive database linked, at a minimum, by location.

### 3.3.3 Priority Setting and Impact Assessment

There is growing concern over the efficiency of investments in agricultural research in developed and developing countries. Key issues include whether benefits accrue to consumers, farmers, or large corporations and whether emphasis on short-term productivity gains causes long-term degradation to the natural resource base. Texas A&M University participates in a project using crop simulation models, GIS, and a suite of spatially adapted methods to assess the economic, environmental, and sociological impact of new technologies. Data and expert opinions are assembled through collaboration with national and regional partners. GIS is used to establish a spatial framework and to organize and analyze spatially explicit data. The simulation models allow decisionmakers to evaluate production and environmental impacts of new technology, while economic sector and farm-level models permit assessing possible socioeconomic consequences. Environmental consequences of technologies adopted by farmers in developing countries are estimated at field, farm, and watershed levels. Methods are being evaluated to estimate the adaptation of new technology to geographically similar zones in areas that were both contiguous and noncontiguous to the locations where the technology was developed. Case studies of research sponsored by USAID serve as platforms for developing and evaluating the methodology.

To date, the project has demonstrated that work in priority setting and impact assessment (in agriculture and natural resource management) benefits greatly from ready access to foundation data and from a flexible, object-oriented SDSS. These serve for sample stratification and site selection, for providing inputs to models, and for ensuring that a quantitative baseline will be available for follow-up studies. Further development of these impact assessment tools is being continued under the Global Project of the SANREM CRSP (SANREM CRSP, 2000).

## 3.4 DISCUSSION

While a GIS can be viewed as a spatial decision support tool, most decision makers will not have the expertise to use a GIS nor the time and resources required to assemble key data. Furthermore, some decisions may require outputs from environmental models that are too complex to run within a GIS. Thus, a spatial decision support system includes elements of a GIS as well as environmental models, and SDSSs are coupled with foundation data that allow decision makers to focus on the problem at hand and not be diverted into data-gathering exercises. Within diverse types of SDSSs, a case can be made that many decisions affecting the environment require participation of multiple stakeholders. Thus there is a particular need for SDSSs that facilitate discussion and minimize debate over underlying assumptions or objectives.

The ACT is a representative multiuse SDSS that draws from a plethora of international efforts at database construction. For example, the climate data originated from individual scientists'—and governments'—efforts to collect and organize meteorological data. These data were then combined to the create data sets of long-term climatic normals. These long-term normals were processed by yet another group (e.g., the Center for Resource and Environmental Studies, Australia National University, for Africa) to produce the climatic surfaces. We used these surfaces (and created other surfaces, following similar methods) and modeled the growing season and extracted characteristics of the season. Other international organizations contributed data on soils, socioeconomic conditions, as well as terrain and demography.

Although designed for agricultural and natural resource management applications, the structure of the ACT SDSS databases and the provision of flexible tools to access, visualize, and explore the data allow for additional benefits to accrue. For example, Lindsay et al. (1998), working on malaria in East Africa, and V. Pietra (personal communication, 2000), in Madagascar, used the foundation databases to examine possible impacts of global climate change on the distribution of malaria and malaria vectors. The transaction costs, not to mention the specialty GIS skills of data generation and assembly, would probably have prevented these health researchers from using spatial tools and data. When properly designed and implemented, the reward from paying the base transaction cost only once (of deploying an SDSS for public use) can continue to grow across disciplines and geographies.

Environmental modeling has two major roles to play in SDSS. The first is in processing of data to create spatial data sets that are more easily accessible to decision makers. In many cases, this involves interpolating sparse or discontinuous data to create more complete data sets. Often use of auxiliary data, such as elevation or transportation access, can significantly improve the interpolations but, in turn, require a prior iteration of modeling to obtain the auxiliary data set.

The second role of environmental modeling is as a decision tool per se. This may range from relatively simple static models to process-based simulations. There still is much discussion about strategies for linking models to SDSSs (Hartkamp et al., 1999). Data quality and error propagation are serious concerns. In some situations, such as highly variable soil conditions, sensitivity analyses can be used to account for uncertainties, producing multiple sets of results for decision makers to review. However, this can create problems both in terms of computation time required and of subsequent interpretation of results.

## 3.5 CONCLUSIONS AND LESSONS LEARNED

Spatial decision support tools require environmental models both as sources of data and as key tools to guide decision making. For many environmental concerns, it seems unlikely that a single model will resolve all possible concerns of decision makers. Often, requirements of multiple stakeholders must be considered, and source data will have sufficient error or uncertainty to permit diverse interpretations. In this context, key objectives of a SDSS should be to provide a foundation that places stakeholders in an equitable position and to offer a flexible set of tools that permit viewing and analyzing the data from multiple perspectives. A corollary objective might also be that the SDSS have an open architecture capable of refinement by interested parties independent of the originators.

We see a large gap between the design and implementation of an SDSS to be run by a the technical team that developed it and the design and implementation of an SDSS for delivery to the potentially nontechnical public. An object model offers a technically sound architecture to support SDSS targeted for the spectrum of potential users and provides a critical design pillar if the target audience is nontechnical. An open structure enables con-truction of controls that become building blocks for continued enhancement of an SDSS, whether by the original design group or by others. An SDSS requires delivery of at least some foundation data, as the expertise needed to create and assemble these data is high and the transaction costs of doing so steep, but the opportunity to pay this cost only once is of enormous benefit to multiple users. Besides providing foundation data and a core tool set, key features of the SDSS include links to auxiliary databases and documentation, and the ability to incorporate additional, application-specific tools, including various types of highly specific, tailored, environmental models.

## 3.6 REFERENCES

Arentze, A. T., Borgers, A. W. J., and Timmermans, H. J. P. 1996. Design of a View-Based DSS for Location Planning. *International Journal of Geographical Information Systems*, 10, no.2, 219–236.

Bouma, J., and Jones, J. W. 2000. An International Collaborative Network for Agricultural Systems Applications (ICASA). *Agricultural Systems* (in press).

Carver, S. J. 1991. Integrating Multi-Criteria Evaluations with Geographical Information Systems. *International Journal of Geographical Systems*, 5, no. 3, 321–339.

Collis, S. N. and Corbett, J. D. 1999. A Methodology for Linking Spatially Interpolated Climate Surfaces with Crop Growth Simulation Models. *Agricultural Ecosystems and the Environment* (In press).

Corbett, J. D., Collis, S. N., Bush, B. R. Muchugu, E. I., O'Brien, R. F., Jeske, R. Q., Burton, R. A., Martinez, R. A., Stone, C. M., White, J. W., & Hodson, D. P. 1999. *East African Country Almanacs. A Resource Base for Characterizing the Agricultural, Natural, and Human Environments of Kenya, Ethiopia, Uganda, and Tanzania*. CD-ROM. Temple, TX: Blackland Research Center.

Corbett, J. D., and Dyke, P. T. 1999. Institutional Adoption of Spatial Analytical Proce-dures: Where Is the Bottleneck? In S. L. Greene and L. Guarino (eds.), *Linking Genetics and Geography: Emerging Strategies for Managing Crop Biodiversity*, CSSA Special Publication no. 27, pp. 101–110. Madison, WI: American Society of Agronomy and Crop Science Society of America.

Corbett, J. D., and O'Brien, R. F. 1997. *The Spatial Characterization Tool—Africa* (vol. 1.0). Texas Agricultural Experiment Station, Texas A&M University, Blackland Research Center Report No. 97-03, December 1997, CD-ROM. Temple, TX: Black-land Research Center.

Dale, V. H. and English, M. R. ( eds.) 1999. *Tools to Aid Environmental Decision Making*. New York: Springer–Verlag.

Daniel, L. 1992. SDSS for Location Planning, or The Seat of the Pants is Out. *GeoInfo Systems*, 2(9): 16–24, December 1992.

Deichmann, U. 1994. *A Medium Resolution Population Database for Africa, Database Documentation and Digital Database.* Santa Barbara, CA: National Center for Geographic Information and Analysis, University of California, Santa Barbara.

FAO and IGAD. 1995. *Crop Production System Zones of the IGAD Sub-Region.* Rome, Italy: FAO.

Graham, I. 1998. *Requirements Engineering and Rapid Development: A Rigorous, Object-Oriented Approach.* Harlow, England: Addison–Wesley Longman Ltd.

Hammer, G. L., Hansen, J. W., Phillips, J. G. Mjelde, J. W., Hill, H., Love, A. and Pottgieteri, A. 2001. Advances in Application of Climate Prediction in Agriculture. *Agricultural Systems* (in press).

Hartkamp, A. D., White, J. W., and Hoogenboom, G. 1999. Interfacing Geographic Information Systems with Agronomic Modeling: A Review. *Agronomy Journal,* 91, 761–772.

Hassan, R. M. (ed.). 1998. *Maize Technology Development and Transfer: A GIS Application for Research Planning in Kenya.* Wallingford, England: CAB International.

Hutchinson, M. F. 1995. Interpolating mean rainfall using thin plate smoothing splines. *International Journal of Geographical Information Systems,* 106, 211–232.

ICIS Project 1999. *The International Crop Information System.* http://www.cgiar.org/icis/ (30 May 2000).

Jaetzold, R., and Schmidt, H. 1983. *Farm Management Handbook of Kenya: Natural Conditions and Farm Management Information.* Nairobi, Kenya. Ministry of Agriculture.

Jones, J. W. 1999. ICASA: Advancing Cooperative Development and Application of Systems Analysis Tools and Crop Models. In M. Donatelli, C. Stockle, F. Villalobos, and J. M. Villar Mir (eds.), *Modeling Cropping Systems*, pp. 25–30. Lleida, Catolonia, Spain: European Soc for Agronomy, University of Lleida.

Lal, H., Fonyo, C., Negahban, B., Boggess, W. G., and Kiker, G. A. 1991. Lake Okeechobee Agricultural Decision Support System (LOADSS). Paper presented at the 1991 International Winter Meeting sponsored by the American Society of Agricultural Engineers, December 17–20, 1991, Chicago, IL.

Leung, Y. 1997. *Intelligent Spatial Decision Support Systems.* Berlin, Germany: Springer-Verlag.

Lindsay, S. W., Parson, L., and Thomas, C. J. 1998. Mapping the Ranges of Relative Abundance of the Two Principal African Malaria Vectors, Anopheles Sambiae Sensu Stricto and An. Arabiensis, Using Climate Data. *Proceedings, Royal Society of London: Biological Sciences*, 265, 847–854.

Luitjen, J. C., Knapp, J. W and Jones, J. W. 2000. A Tool for Community-Based Water Resources Management in Hillside Watersheds. *Agricultural Systems* (in press).

Macro International, Inc. 2000. Headquarters, 11785 Beltsville Drive, Calverton, MD, 20705, Personal communication, July 2000.

Malczeqski, J. 1997. Spatial Decision Support Systems, NCGIA Core Curriculum in GIScience, http//:www.ncgia.ucsb.edu/giscc/units/127/u127.html, posted October 6, 1998.

McConnell, S. 1996. *Rapid Development.* Redmond, WA: Microsoft Press.

Microsoft Corporations. 1995. *The Windows® Interface Guidelines for Software Design.* Redmond, WA: Microsoft Press.

Power, D. J. 1997. What Is a DSS?. *The On-Line Executive Journal for Data-Intensive Decision Support*, October 21, 1, no. 3, http://dssresources.com/papers/whatisadss/index.html. (May 30, 2000).

SANREM CRSP. 2000. Global Projects. http:// www.sanrem.uga.edu:8080/ProGlo.html (30 May 2000).

Scholes, R. J., van Breman, N. , and Corbett, J. D. 1994. Driving Variables of Ecosystem Function and Global Change in Tropical Regions. Paper prepared at the TSBF/GCTE Workshop: The Response of Multi-Species Agricultural Systems to Global Change: The Role of Biodiversity and Complexity, Kenya, May 9–13, 1994.

Shenk, D. 1997. *Data Smog–Surviving the Information Glut*. San Francisco: Harper.

Sombroek, W. G., Braun, H. M., & van der Pouw, B. J. 1982. Exploratory Soil Map and Agro-Climatic Zone Map of Kenya. Exploratory Soil Survey Report No. E1. Nairobi, Kenya: Kenya Soil Survey, Kenya Agricultural Research Institute.

Stuth, J. W. and Lyons, B. G. 1993. *Decision Support Systems for the Management of Grazing Lands. Man and the Biosphere Series*, volume 11. Paris, France: UNESCO, and New York: The Parthenon Publishing Group.

USGS EROS Data Center. 2000. Global Land Cover Characterization. http://edcdaac.usgs.gov/glcc/glcc.html (May 30, 2000).

Williams, S., and C. Kindel. 1994. The Component Object Model: A Technical Overview. http://msdn.microsoft.com/library/techart/msdn_comppr.htm (May 30, 2000).

## 3.7 ABOUT THE CHAPTER AUTHORS

John D. Corbett is a research scientist and assistant professor at the Blackland Research and Extension Center (Temple, Texas). His research interests center around the development and application of spatial characterization tools for environmental assessment.

Jeffrey W. White is head of the GIS and Crop Modeling Laboratory at the International Maize and Wheat Improvement Center (CIMMYT, Texcoco, Mexico). He conducts research on the use of spatial tools and crop simulation models in agriculture and natural resource management.

Stewart N. Collins is an Assistant Research Scientist at the Blackland Research and Extension Center (Temple, Texas). His research interests target the integration of environmental modeling with spatial tools and data in an object-oriented programming environment.

# 4

# *Data Sources and Measurement Technologies for Modeling*

## 4.0 SUMMARY

A heightened awareness of environmental problems has developed over the past several decades and this has spurred a need for reliable geospatial data to enable better understanding of environmental processes and their impacts. Environmental models have also undergone changes and these have created new requirements for geospatial data. Similarly, increased attention in the 1990s to the analytical and environmental modeling capabilities of GIS has generated additional requirements for geospatial data. In view of the critical role data plays in any kind of spatial modeling, this chapter focuses on sources of geospatial data, data capture technologies, data access and distribution methodologies, and critical data issues facing the GIS and environmental modeling communities at the threshold of the next millennium. Emphasis is given to new information-gathering initiatives for remotely sensed data, and to advancements in integrating data from global positioning systems (GPS) and remote sensing (RS) with GIS. Also discussed are data repositories and digital libraries and the role of the internet in providing access to these data resources. Significant changes are on the horizon, including an explosion in the quantity and quality of geospatial data available to the GIS and environmental modeling communities.

## 4.1 ENVIRONMENTAL MODELING, DATA, AND GIS

Passage of the National Environmental Policy Act (NEPA) in 1966 established the first major environmental legislation in the United States. Since then, the country has experienced considerable change in population growth and its distribution, in the nature of environmental pollution and problems, in computer and GIS technology, and in the nature of environmental modeling and data collection efforts. Environmental problems in earlier decades were considered to be specific to air, water, soil, biotic, or human resources, and were typically handled by separate and independent agencies and disciplines. Many environmental problems have evolved spatially, growing from local to regional scales (Foody and Curran, 1995; Varma et al., 1992a, 1992b; Varma, 1996).

Rapid population growth, urbanization, increased mobility, industrial and agricultural development, have resulted in smog, acid rain, oil spills, soil erosion, degraded quality and quantity of surface water and aquifers, and increased pressure from development on coastal zones throughout the world (Harmancioglu et al., 1998; Sinha et al., 1989, 1990a, 1990b; Varma et al., 1992a, 1992b; Varma, 1996). Overuse and misuse of land resources have negatively impacted terrestrial ecology and habitat for biodiversity. The phenomena and consequences of ozone depletion and climate change affect all resources of the Earth in a number of interactive ways. Such pressures and problems have resulted in unsustainable exploitation of living and nonliving resources, and have spread from local levels to regional, international, and global scales. The United Nations Conference on Environment Development (UNCED) Agenda 21, presented in Rio de Janerio in 1992, changed "sustainability" from a philosophy to a management strategy where environmental issues were seen to be integrated with economic and development decision-making (Harmancioglu et al., 1998; United Nations, 1992; Varma, 1996).

Greater public awareness and concern exists today at local, regional, and global levels over the impact of pollution and natural hazards on public health and welfare. Not only environmentalists, project managers, administrators, and regulators, but society itself are emphasizing the need for a better understanding of how environmental processes work under natural and human-made conditions. This need is being partially addressed by Earth System Science, which considers earth as a synergistic physical system of interrelated phenomena, governed by complex processes involving the lithosphere, atmosphere, hydrosphere, and biosphere. Earth system science recognizes and emphasizes that there are fundamental relevant interactions of chemical, physical, biological, and dynamic processes that extend over spatial scales from microns to the magnitudes of planetary orbits, and over time scales ranging from milliseconds to billions of years (http://www.usra.edu/esse/essonline/).

Environmental data are needed for one or more of the following reasons (Harmancioglu et al., 1998; Varma, 1996):
1. To identify the nature, trends, and anomalies in characteristics of environmental processes in order to better understand of these processes
2. To assess the effects of natural and human-made factors upon the general trends in environmental processes

3. To evaluate the effectiveness of pollution control measures;
4. To assess the appropriateness of environmental quality standards;
5. To assure compliance with established quality standards and enforcement of quality control measures
6. To conduct environmental impact assessments
7. To monitor and assess the general quality of the environment at regional to global scales
8. To develop, calibrate, and validate models of environmental processes.

The availability of appropriate and adequate environmental data, and the full extraction of information from collected data, are important concerns. In the past, many data collection efforts were local in nature to meet the objectives of environmental impact assessments for projects, monitoring environmental resources for compliance and enforcement purposes, or for modeling problems at the local level. With the evolution and broadening of environmental problems, the nature and type of data requirements have changed and newer data collection efforts have been undertaken. Moreover, environmental data need to be collected in a way that facilitates understanding the interrelationships and interactions among various earth resources during an environmental disaster (natural or human-made). There is a need to integrate information on all resources of the Earth so that we become "information rich" and not just "data rich".

Multidisciplinary and interorganizational approaches leading to integrated environmental management are necessary to allow consideration of all components and processes in the environment; their spatial, temporal, and human dimensions; their interactions and correlations; coupled with social, economic, political, and legal impacts (Harmancioglu et al., 1997, 1998). Modeling has also experienced significant evolution within the last couple of decades, moving from simple empirical models that simulate individual processes to integrated models of the physical environment (Waters, Chapter 1; Couclelis, Chapter 2; and Mitasova and Mitas, Chapter 8) and human systems (see Grove et al, Chapter 7). Such models are being applied to problems at different spatial, temporal, and human scales. The data input to these models needs to address these variabilities in order to develop appropriately calibrated and validated models (see Giudici; Chapter 5, Thomas & White, 1995). Calibrating and validating 3D models, visualization models, and 4D or temporal models is becoming easier with the availablitiy of high-resolution remote sensing, elevation, and GPS data, or some combination of these. Multidimensional (3D & 4D) and visualization tools are still immature but evolving (see Rogowski and Goyne, Chapter 6; Copsey, Chapter 11; Raper, 1989; Varma, 1997). Moreover, we are still struggling with how to interpret and infer voluminous data sets (in terabytes) derived from remote sensing systems (Lillesand and Kiefer, 1994; Lyon and McCarthy, 1995; Mather, 1995). Organization, retrieval, computer processing, and management of huge data collections are new challenges that scientists, government agencies, and private industry are currently dealing with (Blount et al., 2001; Mather, 1999; Sanchez and Canton, 1999; Short, 2000).

Effective and responsible environmental management and modeling is dependent on reliable and adequate information about how the environment behaves under natural and human-made impacts. Environmental models are needed to develop appropriate decisions and policies regarding the use of earth resources in order to minimize or eliminate risks (associated with air, land, and water pollution) or to adapt to the fury of nature (for example, landslides, earthquakes, hurricanes, tornadoes).  The development of a decision support system (see Corbett and White, Chapter 3; Frysinger, Chapter 9) may provide an appropriate means of using models and geospatial data to address a particular environmental problem.  Questions about the adequacy of information, information systems, and institutional arrangements to meet the requirements of environmental management and decision making were raised in the early 1990s (Harmancioglu et al., 1997, 1998; Varma, 1992a, 1992b; WMO, 1994).  As a result, major efforts were undertaken at regional and international levels to improve the  status of environmental information systems (Clark et al., 1991; Fedra, 1997; Medyckyj-Scott et al., 1996; Mounsey, 1991; Strachan and Stuart, 1996).

Since its inception in the late 1960's, GIS has matured into a powerful tool that can integrate diverse types of spatial data and perform a wide variety of spatial analyses.  This evolution has been driven by significant advances in computer technology and the availability and quality of data (see Longley et al. 1999a, 1999b; Maguire et al., 1991a, 1991b).  From the early 1990's, persistant efforts have been made to advance the integration of GIS and environmental modeling with some notable success (see NCGIA, 1996; Waters, Chapter 1).  Forthcoming developments in both GIS and environmental modeling  are anticipated to be user or use driven.

This chapter deals with different sources of spatio-temporal data, data capture technologies, data access and distribution methodologies, and the associated critical issues facing the GIS and environment modeling communities at the turn of the millennium.  An overview of tradtional data sources and measurement issues is presented in Section 4.3.  Noteable advancements in satellite remote sensing are discussed in Section 4.4 together with issues related to the integration of GIS and remote sensing.  The utility of GPS technology for data capture relative to GIS is presented in Section 4.5.  Issues pertaining to data warehousing and access, as well as the role of the Internet, are explored in Section 4.5. Accurate and efficient data capture, data access, and data distribution from multiple sources in a manner that the accuracy and consistency of data are maintained for the purposes of modeling, analysis, and decision making will be a continuing challenge for the twenty-first century.

## 4.2 DATA TYPES, SOURCES AND ISSUES

GISs and environmental models function with a broad spectrum of geospatial data that are used for diverse applications and spatial analyses at different scales.  These data generally come in different formats and from various sources and measurements.  Data are collected through observation, measurement, and inference.  The examination and organization of data into a useful form produces information, which then enables appropriate analysis and modeling.  Two important concepts that apply to data quality are precision and accuracy.  Data precision refers to the number of decimal places used to represent measurements of such things as geographic coordinates, chemical concentrations in ground water, meteorological observations, etc.  Accuracy is a measure of how closely data represents reality within specified tolerances.  The appropriate level of data precision and accuracy will depend on the type of application and analysis for which the data are being used.  In the following subsections, data types, data sources, and related issues are discussed.

## 4.2.1 Data Types

For discussion purposes, data may be categorized into several different types that include: spatial data, attribute data, temporal data, and metadata. Spatial data come in two types of structure – vector and raster. The vector structure represents spatial features as a series of points (nodes), straight line segments (vectors), and polygons. Vector structures are usually employed to represent linear objects like roads and rivers, or areal features such as a field of corn or a lake. In contrast, raster data are comprised of rectangular cells that are arranged in a gridded array. Common forms of raster data include satellite imagery and scanned maps such as digital raster graphic (DRG) products from the USGS. Table 4.1 compares some major characteristics of these two data structures. Typically, a vector structure is preferred for use by mapping, surveying, and utilities companies, whereas, raster data is preferred for environmental and resource management applications. Fortunately, there are now computer and software systems that work with both types of data structure. Over the years, a variety of data exchange formats have been developed to enable the movement of data between different software programs (see Table 4.2).

**Table 4.1** Vector and Raster: Characteristics, Data Sources, Applications Adapted from Hohl, 1998; Varma, 2000

|  | **Vector** | **Raster** |
|---|---|---|
| Characteristics | More compact but more complex data structure<br>Supports topological relationships efficiently<br>Is better suited to support graphics<br>Representation of high spatial variability is inefficient<br>Manipulation and enhancement of digital images cannot be effectively done in vector domain<br>Overlay operations more difficult to implement<br>Better suited to small amounts of data<br>Time-sensitive data input and management<br>High degree of geometric accuracy<br>High quality graphic representation | Simple but less compact data structure<br>Overlay operations are easily and efficiently implemented<br>High spatial variation is efficiently represented<br>Topological relationships difficult to represent<br>Continuous geographical variation<br>Favored for efficient manipulation and enhancement of digital images<br>Fast data capture, usually directly from data input device<br>Output may be less visually pleasing as it has blocky appearances rather than smooth lines; can be improved but storage requirements are excessive |
| Data Sources | Manual or automated digitizing from hardcopy<br>Coordinate Geometry (COGO)<br>Third party data (DLG, DRG,TIGER/ line)<br>GPS/Surveying<br>Photogrammetric Surveys | Remote Sensing<br>Digital Orthophotographs<br>Scanners<br>CCD Cameras |
| Applications | Planning and Emergency Applications<br>Parcel map based information<br>Linear network analysis and modeling<br>Infrastructure/Asset Management | Environmental & Resource Management<br>Orthophoto mapping<br>Terrain modeling<br>Land Use/Land Cover<br>Crop Production Estimates |

**Table 4.2** Formats for Data Exchange/Interchange. Adapted from (Hohl, 1998).

| Format | Agency/ Vendor | Vector | Raster | Attribute | Comments |
|---|---|---|---|---|---|
| DXF | AutoCAD | Yes | | | Exchange between CAD systems and CAD to GIS |
| e00 | ESRI | Yes | | Yes | Proprietary format but others can read/write e00 formats |
| Arcview Shape Files | ESRI | Yes | | Yes | SHP-main, SHX-index, DBF- dbase tables |
| MIF/MID | MapInfo | MIF | | MID | Proprietary |
| DGN | Microstation | Yes | | No | Openly documented |
| DLG | USGS | Yes | | Only integer | Several varieties |
| TIGER/Line | US Census Bureau | Yes | | Yes | Transferred from Topologically Integrated Geogrpahic Encoding and Referencing (TIGER) database |
| SDTS | USGS US Federal Agencies | Yes | Yes | Yes | Standard format used in US, Australia, and South Korea |
| SAIF | Canada | Yes | Yes | Yes | National data exchange standard |
| NTF | Great Britain | Yes | Yes | Yes | National data exchange standard |
| VPF | NATO Digest standard | Yes | | Yes | Used by the Digital Chart of the World |
| TIFF or TIF | Tagged Image file format | | Yes | | Frequently used for imagery, GeoTIFF is it extension and includes geo-referencing information |
| JPEG or JPG | Joint Photographic Experts Group | | Yes | | Commonly used for imagery |

Attribute data provide critical identification information about the spatial objects and include such things as feature names, crop types, or type of road surface. Temporal data generally exhibit variation and change over time. A good example of temporal data are temperature, wind speed and direction, and precipitaion as recorded by weather stations. Metadata are information about data, and as such, are most useful when data are transferred between users. Typical metadata might include such things as map projection, source material and date acquired, and how the data was processed. The absence of metadata severely limits the utility of any spatial data.

## 4.2.2 Data Sources

Existing data sources may be in analog form (such as traditional printed maps) or as digital files. The following discussion focuses on readily available sources of digital data. Government agencies including the U.S. Bureau of the Census, U.S. Geological Survey (USGS), National Oceanic and Atmospheric Administration (NOAA), National Imagery and Mapping Agency (NIMA), National Aeronautical and Space Administration (NASA), and U.S. Environmental Protection Agency (EPA) have generated huge amounts of digital data over the last decade. An increasing amount of digital data is also being created by State and local government agencies. Similarly, commercial sources of digital data, and a sampling of these include Etak, DeLorme, NAVTECH, LAND INFO International, Geographic Data Technology, Map Express, Earth Satellite Corporation, Space Imaging, EarthWatch, and GlobeXplorer Inc. Web sites for most of these data sources, and many more, are presented in Section 4.9. Many of these sources provide digital data suitable for use in GISs.

The USGS produces a number of data products that include Digital Line Graphs(DLGs), digital raster graphics (DRGs), digital elevation models (DEMs), digitalorthophotoquads (DOQs), and the geographic names information system (GNIS). DLG vector data are available at several scales: 1:24000 (large), 1:100,000 (intermediate), and 1:2,000,000 (small). DRGs are raster data created by the USGS scanning graphic maps and outputting the resulting image files in TIFF format, with GeoTIFF extensions to describe the cartographic and georeferencing information. However, DRGs lack the topological and attribute information that make vector data such as the DLGs more useful for analysis. DEMs produced by the USGS correspond to the format of the topographic quadrangle maps, and those created for the standard seven-and-a-half minute quadrangles have been characterized by 30 m horizontal spacing between elevation points, however, these are rapidly being replaced by DEMs having 10 m elevation postings. One-degree DEM data, available from the USGS on the Internet, correspond to 1 degree by 2 degree (1:250,000 scale) USGS quadrangles and have elevations every three arc seconds (http://mapping.usgs.gov/esic/esic.html). GTOPO30 is a global digital elevation model (DEM) with horizontal spacing of 30 arc seconds (approximately 1 km) that is available from the EROS Data Center at http://edcdaac.usgs.gov/gtopo30/gtopo30.html. Other sources of elevation data include Digital Terrain Elevation Data (DTED) produced by NIMA (http://www.nima.mil/), and ETOPO5 digital 5-minute gridded land and sea floor elevation data produced by NOAA (http://www.ngdc.noaa.gov/mgg/global/etopo5.html).

The U.S. Bureau of the Census developed the Topological Integrated Geographic Encoding and Referencing (TIGER) database beginning with the 1990 census (http://www.census.gov/). The Bureau makes these data available to GIS users through extracts known as TIGER/Line files. The geometric data for major urban areas in TIGER/Line files derive from 1980 Geographic Base File/Dual Independent Map Encoding (GBF/DIME) data, the predecessor to TIGER. There is a statewide prototype cooperative effort underway to enhance TIGER (Sperling, 1999), that demonstrates how the Bureau can use local-level data to improve both positional and attribute accuracy and reduce redundancy of effort among Federal, State, and local government agencies.

The most comprehensive worldwide vector data set for use in GIS systems is the Digital Chart of the World (DCW) (http://www.lib.ncsu.edu/stacks/gis/dcw.html). DCW is a 1:1,000,000-scale map of the world based on the NIMA Operational Navigation Chart series. The DCW is comprised of seventeen information layers including political boundaries, cities, transportaion networks, rivers and lakes, and coastlines, to name a few, and is distributed in vector product format (VPF) on CD-ROM media.

Global, national and regional environmental data are accessible from numerous web sites. Both global and regional biogeographical data sets are hosted by NOAA (http://www.ngdc.noaa.gov/seg/eco/ged_toc.shtml) in support of global change characterization and modeling. Water quality data in the form of maps and tables, are available at EPA's "Surf Your Watershed" web site (http://epa.gov/surf/), that also lists hundreds of URL's for other environmental information and can be searched by state or by drainage basin. The USGS posts readings taken every 15 minutes from the National Stream Gauge Network regarding stream flows and water quality (http://water.usgs.gov/). Information about wetlands is available at the National Wetlands Inventory web site (http://www.nwi.fws.gov/). NOAA's National Climatic Data Center posts a variety of weather and climatic data sets (http://weather.ncdc.noaa.gov/). Other web sites that are good sources of environmental and GIS data or information are listed at the end of this chapter.

The Federal Geographic Data Committee (FGDC) has established the National Geospatial Data Clearinghouse, as well as standards for the minimum set of metadata necessary to adequately describe geospatial data, thereby facilitating its use by others (http://wwww.fgdc.gov/). Invariably these data have to be processed and manipulated to make them amenable for use in modeling and analysis. Integrating data from diverse sources requires that attention be given to format, scale, projection, coordinate reference system, resolution, precision, and accuracy. Incompatibility may propagate error in GIS and environmental modeling applications (Carlotto, 1995; Heuvelink, 1999). Advances in CAD and GIS systems have eased data transfer between these systems (Koch et al., 1999). More important is the need to advance GIS from a computer-aided mapping tool to one that can routinely perform spatio-temporal analyzes (Levinsohn, 2000).

### 4.2.3 Data Collection, Conversion, and Integration

Spatio-temporal data are collected primarily by means of field surveying, instrumentation, photogrammetry, and remote sensing. Field data collection using sampling, surveying and/ or GPS instruments may be required for gathering geospatial information. Such field work can help fill gaps in existing data, provide ground reference information for use in calibrating remote sensing instruments, for processing and interpreting remote sensing data, and in the establishment of ground control (see Pickles, 1995). Field surveying techniques involve the use of precise electronic distance measuring equipment and/or GPS in conjunction with an existing geodetic control network (GCN), to accurately locate features in the landscape. Instrumentation placed in the field to measure environmental phenomena are increasingly self-supporting with their own solar power and ability to automatically transmit data readings via satellite relay to a reception point for processing and archiving. Weather instruments, seismographs, and stream gauges are examples of telemetered equipment. Photogrammetry, the process of using remote sensing imagery to create a precise stereo model and compile geometrically accurate map products, has evolved into a digital process that uses soft-copy input. Remote sensing includes airborne and satellite imaging systems, both passive and active. Aerial photography is a passive airborne source of remote sensing data, whereas, light detection and ranging (LIDAR) and interferometric synthetic aperature radar (IFSAR) technologies employ an active signal to obtain information about the environment. Similarly, multispectral scanners on-board satellites are passive sensors, however, radar systems are an active source of data.

Digital data is also created by the process of converting existing maps or graphic documents into an appropriate digital form using digitizing or scanning methods (see Clarke, 1997; Hohl, 1998; Jackson & Woodsford, 1991 for details on digitizing and scanning). Data conversion is still an important task in GIS projects and constitutes a major portion of the cost and time of projects. Map conversion projects need to pay attention to scale, level of generalization, datum, coordinate system, projection, date of source, and accuracy (see Fisher, 1991 for spatial data sources and data problems; Flowerdew, 1991, for spatial data integration; Hohl, 1998, for discussion of data conversion, and Muller et al., 1995; and Weibel and Dutton, 1999, for generalization issues).

One of the strengths of GISs is the ability to integrate data from diverse sources. An environmental application or modeling effort may require the use of traditional maps containing point, line, and polygon features, as well as grids, satellite images, digital elevation models (DEMs), digital orthophotos, and scanned data files. Integrating images (raster data) with feature coverages (vector data) has become increasingly common thanks to improvements in GIS data handling capability.

## 4.3 REMOTELY SENSED DATA AND DATA CAPTURE

This section discusses a number of important considerations involved in obtaining remote sensing data and subsequent derived geospatial information (Barnsley, 1999; Campbell, 1996; Dowman, 1999; Estes and Loveland, 1999; Varma, 2000). Remote sensing includes aerial and satellite sensor systems that capture emitted or reflected energy from specific portions of the electromagnetic spectrum. The segment of the electromagnetic spectrum that is most commonly sensed for environmental applications extends over the ultraviolet, visible, near infrared, and thermal wavelengths (Table 4.3). An increasing number of satellites and sensors are used to monitor different aspects of the environment. Weather satellites observe atmospheric conditions globally day and night to monitor cyclonic systems and predict daily weather, track hurricanes, typhoons, and tornadoes, and document volcanic eruptions and their emissions. Similarly, there are an increasing number of satellites and sensor systems for observing land and/or sea (http://eospso.gsfc.nasa.gov/eos_homepage/images.html) that have been developed for various environmental applications (Table 4.4). Most satellite imaging systems are electromechanical scanners, linear array devices, or imaging spectrometers that operate in either a "sweep" or "push broom" mode (Estes and Loveland, 1999).

The five main sources of information that can be exploited by remote sensing systems relate to variations in the recorded signal as a function of spectral wavelength, directional angle, spatial location, temporal variations, and wave polarization of solar radiation reflected from the Earth's surface (see Barnsley, 1999). Another important characteristic is spatial resolution, which varies with different satellite sensor systems (Table 4.5).    Typically, different spatial resolutions are required for particular environmental applications (Table 4.6).  Other considerations include accuracy and scene size.    Additionally, the varied satellites carry sensor systems characterized by distinct spectral resolutions.  Table 4.7 compares the spectral resolutions of the Landsat TM 7 and SPOT 4 satellites.  Different environmental applications require data at different spectral resolutions (see Table 4.8).  Characteristics of selected satellites appear in Table 4.9.

Remote sensing data are usually available at different levels of processing. Raw data are as they were received from ground receiving stations and have not undergone any additional processing.  However, raw data contains geometric errors due to the Earth's rotation and sensor orientation.  Geometric distortion can be removed by selecting map coordinates common to the image, and registering the image to the map.  Use of GPS derived ground control provides a more rigorous registration.   When a DEM is incorporated into the process of eliminating geometric distortion, the result is an ortho-rectified image that meets national map accuracy standards.

**Table 4.3** Electromagnetic Spectrum

| Spectral Band | Wavelength (micrometer,mm) |
|---|---|
| Ultraviolet | 0.1 to 0.4 |
| Visible blue | 0.4 to 0.5 |
| Visible green | 0.5 to 0.6 |
| Visible red | 0.6 to 0.7 |
| Near infrared | 0.7 to 1.2 |
| Thermal—short wave infrared (SWIR) | 1.2 to 3.0 |
| Thermal—mid wave infrared (MWIR) | 3.0 to 5.0 |
| Thermal—long wave infrared (LWIR) | 5.0 to 14.0 |

**Table 4.4** Selected Satellite Systems and Appropriate Environmental Applications.
Adapted from (Varma, 2000)

| Satellite systems/ applications | Landsat 7 | SPOT-4 | IRS-1C&1D | ERS-1&2 | Radarsat-1 | JERS-1 | SeaWiFS | AVHRR | MODIS | ASTER | EROS-A1 | IKONOS |
|---|---|---|---|---|---|---|---|---|---|---|---|---|
| Land use | X | X | X | | | | | X | X | X | X | X |
| Flood extent | X | X | X | X | X | X | | | X | | | X |
| Environmental monitoring | X | X | X | X | | | | | X | X | | X |
| Damage assessment | X | X | X | | | | | | X | X | | X |
| Crop identification | X | X | X | | | | | X | X | X | | X |
| Crop development | X | X | X | | | | | X | X | X | | X |
| 3–dimensional mapping | | X | | | X | X | | | | X | X | X |
| Oil spill detection | X | | X | X | X | X | | | X | | | |
| Regional environmental mapping | X | | | | | | | X | X | X | | |
| Monitoring of coastal zones | X | X | X | X | X | X | X | | X | X | | X |
| Drought | X | X | X | | | | | X | X | X | | X |
| Fires | X | X | X | | | | | | X | X | | X |
| Night coverage | X | | | X | X | X | | | X | X | | |
| Ocean pollution monitoring | | | | | | | X | | X | X | | |
| Algae detection | X | X | X | | | | X | X | X | X | | X |
| High-resolution mapping | | | | | | | | | | | X | X |
| Infrastructure identification | X | X | X | X | X | X | | | X | X | | X |
| Terrain analysis | X | X | X | X | X | X | | | X | X | | X |

**Table 4.5** Spatial Resolution of Selected Remote Sensing Systems.  Adapted from Varma, 2000.

| Sensor | Resolution (m) | Area (m2) |
|---|---|---|
| Landsat TM 7 | 30 x 30 | 900 |
| Landsat TM 7 (Panchromatic) | 15 x 15 | 225 |
| SPOT 4 (Multispectral) | 20 x 20 | 400 |
| SPOT 4 (Panchromatic) | 10 x 10 | 100 |
| IKONOS (Multispectral) | 4 x 4 | 16 |
| IKONOS (Panchromatic) | 1 x 1 | 1 |

**Table 4.6** Spatial Resolution and Applications. Adapted from Varma, 2000.

| Spatial Resolution | Map Scale | Applications |
|---|---|---|
| 1 meter | 1: 2,000 | Land parcels<br>Sidewalks, vehicles, utilities |
| 10 meters | 1: 24,000 | Locate/map buildings/streets |
| 20/30 meters | 1: 62,500 | Land surface properties |
| 80 meters | 1: 250,000 | Regional mapping of geology, vegetation, and land-use |

**Table 4. 7** Spectral Bands (in micrometers) for Two Satellites.  Adapted from Varma, 2000.

| Landsat 7 | Color | SPOT 4 |
|---|---|---|
| Band 1 (0.45–0.52) | Blue | |
| Band 2 (0.52–0.60) | Green | Band XS1 (0.50–0.59) |
| Band 3 (0.63–0.69) | Red | Band XS2 (0.61–0.68) |
| Band 4 (0.79–0.90) | NearIR | Band XS3 (0.79–0.89) |
| Band 5 (1.55–1.75) | SWIR | Band XS4 (1.55–1.75) |
| Band 7 (2.08–2.35) | SWIR | |

**Table 4.8** Applications of Spectral Wavelengths

| Color/Spectrum | Applications |
|---|---|
| Visible blue | Map shallow water, discriminate soil vs. vegetation |
| Visible green | Vegetation health |
| Visible red | Vegetation species |
| Near infrared | Vegetation health/stress, vegetation classification |
| Mid-infrared | Land/water boundary, moisture in soils |
| Thermal infrared | Heat sources |
| Microwave (radar) | See at night, through clouds |

**Table 4.9** Characteristics of Selected Satellites.
(Adapted from Short, 2000; Turner and Hansen, 2000; Varma, 2000)

| Satellite | Operator | Launch Status | Spatial Resolution (m) P-panchromatic, M-multispectral | Spectral Resolution | Repeat Cycle (days) | Footprint (km) |
|---|---|---|---|---|---|---|
| SPIN-2 | Russia | Periodic | 2 (P) 10 (P) | Panchromatic cameras | 8 | 180 x 180 200 x 200 |
| IRS/1C-D | ISRO India | 1995/97 | 5 (P) 20 (M) | 1 (P) 3 (M) | 5 (P) 24 (M) | 70 x 70 (P) 150 x 150 (M) |
| Landsat 7 | NASA/ USGS | April 15, 1999 | 15 (P) 30(M) 60 (T) | 1 (P) 7 (M) | 16 | 185 x 185 |
| IKONOS | Space Imaging Corp | Sept. 24, 1999 | 1 (P) 4 (M) | 1 (P) 3 (M) | 3.5 - 5 | 11 x 11 |
| EROS A1 | ImageSat | 2000 | 1 (P) | 1 (P) | 90 minutes | 12.5 x 12.5 |
| Quickbird 2 | Earthwatch Corp | 2001 | 0.6 (P) 2.5 M) | 1 (P) 4 (M) | 1.5 - 4 | 16.5 x 16.5 |
| Terra's ASTER | JPL/NASA | December 16, 1999 | VNIR: 15 IR: 30-90 | 14 (M) | 16 | Variable |
| Orbview4 | Orbital Imaging Corp | 2001 | 1 (P) 4 (M) 8 (H) | 1 (P) 4 (M) 200 (H) | 3 | 8 x 8 (P,M) 5 x 5 (H) |
| SPOT 5 | SPOT Image Corp | 2002 | 2.5 (P) 10 (M) 20 (Mid IR) | 1 (P) 3 (M) | 1-4 | 60 x 60 |

### 4.3.1 Airborne Remote Sensing Systems

Remote sensing has its origins in traditional aerial photography, which has been project-based and usually produces black and white (panchromatic), natural color, or color-infrared imagery. The USGS National High Altitude Photography (NHAP) program operated between 1980 and 1987 and produced either black and white (B/W) or color infrared (CIIR) coverage for the lower 48 states with a 5-year acquisition cycle. The B and W photos are at 1:80,000 scale (1" = 1.26 miles) and the CIR photos are at 1:58000 scale (1"= 0.9 miles). NHAP was succeeded by the National Aerial Photography Program (NAPP) in 1987, which continues to the present. NAPP has a 7-year acquisition cycle, and receives joint funding from US Federal and State government agencies for the acquisition of B/W and CIR photos at 1:40,000 scale (1" = 0.6 miles). The NAPP photography is used by the USGS to produce Digital orthophoto quarter quadrangles (DOQQs), which are map images at 1:12,000 scale and cover ¼ of a standard 1:24000 topographic quadrangle map. The horizontal resolution for DOQQs is 1 meter, which meets national map accuracy standards (see Fowler, 1999 for a primer on effective use of Digital Orthophotos, and Murphy, 2000, for a discussion of digital cartographic data). DOQQs are not currently available for the entire USA. Web-compatible (JPEG-files) versions of DOQQ's may be downloaded from Microsoft's TerraServer site (http://terraserver.homeeadvisor.msn.com/).

Several new airborne remote sensing systems have become available in recent years. One of these is the airborne visible infrared imaging spectrometer (AVIRIS) developed by NASA. AVIRIS is a hyperspectral system that uses 224 contiguous spectral channels to sense in the visible to near-infrared (400 to 2,500 nm) portion of the electromagnetic spectrum. The spatial resolution of AVIRIS data can be as high as 4 m$^2$. With proper callibration and correction for atmospheric effects, this instrument is well suited to monitoring land, air, and water environments (http://makalu.jpl.nasa.gov/html).

Two other remote sensing systems that deserve mentioning use active sensors to generate high-resolution elevation data. light detection and ranging (LIDAR) systems utilize laser technology to accurately sense the surface of the Earth to about 15 cm vertical resolution under favorable conditions. Because it uses a laser signal, LIDAR is unable to sense through clouds. In contrast, interferometric synthetic aperture radar (IFSAR) transmits a radar signal that is able to pass through clouds and can capture elevation data to a vertical resolution of 1 to 2 m. Both LIDAR and IFSAR signals have difficulty penetrating the vegetation canopy. Precise ground reference information is needed in order to verify the accuracy of the data produced by these systems.

### 4.3.2 Remote Sensing Satellite Systems

Satellite remote sensing technology has advanced significantly since the launch of the first Landsat satellite in 1972. The new era in space-based remote sensing observations of the Earth's land-ocean-atmosphere system has a broad suite of sensors being sponsored by NASA through its Earth Observing System program. Concurrently, commercial vendors are in the process of placing in orbit satellites that provide sub-meter spatial resolution.

**Earth Observing System (EOS) Satellites.** Designed by NASA with international partners in many cases, the EOS satellites are intended to monitor nearly all aspects of the Earth system. Terra, the flagship of EOS, was launched December 18, 1999, and began collecting data February 24, 2000, from its five instruments: (1) the Moderate Resolution Imaging Spectroradiometer (MODIS), (2) the Advanced Spaceborne Thermal Emission and Reflection Radiometer (ASTER), (3) the Multi-angle Imaging Spectroradiometer (MISR), (4) the Clouds and the Earth's Radiant Energy System (CERES), and 5) the Measurement Of Pollution in the Troposhrere (MOPITT). These five instruments provide simultaneous, geolocated measurements on the state of earth's atmosphere, land, and oceans. Scheduled to join Terra in orbit about the earth are two additional EOS platforms. Aqua is planned for launch in July 2001 and will carry six instruments for monitoring cloud properties, radiative energy flux, land surface wetness, sea ice, snow cover, sea surface temperature, and sea surface wind fields. In June 2003, Aura is expected to be launched with two sensors that will provide data on the chemistry and dynamics of earth's atmosphere from the ground to the mesosphere. At a date still to be determined, NASA plans to launch the Vegetation Canopy LIDAR (VCL) satellite to provide the first global inventory of tree height, forest canopy structure, and biomass. More information about these and other NASA satellites and the sensor systems they carry can be found at http://eospso.gsfc.nasa.gov/eos_homepage.html

**Other New and Planned Satellites.** Launched into orbit April 15, 1999, Landsat 7 is the newest in this series of NASA satellites. New features on Landsat 7 include a panchromatic band with 15 meter spatial resolution, and a thermal infrared channel with 60 meter resolution. SPOT Image is expected to launch their newest satellite, SPOT 5, in 2002 with a 5 meter resolution panchromtic band, a 10 meter resolution multispectral band, and a 20 meter near infrared band. Other satellites planned for launch in the near future include the QuickBird 2 and the IKONOS 2 satellites with sub-meter spatial resolution in the panchromatic band, and the Orbview4 satellite with an 8 meter hyperspectral sensor (see Estes and Loveland, 1999; Short, 2000; Varma, 2000).

**Active Satellite Sensors.** Space borne RADAR systems have also been in use for many years. Those currently operational include ERS-1 and ERS-2, JERS-1, and Radarsat-1. System specifications, analysis capabilities, and applications of these radar sensors vary (Short, 2000; Varma, 2000).

The newest radar system to provide broad scale coverage of earth is the Shuttle Radar Topography Mission (SRTM) that was launched into space from the Space Shuttle Endeavor on February 11, 2000 (Chien, 2000; GeoInfo Systems, 2000; Hughes, 2000; Varma, 2000). SRTM acquired enough data during 10 days of operation to obtain the most complete near-global, 30 meter high-resolution database of earth's topography ever produced. The previous best global elevation data coverage was one kilometer resolution. The data are being processed continent by continent, then sent to NIMA to assess data quality, and finally to USGS's EROS Data Center (EDC) for archiving and distribution (more details at http://edcdaac.usgs.gov/). This data will prove valuable as input to hydrological models at global and regional scales.

### 4.3.3 Remote Sensing–GIS Integration Issues

Remote sensing technology is capable of providing spatial data that can quantitatively describe an environmental process with some degree of accuracy (Danson and Plummer, 1995; Lyon and McCarthy, 1995; Mather, 1995; Singh 1989; Varma, 2000). By integrating remote sensing with GIS, an even greater potential for environmental applications is achieved (Ehlers et al., 1989; Shelton and Estes, 1980). Debate has focused on different ways the integration may be accomplished (Bethel, 1995; Davis and Simonett, 1991; Hinton, 1996; Wilkinson, 1996). Historically, database, cartographic, and image processing systems were separate with facilities to transfer data between them. Attempts have been made to run GIS and image processing modules with a common user interface and simultaneous display. This has benefitted GIS by allowing geometric and radiometric correction and image classification, whereas, remote sensing has been a very good source of geocoded digital data for GIS and has allowed line/feature extraction, map revision, interpretation of land use, as well as format conversion. Synergy between two technologies has been the motivating factor for integration.

Environmental, urban, population, precision farming, and agriculture are applications that have benefitted from this integration (Abiodum, 2000; Bibby and Sheppard, 1999; Conitz, 2000; Couloigner and Ranchin, 2000; Dobson et al., 2000; Larsen, 1999; Lyon and McCarthy, 1995; Nogales, 2000; Pratt, 1999, 2000; Quattrochi and Luvall, 1999; Tappan et al., 2000; Thenkabail et al., 2000; Thompson, 1999; Wilson, 1999; Wilson et al., 2000). Appropriate combinations of satellite and ancillary data result in more informative products and analyses (Stewart, 1998).

Hydrologically useful data can be obtained from NOAA, SPOT, and Landsat satellites. Such data can be used to better define soil and land cover over a watershed, and thus can be useful in determining infiltration, evapotranspiration, and runoff. Since RS measures data over space rather than at point, it can be used to correct errors in input data based on point measurements (e.g. in case of rainfall, evaporation). By far, the most important data produced by remote sensing are on soil moisture and evapotranspiration and are attained through satellite thermal infrared images. These data can be employed to model water exchange between land surface and atmosphere.

RS data are useful in obtaining temperature estimates (using thermal infrared), soil moisture (using microwave), and groundwater resource evaluation. Mapping snow and ice conditions and observing water quality parameters such as algae, chlorophyll, and aquatic life parameters, as well as thermal pollution and oil spills, can be enhanced considerably using RS. NOAA infrared satellite data have been used by tightly coupled models for the computation of monthly runoff. Distributed or multiple coupled real-time flood forecasting models have used remote sensing data as input (e.g., radar rainfall measurements). Multiple-coupled hydrological models have been developed based on digital terrain models and GIS, and incorporating Landsat data. Macroscale hydrological models have been coupled with atmospheric models using satellite data. Multispectral Landsat satellite data have been used for model parameter estimation. In addition, RS data have been used as an input or as an integrator for applications such as weather forecasting, watershed management for conservation planning, drought assessment/forecasting, mapping of groundwater potential to support conjunctive use of surface and ground waters, inventorying of coastal processes and marine process environment, flood-damage assessment, and the development of a resource information matrix for irrigation development (Singh, 1995).

## 4.4 GLOBAL POSITIONING SYSTEMS

The Navigation Satellite and Ranging Global Positioning System (NAVSTAR GPS) is a constellation of twenty-four dedicated satellites orbiting the Earth at a distance of 12,600 miles that comprise a radio-navigation system permitting accurate determination of locations  worldwide, 24-hours a day, in three dimensional (latitude, longitude, and elevation) space  (Lange and Gilbert, 1999).  The 24 satellites are deployed in six evenly spaced orbits and complete a single revolution every 12 hours,  so that several are "visible" from any point on Earth at any given time.  The first satellite was launched on February 22, 1978, and by mid-1994, all 24 were broadcasting.  The U.S. Department of Defense (DOD) is responsible for building, and launching the orbital positioning and monitoring of NAVSTAR.  Steadily decreasing size and cost of  GPS equipment has led to a dramatic increase in civilian users.

### 4.4.1 GPS Technology and Data Collection

GPS receivers are able to provide information about absolute positioning (where am I?), relative positioning (where am I with respect to you?), orientation (where I am heading?), and timing (what time is it?) very efficiently.  The technology has evolved in capability from static GPS systems, to differential GPS, to real-time kinetic GPS (see Falkner, 1995; Kennedy, 1996; Leick, 1995; Sickle, 1996; Smith and Rhind, 1999 for details on GPS technology).  The position of a point is determined by measuring the length of time it takes a signal to travel from a satellite to a GPS receiver.  To obtain a two-dimensional location reading (latitude and longitude), signals must be received from a minimum of three satellites. A three-dimensional position (latitude, longitude, and elevation) requires receipt of signals from at least four satellites.  For a long time, GPS system accuracy was affected by intentional garbling of the GPS signal by DOD, a process termed "selective availability" (SA).  By executive order on May 2, 2000, the Clinton administration instructed DOD to remove SA, thereby impoving the accuracy of GPS positions from 100 meters to about 15, with a similar improvement in elevation values (Divis, 2000).

Further improvement in accuracy–from 5 meter to submeter–can be achieved through differential correction.  Differential GPS utilizes two receivers, one placed at a known location such as a surveyed point like a USGS benchmark. This stationary receiver serves as a base station, whereas the second receiver is transported around and used to record location coordinates in the field.  Comparing the known base station location to where the signals indicate it is, yields an error value that is valid for correcting the locational readings of the roving receiver.  Differential corrections can be applied in the field as the data is collected in real-time, or after the data has been collected by means of postprocessing.  Although postprocessing usually produces more accurate results, real-time differential correction is more suitable for applications involving navigation.

Regional implementations of real-time differential correction systems are provided by a number of sources. To facilitate navigation of United States waterways, the U.S. Coast Guard has established Continuously Operating Reference Stations (COREs) in coastal areas and along navigable reaches of rivers. CORE stations transmit real-time differential correction for GPS receivers. There is no charge for accessing these signals, however, most GPS receivers require a separate antenna to capture transmissions from these beacons. Similarly, the Federal Aviation Administration's Wide Area Augmentation System (WAAS) provides real-time differential correction support for airplanes. Various companies also provide differential correction services (http://gauss.gge.unb.ca/). Seamless global real-time positioning at 20 cm vertical accuracy and 10 cm horizontal accuracy is being achieved through NASA's new Global Differential GPS (GDGPS). This system is designed for dual-frequency receivers, which are expected to become the norm with the promise of two new civilian GPS frequencies (Muellerschoen et al., 2001). Access to GDGPS is currently limited, but this capability will soon becomes ubiqitus.

### 4.4.3 GPS–GIS Integration

When importing GPS data into a GIS, special attention must be paid to coordinate systems and map projections to insure compatibility (Maling, 1991; Seeger, 1999). Current software packages make the necessary transformations relatively easy. GPS data is integrated with GIS, using one of three approaches: *Data-focused integration*–least expensive with data downloaded from a GPS data logger into a computer for processing and eventual export to a GIS database; *Position-focused integration*–positions transferred through NEMA-0183 open protocol messages; and *Technology-focused integration*– through use of component technologies such as Active X that enable GIS and GPS components to be included within the same code (Harrington, 1999). Data-focused systems have entailed one-way data flow from the GPS product to the GIS, however, the ability to take existing GIS data back into the field for revision using GPS is providing impetus for two-way data flow. Table 4.10 lists some of the GPS-GIS integration products that can export vector and raster data to CAD/GIS systems on a real- time basis, allowing development of real-time maps (Harrington, 2000).

GPS provides enhanced capabilities when integrated with video or digital cameras, with airborne and satellite remote sensing systems, and when used in conjunction with GIS. One of the key advancements has been terrestrial/infrastructure imaging using vehicle-based systems. State-of-the-art vans equipped with GPS, digital imaging, and other sensors, are used to make single-pass data collections at highway speed (Varma, 1997, 2000; see Pfister, 1999 for a discussion of video mapping). Integrating GIS and GPS data has improved environmental characterization, modeling, and decision support (Cornelius and Sear, 1995), emergency management during natural disasters (Giles and Speed, 2000) and archaeology (Corbley, 1999; Michelsen, 1999; Stevenson and Robins, 2000). Interferometric synthetic-aperture radar (IFSAR), LIDAR, and real-time kinematic surveys based on mobile GPS (see Lowe, 2000) have provided higher levels of detail (1 to 2 m resolution) and vertical accuracy (about 15 cm) for terrain mapping. Greater levels of detail, in turn, improve our ability to analyze and predict the movement of water and related contaminants in natural and anthropogenic landscapes (Wilson et al., 2000).

## 4.5 DATA ACCESS AND DISTRIBUTION

Solutions to environmental problems often require data exchange at local, national, and global (international) levels. Such exchanges may be needed for data of the same type (e.g., water or air quality data collected by different methods), for data of different types of one discipline (e.g., marine physical, chemical, biological, and other oceanographic data types), or for data of different disciplines (e.g., oceanographic, meteorological, geophysical, or demographic data). The need for integrating multiple concerns in environmental modeling and related data exchange imposes significant demands on our capacity to handle environmental data so that information flow can be properly realized at local, regional, and global levels. These demands imply the need for integrated approaches to data handling, involving integration of data, information, information systems, and institutional practices. Computers and communication technologies have fundamentally changed the way in which data and information are managed and made accessed. However, institutional practices and arrangements need to improve to facilitate data exchange and related decision making.

A need exists for the development of integrated and interactive environmental information systems (Bouma and Groenigen, 1995; Egenhofer and Kuhn, 1999; Foody and Curran, 1995; Hallet et al., 1996; van Oosterom and Schenkelaars, 1995). Fedra (1997) describes two integrated environmental information systems with easy-to-use and easy-to-understand multimedia interface. One of the systems is used for local air quality assessment and management (AirWare), and the other focuses on regional water resources management (WaterWare). Both include interactive information systems equipped with GIS, simulation models, optimization models, and expert systems. AirWare includes a GIS with background maps and utilizes a DEM and satellite imagery. In addition, it has object databases containing time series air quality observations, meteorological data, emission inventories, and model scenarios. The expert system within AirWare is used to estimate point and area source emissions. WaterWare provides advanced tools to analyze environmental impacts and constraints of water resources management options. The system uses and integrates a set of databases, GIS, simulation models, and expert systems. It is developed in a modular, easy-to-use, interactive framework and has an open, object-oriented structure. This structure involves a hybrid GIS with hierarchical map layers, object databases, time series analysis, expert system, reporting functions, and a multimedia user interface with analysis for water quantity and quality problems.

Use of environmental information systems and environmental models requires data from multiple sources, which may be in different formats, structures, projections, and at different scales. Data access and sharing, as well as data currency and consistency, are important concerns. Sharing data has legal implications that users need to be aware of (see Onsrud, 1999). The Open GIS Consortium, with representation from industry leaders, is currently looking at ways of facilitating data transfer among GIS systems (http://www.opengis.org). Other entities addressing data access issues include the U. S. National Center for Geographic Information and Analysis at the University of California, Santa Barbara, the University Consortium for GIS (http://www.ucgis.org), the European Science Foundation's GISDATA program (http://www.shef.ac.uk/uni/academic/D-H/gis/gisdata.html), and the Alexandria Digital Library project (http://alexandria.sdc.ucsb.edu).

**Table 4.10** GPS/GIS Integration Products. Adapted from Graham, 2000.

| Product | GPS Type | Real-Time/Back-ground Maps; Feature Types | GIS/CAD Export & Comment |
|---|---|---|---|
| Reliance, Reliance RT, Reliance PenMap www.astech.com | 12-channel L1 or L1/L2 RT Capable or beacon   RT integrated | Yes vector & raster; Point, line & polygon with user-defined attributes | Yes; Includes interfaces to external sources |
| GPS Workhorse www.bakergeore search.com | 8-channel L1 (Motorola Oncore), RT capable | Yes vector & raster; Point, line & polygon with user-defined attributes | Yes, also GPS data saved directly into ESRI shape files; Add- on modules  support external devices; digital cameras, video and real-time sketching available |
| ALTO-G12, HP-GPS-L4, PC-GPS 3.6 www.cmtinc.com | 12-channel L1/L2 ALTO L1; HP-GPS-L4, RT capable | Yes vector & raster; Point, line & polygon with user-defined attributes | ODBC database can be joined to or exter nally converted to GIS formats; Digital camera interface; interface to external sensors |
| SR510, SR520, SR530 www.leciageosys tems.com | 12- or 24-channel L1 or L1/L2 RT capable or integrated | yes vector; Point, line & polygon with user-defined attributes | Yes; GS50 can be connected to almost all Lecia receivers; SKI-Pro processing software also available. |
| Axis GPS, Midas GIS www.sokkia.com | DGPS 12-channel L1 receiver with internal dual-channel beacon receiver | Yes vector & raster; Point, line & polygon with user-defined attributes using Database Builder | Yes; Midas I/O module for importing vector and raster maps |
| GeoExplorer, 3 Path finder Pro XR/XRS www.trimble.com | 12-channel L1 DGPS capable with BoB; 12-channel L1, RT capable or beacon RT integrated | Yes vector or vector/ Raster with Aspen; Point, line and polygon with attributes | Yes, through path finder office and/or Aspen; Includes Pathfinder office software for data processing and export to CAD/GIS; Aspen can be purchased separately |

The Alexandria Digital Library project is an effort to provide the services of a map and imagery library over the Internet, and to exploit the power of geographic location as a means of organizing information (Goodchild, 1996). A Spatial Data Transfer Standard (SDTS) has been established by the USGS to facilitate the transfer of spatial vector data and many commercial vendors have implemented it (http://mcmcweb.er.usgs.gov/sdts). Similarly, there are efforts to develop metadata catalogs and data transfer frameworks and standards at different levels and scales (Guptill, 1991, 1999; National Research Council, 1994, 1999; Rhind, 1999; Salge, 1999).

Public demand for access to information is increasing, and this has even led to court challenges (Dansby, 2000). To help meet the demand for data, many government agencies have created web access to catalogs of their holdings and mechanisms for placing orders. A good example is the USGSs Earth Explorer web site that provides access to millions of land-related products including satellite imagery (Landsat, AVHRR, Corona), aerial photography, digital cartographic data, and paper maps (http://earthexplorer.usgs.gov/).

GIS in networked environments pose challenges of their own (Coleman, 1999). Internet use has become common in all sectors of business, as well as the home (Cohen, 2000). It influences the way we communicate and provides access around the world. It is believed that there are over 80 million Internet users now and the number is growing rapidly. New browsers have been developed, and they are taking on the role of operating systems in some cases. This has the potential to change the landscape of computing from desktop to wireless. Similarly, map servers are being created (http://mapweb.parc.xerox.com/map) to produce maps on the web (Harder, 1998; Plewe, 1997). New developments in data storage devices mean significant increases in storage capacity (Hariharan and Kobler, 2000).

## 4.6 CONCLUSIONS

From the perspective of GIS, environmental modeling, and spatio-temporal data, this is an exciting and dynamic time. GIS has transitioned from a technology-driven phase to a data-driven phase, and is now entering a use-driven phase. Progress has also been made with respect to integrating GIS and environmental modeling, however, there is no consensus whether the coupling should be tight, loose, or embedded. Each of these approaches is appropriate depending on the environmental problem being addressed and the socioeconomic, institutional, and technology context in which the tools are being used. Whereas one solution may not be appropriate for all situations, and each approach has its own unique data and data exchange requirement, in the future, users should expect to find modeling modules integrated directly into GIS packages.

Accessibility to spatio-temporal data has improved considerably over the past decade. The internet with its global connectivity has been a major driver in this regard. Data providers at all levels of government, private industry, and NGO's, are advertising and distributing data over the internet. Seamless data sets are now being distributed by the USGS from its website at http://edcnts14.cr.usgs.gov:81/Website/seamless.htm. The ability to extract seamless data by areal units such as counties, drainage basins, or map extents will be forthcoming. An even more exciting initiative by the USGS is "Gateway to the Earth," a single entry point that will enable users to find and acquire earth science information from any source in the Bureau. Other data providers are certain to provide similar capabilities in the near future.

Both the GIS and environmental modeling communities have benefitted from the development of data exchange formats, advancements in GPS technology, and the increasing number of remote sensing satellite systems, acquiring new information about environmental trends and processes. Anticipated developments related to these subjects include: (1) miniaturization and integration of GPS technology with cell phones, watches, cars and other forms of transportation; (2) sub-meter spatial data resolution from satellite platforms; (3) satellite based hyperspectral scanner systems; (4) satellite based LIDAR and IFSAR high-resolution elevation data systems; and (5) dedicated satellites for monitoring natural and human-made hazards. On one hand, we collect data at a local level and for a shorter discrete time frame. For example, we have water and air quality monitoring programs for purposes of enforcement and assuring compliance with environmental standards. We have cost implications and suffer from inadequate and missing data. Converting these data into information is a challenge. On the other hand, remote sensing has allowed us to capture data continuously over regional and global scales. We have giga and terra bytes of data. Storage and organization of such data and making inferences from such data to extract information correctly, is a challenge. The issues of accuracy, uncertainties, and appropriate use of such information will be crucial.

The past decade has also witnessed incredible advances in technology. Today's personal computers are more powerful and far cheaper than the computer servers used for processing scientific data in 1990. An emerging technological development of interest to the modeling community is the ability to cluster low cost personal computers into larger parallel computing systems that create cost-effective, high performance computing environments. Computing is moving from a distributed and fixed desktop environment to an emerging and more mobile wireless society. The internet is expanding rapidly and has reached the far corners of the globe, and for many, has become an indispensable form of communication. To an increasing degree, the internet is democratizing access to data and information, environmental models, and geospatial data processing and analysis.

A number of important issues pertaining to data production, ownership, certification, and liability have been raised and are being discussed, however, it is unlikely they will be resolved in the near future. In general, private industry believes government agencies should cease being data producers and become data purchasers; that the government's role should focus on the development of standards; that government should assume the role of certifying the quality of commercial data products and the associated liability; and underwrite the cost. On the other side, government agencies are skeptical that private industry will address unique data requirements or products that have limited markets; fear industry control of data; and question why data produced with public money should be licensed by private industry for resale to the public (Tosta, 2001). Another concern is the long-term archival preservation of data and its continued accessibility – something private industry has shown little interest in. However, those issues that are resolved will have a significant impact on the GIS and environmental modeling communities.

## 4.7 REFERENCES

Ade Abiodun, A. 2000. Development and Utilization of Remote Sensing Technology in Africa. *Photogrammetric Engineering & Remote Sensing*, 66, no. 6, 674–686.

Barnsley, M. 1999. Digital Remotely-Sensed Data and Their Characteristics. In Longley, P. A., Goodchild, M. F., Maguire, D. J., and Rhind, D. W. (eds.) *Geographical Information Systems, Volume 1: Principles and Technical Issues*, 451–466. New York: Wiley.

Bethel, J. S. 1995. Photogrammetry and Remote Sensing. In Chen, W. F. (ed.), *The Civil Engineering Handbook,* 1964–2001, New York: CRC Press, Inc.

Bibby, P., and Shepherd, J. 1999. Monitoring Land Cover and Land-Use for Urban and Regional Planning. In Longley, P. A., Goodchild, M. F., Maguire, D. J., and Rhind, D. W. (eds.), *Geographical Information Systems, Volume 2: Management Issues and Applications,* 953–965. New York: Wiley.

Bouma, J., and van Groenigen, J. W. 1995. Interactive GIS for Environmental Risk Assessment. *International Journal of Geographical Information Systems,* 9, no. 5, 509–526.

Campbell, J. B. 1996. *Introduction to Remote Sensing,* 2nd ed. New York: Guilford.

Carlotto, M. J. 1995. Text Attributes and Processing Techniques in Geographical Information Systems. *International Journal of Geographical Information Systems,* 9, no. 5, 621–635.

Chien, P. 2000. Endeavor Maps the World in Three Dimensions. *GeoWorld,* 13, no. 4, April, http://www.geoplace.com/gw/2000.

Clark, D. M., Hastings, D. A., and Kineman, J. J. 1991. Global Data Bases and Their Implications for GIS. In Maguire, D. J., Goodchild, M. F., and Rhind, D. W. (eds), *Geographical Information Systems: Principles and Applications, Volume 2: Applications,* 217–231. Essex, England: Longman.

Clarke, K. C. 1997. *Getting Started with Geographic Information Systems.* Englewood Cliffs, NJ: Prentice Hall.

Cohen, J. 2000. Making Most of the Internet. *Planning,* 66, no. 7, 20–23, American Planning Association.

Coleman, D. J. 1999. GIS in Networked Environments. In Longley, P. A., Goodchild, M. F., Maguire, D. J., and Rhind, D. W. (eds.), *Geographical Information Systems, Volume 1: Principles and Technical Issues*, 317–329. New York: Wiley.

Conitz, M. 2000. GIS Applications in Africa. *Photogrammetric Engineering & Remote Sensing,* 66, no. 6, 672–673.

Corbley, K. P. 1999. Pioneering Search for a Primitive City. *GeoInfo Systems,* 9, no. 6, 30–34.

Cornelius, S., and Sear, D. 1995. Evaluating Field-Based GIS for Environmental Characterization, Modeling and Decision Support. *International Journal of Geographical Information Systems,* 9, no. 4, 475–485.

Couloigner, I., and Ranchin, T. 2000. Mapping of Urban Areas: A Multiresolution Modeling Approach for Semi-Automatic Extraction of Streets. *Photogrammetric Engineering & Remote Sensing,* 66, no. 7, 867–874.

Dansby, H. B. 2000. Access to Government Data Reaches Highest Court. *GeoWorld*. 13, no. 2, February, http://www.geoplace.com/gw/2000.

Danson, F. M., and Plummer, S. E. (eds.) 1995. *Advances in Environmental Remote Sensing*. New York: Wiley.

Davis, F. W., and Simonett, D. S. 1991. GIS and Remote Sensing. In Maguire, D. J., Goodchild, M.F. and Rhind, D.W. (eds.) *Geographical Information Systems: Volume 1: Principles,* 191–213, Essex, England: Longman.

Divis, D. A. 2000. SA No More: GPS Accuracy Increases 10 Fold. *Geospatial Solutions, 10,* no. 6, 18–20.

Dobson, J. E., Bright, E. A., Coleman, P. R., Durfee, R. C., and Worley, B. A. 2000. LandScan: A Global Population Database for Estimating Populations at Risk. *Photogrammetric Engineering & Remote Sensing*, 66, no. 7, 849–857.

Egenhofer, M. J., and Kuhn, W. 1999. Interacting with GIS. In Longley, P. A., Goodchild, M. F., Maguire, D. J., and Rhind, D. W. (eds.) *Geographical Information Systems, Volume 1: Principles and Technical Issues,* 401–412. New York: Wiley

Ehlers, M., Edwards, G., and Bedard, Y. 1989. Integration of Remote Sensing with Geographic Information Systems: A Necessary Evolution. *Photogrammetric Engineering and Remote Sensing,* 55, no. 11, 1619–1627.

Estes, J. E., and Loveland, T. R. 1999. Characteristics, Sources, and Management of Remotely-Sensed Data. In Longley, P. A., Goodchild, M. F., Maguire, D. J., and Rhind, D. W. (eds.), *Geographical Information Systems, Volume 2: Management Issues and Applications,* 667–675. New York: Wiley

Falkner, E. 1995. *Aerial Mapping: Methods and Applications*. Boca Raton, FL: Lewis Publishers.

Fedra, K. 1997. Integrated Environmental Information Systems: From Data to Information. In Harmancioglu, N. B., Alpaslan, M. N., Ozkul, S. D., and Singh, V. P. (eds.), *Integrated Approach to Environmental Data Management Systems,* 367–378, Dordrecht: Kluwer Academic Publishers.

Fisher, P. F. 1991. Spatial Data Sources and Data Problems. In Maguire, D. J., Goodchild, M. F. and Rhind, D. W. (eds.), *Geographical Information Systems: Volume 1: Principles,* 175–189. Essex, England: Longman.

Flowerdew, R. 1991. Spatial Data Integration. In Maguire, D. J., Goodchild, M. F. and Rhind, D. W. (eds.), *Geographical Information Systems: Volume 1: Principles,* 375–387. Longman Scientific and Technical.

Foody, G. M., and Curran, P. J. (eds.) 1995. *Environmental Remote Sensing from Regional to Global Scales*. New York: Wiley.

Fowler, R. 1999. Digital Orthophoto Concepts and Applications: A Primer for Effective Use. *GeoWorld,* 12, no. 7, 42–46.

Geo Info Systems (2000). SRTM Maps the World. *GeoInfo Systems,* 10, no. 4, p.12.

Giles, J., and Speed, V. 2000. GPS/GIS Mapping for Emergency Management. *Geospatial Solutions,* 10, no. 6, 36–40.

Goodchild, M. F. 1996. Directions in GIS. In NCGIA, Third International Conference/ Workshop on Integrating GIS and Environmental Modeling CD-ROM. www.ncgia.ucsb.edu/conf/SANTA_FE_CD-ROM/santa_fe.html

Graham, L. A. 2000. GPS/GIS Integration—New Products Ease Data Transfer from the Field to the Desktop. *GeoWorld.* 13, no. 1, http://www.geoplace.com/ gw/2000.

Guptill, S. C. 1991. Spatial Data Exchange and Standardization. In Maguire, D. J., Goodchild, M. F. and Rhind, D. W. (eds.) *Geographical Information Systems: Volume 1: Principles,* 515–530. Essex, England: Longman.

Guptill, S. C., 1999. Metadata and Data Catalogues. In Longley, P. A., Goodchild, M. F., Maguire, D. J., and Rhind, D. W. (eds.) *Geographical Information Systems, Volume 2: Management Issues and Applications,* 677–692. New York: Wiley.

Hallet, S. H., Jones, R. J. A., and Keay, C. A. 1996. Environmental Information Systems for Planning Sustainable Land Use. *International Journal of Geographical Information Systems,* 10, no. 1, 47–64.

Harder, C. 1998. *Serving Maps on the Internet—Geographic Information on the World Wide Web.* Redlands, CA: Environmental Systems Research Institute, Inc.

Hariharan, P.C., and Kobler, B. 2000. Data Storage Choices Abound. *GeoWorld.* 13, no. 7, July, http://www.geoplace.com/gw/2000.

Harmancioglu, N. B., Alpaslan, M. N., Oksul, S. D., and Singh, V. P. (eds.) (1997). *Integrated Approach to Environmental Data Management Systems.* Dordrecht: Kluwer Academic Publishers.

Harmancioglu, N. B., Singh, V. P., and Alpaslan, M. N. (eds.) 1998. *Environmental Data Management.* Dordrecht: Kluwer Academic Publishers.

Harrington, A. 1999. What Is GPS/GIS Integration? *GeoWorld.* 12, no. 11, p. 26.

Harrington, A. 2000. GPS/GIS Integration: Consider the Differences Among GPS Integration Technologies. *GeoWorld,* 13, no. 2, http://www.geoplace.com/gw/2000.

Healey, R. G. 1991. Data Base Management Systems. In Maguire, D. J., Goodchild, M. F., and Rhind, D. W. (eds), *Geographical Information Systems: Principles and Applications, Volume I: Principles,* 251–267. Essex, England: Longman.

Heuvelink, G. B. 1999. Propagation of Error in Spatial Modeling with GIS. In Longley, P. A., Goodchild, M. F., Maguire, D. J., and Rhind, D. W. (eds.) *Geographical Information Systems, Volume 1: Principles and Technical Issues,* 207–217. New York: Wiley.

Hinton, J. C. 1996. GIS and Remote Sensing Integration for Environmental Applications. *International Journal of Geographical Information Systems,* 10, no. 7, 877–891.

Hohl, P. 1998. *GIS Data Conversion: Strategies, Techniques, Management.* Santa Fe, NM: Onword Press.

Hughes, J. R. 2000. SRTM Data Processing is Under Way! *GeoWorld.* 13, no. 4, April, http://www.geoplace.com/gw/2000.

Jackson, M. J., and Woodsford, P. A. 1991. GIS Data Capture Hardware and Software. In Maguire, D. J., Goodchild, M. F. and Rhind, D. W. (eds.), *Geographical Information Systems: Volume 1: Principles,* 239–249. Essex, England: Longman.

Kennedy, M. 1996. *The Global Positioning System and GIS: An Introduction.* Chelsea, MI: Ann Arbor Press, Inc.

Koch, M., Greene, S., and Thomas, L. 1999. Choose the Right Road to CAD/GIS Integration. *GeoWorld.* 12, no. 6, 46–50.

Lange, A., and Gilbert, C. 1999. Using GPS for GIS Data Capture. In Longley, P. A., Goodchild, M. F., Maguire, D. J., and Rhind, D. W. (eds.) *Geographical Information Systems, Volume 1: Principles and Technical Issues,* 467–476. New York: Wiley.

Larsen, L. 1999. GIS in Environmental Monitoring and Assessment. In Longley, P. A., Goodchild, M. F., Maguire, D. J., and Rhind, D. W. (eds.), *Geographical Information Systems, Volume 2: Management Issues and Applications,* 999–1007. New York: Wiley.

Leick, A. 1995. *GPS Satellite Surveying* 2nd ed., New York: John Wiley.

Levinsohn, A. 2000. GIS Moves from Computer-Aided Mapping to Spatial Knowledge Representation. *GeoWorld.* 13, no. 2, http://www.geoplace.com/gw/2000.

Lillesand, T.M., and Kiefer, R.W. 1994. *Remote Sensing and Image Interpretation.* New York: Wiley

Longley, P. A., Goodchild, M. F., Maguire, D. J., and Rhind, D. W. (eds.) 1999a. *Geographical Information Systems, Volume 1: Principles and Technical Issues,* 451–466. New York: John Wiley.

Longley, P. A., Goodchild, M. F., Maguire, D. J., and Rhind, D. W. (eds.) 1999b. *Geographical Information Systems, Volume 2: Management Issues and Applications.* New York: Wiley.

Lowe, J. W. 2000. Maps in Motion: Spatial Data on Mobile Devices. *Geospatial Solutions,* 10, no. 6, 42–44.

Lyon, J. G., and McCarthy, J. (eds.) 1995. *Wetland and Environmental Applications of GIS.* Boca Raton, FL: CRC.

Maguire, D. J., Goodchild, M. F., and Rhind, D.W. (eds.) 1991a. *Geographical Information Systems: Volume 1: Principles.* Essex, England: Longman.

Maguire, D. J., Goodchild, M. F., and Rhind, D. W. (eds.) 1991b. *Geographical Information Systems: Volume 2: Applications.* Essex, England: Longman.

Maling, D. H. 1991. Coordinate Systems and Map Projections for GIS. In Maguire, D. J., Goodchild, M. F. and Rhind, D. W. (eds.) *Geographical Information Systems: Volume 1: Principles,* 135–146. Essex, England: Longman.

Mather, P. M. (ed.) 1995. *TERRA 2 Understanding the Terrestrial Environment: Remote Sensing Data Systems and Networks.* New York: Wiley.

Mather, P. M. 1999. *Computer Processing of Remotely-Sensed Images.* New York: Wiley.

Medyckyj-Scott, D., Cuthbertson, M., and Newman, I. 1996. Discovering Environmental Data: Metadatabases, Network Information Resource Tools, and the GENIE System. *International Journal of Geographical Information Systems,* 10, no. 1, 65–84.

Michelsen, M. W. 1999. The Riddle of the Ruins: GPS/GIS Helps Tribal Leaders Solve Land-Management Problems. *GeoWorld,* 12, no. 2, 54–56.

Muellerschoen, R. J., Bar-Sever, Y. E., Bertiger, W. I., and Stowers, D. A. 2001. NASA's Global DGPS for High-Precision Users. *GPS World,* 12, no. 1, 14–20.

Mounsey, H. M. 1991. Multisource, Multinational Environmental GIS: Lessons Learnt from CORINE. In Maguire, D. J., Goodchild, M. F. and Rhind, D. W. (eds.) *Geographical Information Systems: Volume 2: Applications,* 185–200. Longman.

Muller, J. C., Lagrange, J., and Weibel, R. (eds.) 1995. *GIS and Generalization: Methodology and Practice.* Bristol, PA: Taylor and Francis Ltd.

Murphy, C., 2000. Digital Cartographic Data: The Bottom Line. *CE News.* 12, no. 6, 52–56.

NCGIA. 1996. Third International Conference/Workshop on Integrating GIS and Environmental Modeling CD-ROM. MAIN MENU. Proceedings on CD-ROM. www.ncgia.ucsb.edu/conf/SANTA_FE_CD-ROM/santa_fe.html

National Research Council (NRC) (1999). *Distributed Geolibraries: Spatial Information Resources.* Washington, DC: National Academy Press.

National Research Council, 1994. *Promoting the National Spatial Data Infrastructure Through Partnerships.* National Research Council. National Academy Press, Washington, DC.

Nogales, M. G. 2000. Environmental Impact: Ecuador's Geospatial Mission. *GeoInfo Systems,* 10, no. 4, 24–28.

Onsrud, H. J. 1999. Liability in the Use of GIS and Geographical Datasets. In Longley, P. A., Goodchild, M. F., Maguire, D. J., and Rhind, D. W. (eds.) *Geographical Information Systems, Volume 2: Management Issues and Applications*, 643–651. New York: Wiley

Pfister, B. 1999. Video Mapping: Another Candidate for Next Great Technology. *GeoWorld.* 12, no. 10, 36–37.

Pickles, J. 1995. *Ground Truth: The Social Implications of Geographic Information Systems.* New York: Guilford.

Plewe, B. 1997. *GIS Online: Information Retrieval, Mapping, and the Internet.* Santa Fe, NM: OnWord Press.

Pratt, T. 2000. Dealing with Disaster. *GeoWorld,* 13, no. 7, July, http://www.geoplace.com/gw/2000.

Pratt, T. 1999. In the Wake of a Disaster. *GeoWorld,* 12, no. 2, 44–48.

Quattrochi, D. A., and Luvall, J. G. 1999. Urban Sprawl Urban Pall: Assessing the Impacts of Atlanta's Growth on Meteorology and Air Quality Using Remote Sensing and GIS. *GeoInfo Systems,* 9, no. 5, 26–33.

Raper, J. (ed.) 1989. *Three Dimensional Applications in Geographical Information Systems.* New York: Taylor and Francis.

Rhind, D. W. 1999. National and International Geospatial Data Policies. In Longley, P. A., Goodchild, M. F., Maguire, D. J., & Rhind, D. W. (eds.) *Geographical Information Systems, Volume 2: Management Issues and Applications,* 767–787. New York: Wiley

Salge, F. 1999. National and International Data Standards. In Longley, P. A., Goodchild, M. F., Maguire, D. J., and Rhind, D. W. (eds.) *Geographical Information Systems, Volume 2: Management Issues and Applications* , 693–706. New York: Wiley

Sanchez, J., and Canton, M. P. 1999. *Space Image Processing.* Boca Raton: CRC Press.

Seeger, H. 1999. Spatial Referencing and Coordinate Systems. In Longley, P. A., Goodchild, M. F., Maguire, D. J., and Rhind, D. W. (eds.) *Geographical Information Systems, Volume 1: Principles and Technical Issues* , 427–436. New York: Wiley.

Shelton, R. L. and Estes, J. E. 1980. Remote Sensing and Geographic Information Systems: An Unrealized Potential. *Geo-processing,* 1, no. 4, 395–420.

Short, N. M. 2000. *Remote Sensing Tutorial.* NASA. http://rst.gfc.nasa.gov/Front/tofc.html

Sickle, J. V. 1996. *GPS for Land Surveyors.* Ann Arbor Press, Inc., Chelsea, MI.

Singh. V. P. 1995. Watershed Modeling. In Singh, V. P. (ed.) *Computer Models of Watershed Hydrology*, 1–22. Highlands Ranch, CO: Water Resources Publications.

Sinha, K. C., Varma, A., and Faiz, A. 1990a. Environmental Issues in Developing Countries. *Proceedings of PTRC on Transportation and the Environment in Developing Countries,* September 10–14, 1990, Sussex, England, 37–45.

Sinha, K. C., Varma, A., Souba, J., and Faiz, A. 1989. *Environmental and Ecological Considerations in Land Transport: A Resource Guide,* Technical Paper, Report INU 41, The World Bank, Washington, D.C.

Sinha, K. C., Varma, A., and Walsh, M. P. 1990b. *Land Transport and Air Pollution in Developing Countries,* Technical Paper. Report INU 60. The World Bank, Washington, D.C.

Smith, N. S., and Rhind, D. W. 1999. Characteristics and Sources of Framework Data. In Longley, P. A., Goodchild, M. F., Maguire, D. J., and Rhind, D. W. (eds.) *Geographical Information Systems, Volume 2: Management Issues and Applications,* 655–666. New York: Wiley.

Sperling, J., and Sharp, S. 1999. A Prototype Cooperative Effort to Enhance TIGER. *URISA Journal,* 11, no. 2, 35–42.

Stevenson, M., and Robins, N. 2000. The Mojave Desert: Future Tense. *Planning,* 66, no. 7, 10–15, American Planning Association.

Stewart, J. S. 1998. *Combining Satellite Data With Ancillary Data to Produce a Refined Land-Use/ Land-Cover Map.* Water-Resources Investigations Report 97-4203. Denver: USGS.

Stojic, M. 2000. Digital Photogrammetry—Unleash the Power of 3-D GIS. *GeoWorld,* 13, no. 5, May, http://www.geoplace.com/gw/2000.

Strachan, A. J., and Stuart, N. 1996. UK Developments in Environmental GIS. *International Journal of Geographical Information Systems,* 10, no. 1, 17–20.

Tappan, G. G., Hadj, A., Wood, E.C., and Lietzow, R. W. 2000. Use of Argon, Corona, and Landsat Imagery to Assess 30 Years of Land Resource Changes in West-Central Senegal. *Photogrammetric Engineering & Remote Sensing,* 66, 6, 727–736.

Thenkabail, P. S., Nolte, C., and J. G. Lyon, 2000. Remote Sensing and GIS Modeling for Selection of a Benchmark Research Area in the Inland Valley Agroecosystems of West and Central Africa. *Photogrammetric Engineering & Remote Sensing.* 66, no. 6, 755–768.

Thomas, B. D., and White, S. G. 1995. Calibration and Validation Data for Global Biosphere Models. In Mather, P. M. (ed.) *TERRA 2 Understanding the Terrestrial Environment: Remote Sensing Data Systems and Networks,* 203–212. New York: Wiley.

Thompson, K. 1999. Amber Waves of Grain: The Age of Precision Agriculture Has Arrived. *GeoWorld.* 12, no. 10, 46–48.

Tosta, N. 2001. Licensing Data and People. *Geospatial Solutions,* 11, no. 2, 24–29.

Turner, A. K., and Hansen, J. H. 2000. *Advances in Remote Sensing and Data Capture Technologies for Transportation Applications.* Remote Sensing Workshop at Transportation Research Board Annual Meeting, January 9, 2000, Washington, DC.

United Nations 1992. *Agenda 21: Programme of Action for Sustainable Development.* New York: United Nations.

van Oosterom, P., and Schenkelaars, V. 1995. The Development of an Interactive Multi-scale GIS. *International Journal of Geographical Information Systems.* 9, no. 5, 489–508.

Varma, A. 1996. *Transportation and Water Environment: Sustainable Relationships and Appropriate Technologies.* Report CE-TRA-2-96, Dept. of Civil Engineering, North Dakota State University.

Varma, A. 1997. *Integrating CAD, GPS, GIS, and Imaging Technologies for Infrastructure Management Report No. NSF-CE-TRA-03-1997,* Department of Civil Engineering, North Dakota State University.

Varma, A. 2000. *Remote Sensing Technology and Applications- A Synthesis.* Report No. NASA-CE-TRA-2-2000. Department of Civil Engineering.

Varma, A., Souba, J., Sinha, K. C., and Faiz, A. 1992a. Environmental Considerations of Land Transport in Developing Countries. Part 2. *Transport Reviews,* 12, no.3, 187–198, Taylor and Francis Ltd., London.

Varma, A., Souba, J., Sinha, K. C., and Faiz, A. 1992b. Environmental Considerations of Land Transport in Developing Countries. Part 1. *Transport Reviews.* 12, no. 2, 101–113, Taylor and Francis Ltd., London.

Weibel, R., and Dutton, G. 1999. Generalising Spatial Data and Dealing with Multiple Representations. In Longley, P. A., Goodchild, M. F., Maguire, D. J., and Rhind, D. W. (eds.) *Geographical Information Systems, Volume 1: Principles and Technical Issues ,* 125–155. New York: Wiley

Wilkinson, G. G. 1996. A Review of Current Issues in the Integration of GIS and Remote Sensing Data. *International Journal of Geographical Information Systems,* 10, no. 1, 85–102.

Wilson, J. P. 1999. Local, National, Global Applications of GIS in Agriculture. In Longley, P. A., Goodchild, M. F., Maguire, D. J., and Rhind, D. W. (eds.) *Geographical Information Systems, Volume 2: Management Issues and Applications ,* 981–998. New York: Wiley

Wilson, J. P., Mitasova, H., and Wright, D. 2000. Water Resource Applications of Geographic Information Systems. *URISA Journal,* 12, no. 2, 61–81.

WMO, 1994. *Advances in Water Quality Monitoring—Report of a WMO Regional Workshop (Vienna, March 7–11, 1994).* Technical Reports in Hydrology and Water Resources, No. 42, WMO/TD-NO 612. Geneva: World Meteorological Organization..

## 4.8 WEB SITES FOR ENVIRONMENTAL AND GIS DATA

The list of web sites that follows is not intended to be exhaustive, but indicative of the breadth of data and information that is accessible over the Internet.

### Global Data Sources
Agriculture World Census—http://waffle.nal.usda.gov/agdb/awc90un.html
Australian Surveying and Land Info Group (AUSLIG)—http://www.auslig.gov.au/welcome.html

Biodiversity, Systematics and Collections (NBII)—http://www.nbii.gov/biodiversity/
index.html

Center for Disease Control and Prevention—http://www.cdc.gov

Consortium for Intl Earth Sci info Network (CIESIN)—http://www.ciesin.org/

Department of Energy—http://www.eia.doe.gov

DOE/ORNL—http://www.epm.ornl.gov/chammp/

DTN Weather Services—http://dtnweather.com/gis

Encyclopedia of Life Support Systems–UNESCO—http://www.eolss.co.uk/

EPA—http://www.epa.gov/globalwarming/home.htm

EPA Envirofacts Warehouse—http://www.epa.J. Wileygov/enviro/index_java.html

FAO Digital Soil Map ofthe World—http://edcwww.cr.usgs.gov/glis/hyper/guide/fao

GCMD NASA—http://gcmd.gsfc.nasa.gov/pointers/

Geography Network—http://www.geographynetwork.com/

Global Assessment of Soil Degradation—http://grid2.cr.usgs.gov/data/glasod.html

Global Change Data Center—http://www-tsdis.gsfc.nasa.gov/gcdc/gcdc.html

Global Change Data Information System—http://www.gcdis.usgcrp.gov/

Global Land Cover Database—http://edcwww.cr.usgs.gov/landdaac/glcc/globdoc1_2.html

Global Land 1 km AVHRR—http://edcwww.cr.usgs.gov/landdaac/1KM/
1kmhomepage.html

Global Population Database  (CIESIN)—http://www.ciesin.org/datasets/cir/gpopdb-
home.html

Global Pop. 1 Deg Database—http://www.usra.edu/esse/essonline/ *http://
grid2.cr.usgs.gov/globalpop/index.html

Human Health and Global Environmental Change—http://www.ciesin.org/TG/HH/hh-
home.html

Live Access to Derived Data Sets (NOAA)—http://las.saa.noaa.gov/

NASA-Land Data Assimilation System (LDAS)—http://ldas.gsfc.nasa.gov/index.shtml

National Biological Information Infrastructure—http://www.nbii.gov/

National Climate Data Center—http://www.ncdc.noaa.gov/

National Wetlands Research Center (USGS)—http://www.nwrc.gov/about/sab
/decision.html

NOAA Global Ecosystems Database (GED) —http://www.ngdc.noaa.gov/seg/eco/
ged_toc.shtml

NOAA Global Elevation Data (etopo5)—http://www.ngdc.noaa.gov/mgg/global/
etopo5.html

NOAA - http://www.ncdc.noaa.gov/pw/cg/decadal.html

NRL Monterey Satellite Meteorology—http://www.nrlmry.navy.mil/sat_products.html

Ozone Depletion and Global Environmental Change—http://www.ciesin.org/TG/OZ/oz-
home.html

Stratospheric Ozone and Human Health—http://sedac.ciesin.org/ozone/

The Earth Observing System Data Info Sys (EOSDIS) NASA—http://
    spsosun.gsfc.nasa.gov/New_EOSJ. WileyDIS.html
UNIDATA (Atmospheric and Related Data)—http://www.unidata.ucar.edu/
US Global Change Research Information Office—http://www.gcrio.org/
US National Assessment of Climate Variability and Change—http://
    www.nacc.usgcrp.gov/
UN Framework Convention on Climate Change—http://www.unfccc.de/index.html
Universities Space Research Association's (USRA)—http://www.usra.edu/esse/essonline/
Virtual Library of Ecology, Biodiversity, and the Environment—http://
    www.conbio.rice.edu/vl/
WRI Biodiversity Links—http://www.wri.org/biodiv/biolinks.html
WRI Climate Protection Links—http://www.wri.org/cpi/cpi-net.html
WRI Forest Resources—http://www.wri.org/
WRI Sustainable Agriculture Links—http://www.wri.org/sustag/aglinks.html


**Public Source**

Digital Chart of the World and Data Quality Project—http://www.nlh.no/ikf/gis/dcw/
DOQ—http://edcwww.cr.usgs.gov/glis/hyper/guide/usgs_doq
EROS Data Center home page—http://edcwww.cr.usgs.gov/
Federal Geographic Data Committee—http://www.fgdc.gov
Federal Emergency Management Agency (FEMA)—http://www.fema.gov/
GTOPO30–Global Topographic Data—http://edcdaac.usgs.gov/gtopo30/gtopo30.html
Manual of Federal Geographic Data Products–http://www.fgdc.gov/FGDP/title.html
National Atlas of the U.S.—http://nationalatlas.gov
National Wetlands Inventory Homepage—http://www.nwi.fws.gov/
TIGER/Line—US Census Bureau Home Page–http://www.census.gov
2MIL—http://edcwww.cr.usgs.gov/glis/hyper/guide/2mil
USGS Earth Explorer (millions of land-related products)—http://earthexplorer.usgs.gov/
USGS Home Page (water resources, geology, mapping, biology)—http://www.usgs.gov
USGS Node of NSDI Clearinghouse (digital products)—http://nsdi.usgs.gov/pages/
    nsdi005.html
USGS Seamless Data—http://edcnts14.cr.usgs.gov:81/Website/seamless.htm


**Commercial Data Sources**

Beartooth Mapping Inc.—http://www.beartoothmaps.com/
Claritas International—http://www.claritas.com/
Datamocracy Home—http://www.datamocracy.com
DeLorme—http://www.delorme.com/
Digital Globe—http://www.digitalglobe.com
eMapsPlus Warehouse—http://www.emapsplus.com/
ESRI Data for Your GIS—http://www.esri.com/data/index.html

Etak, Inc.—http://www.etak.com

Geographic Data Technology Inc.—http://www.geographic.com/

GeoPlace.com (The World's Leading Provider of Geospatial Information)—http://www.gisworld.com/

GISLinx (Over 1,700 Categorized GIS Links!)—http://www.gislinx.com

GIS Data Depot—http://www.gisdatadepot.com/data/catalog/index.html

International Coomputer Works—http://icwmaps.com

LAND INFO International—http://www.landinfo.com/

Macon AG—http://www.globalmaps.com

Microsoft TerraServer—http://terraserver.homeadvisor.msn.com/

NAVTECH——http://www.navtech.com/

Virtual Landscapee Technologies, Inc.—http://www.geowarehouse.com


## Remote Sensing and Satellite Images/Data

Aerial Photographs (USGS)—http://mapping.usgs.gov/esic/aphowto.html

ASPRS: The Imaging and Geospatial Information Society—http://www.asprs.org

Earth From Space—http://earth.jsc.nasa.gov

EarthWatch, Inc. (QuickBird)—http://www.digitalglobe.com

EOS Earth Data—http://eospso.gsfc.nasa.gov/eos_homepage/images.html

ImageSat International (EROS A1)—http://www.imagesatintl.com/

ISTAR Americas Inc.——http://www.istar.com/

Landsat TM 7 Satellite Data (USGS/EROS Data Center)—http://landsat7.usgs.gov

NASA AVIRIS Hyperspectral Instrument—http://makalu.jpl.nasa.gov/html

NASA Shuttle Radar Topography Mission—http://www.jpl.nasa.gov/ srtm/

NASA-The Remote Sensing Tutorial—http://rst.gsfc.nasa.gov

NOAA (National Environmental Satellite, Data, and Information Service)—http://www.nesdis.noaa.gov/

ORBIMAGE (OrbView)—http://www.orbital.com

RADARSAT (Radarsat 1)—http://www.rsi.ca/

Space Imaging (IKONOS, IRS-1C, IRS-1D, Landsat 5)—http://www.spaceimaging.com

SPIN (all) Satellite Data (Russia) —http://www.terraserver.com

Spot Image (SPOT 4)—http://www.spot.com

TERRA (EOS AM-1)—http://terra.nasa.gov

USDA-NRCS-Data Resources—http://www.ncg.nrcs.usda.gov/nsdi_node.html

Visualization of Remote Sensing Data—http://rsd.gsfc.nasa.gov/rsd/l

## 4.9 ABOUT THE CHAPTER AUTHOR

Amiy Varma, Ph.D., P.E., A.I.C.P., is Associate Professor in Department of Civil Engineering at North Dakota State University. He teaches and pursues research in areas of transportation planning and engineering, systems analysis, infrastructure management, urban planning, and applications of CAD, GIS, GPS, remote sensing, and visualization technologies for transportation, urban planning, and environmental analysis. He has worked on research sponsored by National Science Foudation, National Science Foundation Experimental Program to Stimulate Competitive Research (NSF-EPSCoR), NASA Experimental Program to Stimulate Competitive Research (NASA-EPSCoR), South Dakota Department of Ttransportation (SDDOT), Indiana Department of Transportation (INDOT), The World Bank, and others.

## 4.12 ACKNOWLEDGMENTS

I thank the editors, especially Mike Crane, for their patience and encouragement, for forming the group of contributors for this book, and for letting me be part of the endeavor. Funding from NSF, EPSCoR, and other sources allowed me to be involved with GIS and related technologies and applications. Finally, I acknowledge the considerable support, sacrifice, and understanding of my loving wife, Jaya, and my wonderful sons, Ashish and Anurah, in providing the impetus to complete this chapter and other research work.

# 5

# *Development, Calibration, and Validation of Physical Models*

*The philosophy is written in this huge book (I mean the universe) which is continuously open in front of our eyes but which cannot be understood if one does not learn the language and does not know the characters with which it is written. It is written in mathematical language and the characters are triangles, circles and other geometrical figures. Without these means a man cannot understand a single word; without these means it is a useless walk in a dark labyrinth.*

*Galileo Galilei*, Il Saggiatore, 1623

## 5.0 INTRODUCTION

Models are tools to simulate the behavior of physical systems. They can predict the future evolution of the systems, they can be used as interpretative tools in order to study system dynamics, and they can give hints for data collection and design of experiments. Models are sometimes used to examine the evolution of generic natural systems, without a specific application to a definite site or population.

A model is developed through several steps, which include defining the goals of the model; determining the structure of the model; formulating the model equations (from continuous physicomathematical equations to discrete equations; writing the computer code to solve these equations; and, finally, calibrating and validating of the model. A logical sequence can be drawn through these steps, but the actual progress of model development is not sequential; in fact, every step depends on all the previous steps, and when the results of a step are not successful, all the previous steps must be analyzed and repeated.

This chapter is devoted to a discussion of these steps and, in particular, to the calibration and validation of models. The general concepts discussed in this chapter should encourage thought about all the activities involved in the development of an environmental model and raise awareness of the difficulties that are sometimes hidden in this process. Among the key points that are stressed throughout the chapter are the importance of proper

100

collection and use of data about the real system; the interconnection between all the afore-mentioned steps; and a view of model development as an open process.

The arguments of this chapter are discussed within a general framework, without explicit reference to specific inverse models, as implemented in commercial codes. However, in order to avoid an abstract and obscure explanation, continuous reference is made to numerical models of groundwater flow.

Modeling efforts cannot give positive answers if data about the physical system being modeled are not properly taken into account at every stage of the model development. Moreover, model development requires the cooperation of scientists and professionals with differing areas of expertise and the ability to work with different kinds of data.

Modeling and data collection are both necessary for the study of natural systems. Modeling efforts that are not supported by a complete database are limited to simple problems or to generic studies. On the other hand, measurements that are not interpreted through a conceptual and a mathematical model are of limited value; they can be used only for the purposes of monitoring the system or developing simple strategies of hazard warning. Available data are often sparse, are usually clustered in relatively small areas, and do not provide an adequate base to evaluate natural variability.

It follows that Geographic Information Systems (GIS) techniques are essential to perform a complete modeling effort in an efficient way. Therefore, although GIS techniques are seldom mentioned in the rest of the chapter, readers will clearly recognize the importance of the connection between GIS and modeling techniques, as described in Chapter 8.

## 5.1 FROM THE REAL WORLD TO ABSTRACT MODELS

Natural systems (for example, biological, geological, and ecological systems) and their relevant phenomena are very complex; therefore, it is practically impossible to model such systems perfectly, and any model of these systems requires simplification and approximation. The complexity of the natural systems and phenomena is reflected in the complexity of the model development and application that is described in this chapter with reference to the schema of Figure 5.1.

Without addressing fundamental philosophical and epistemological problems, it is notable that a physicomathematical model is a tool that permits us to describe the behavior of a complex natural system with powerful mathematical techniques. Therefore, the modeler attempts to link the real world with its physical and mathematical abstraction. This process evolves through the steps shown in Figure 5.1, where the arrows show the standard path through the process. In the following sections of this chapter, each topic shown in Figure 5.1 is discussed . At the end of the chapter it will be clear that the path indicated by the arrows is an ideal case only. Reality is far more complex, and model development and application is not a straightforward process, limited to following the arrows of Figure 5.1, but is an open and evolving process, during which the previous steps have always to be taken into account.

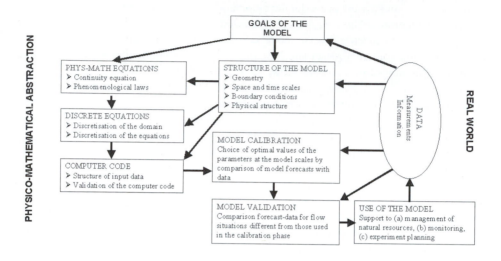

**Figure 5.1** Schema of the process of development and application of a physically based model. Square boxes correspond to different phases of the process; arrows show the standard way followed by the modeler.

## 5.1.1 The Role of Data in the Development and Application of a Model

Before entering into the details of each modeling step, it is important to stress one of the main differences between the schema of Figure 5.1 and similar descriptions proposed by others (see, e.g., Anderson and Woessner, 1991, p. 7). The protocol of model development and application is usually represented by a sequential flow chart, where only a few steps involve comparison of model outcome with data. If one of the steps gives bad results, the sequence is moved back to the previous steps.

On the other hand, Figure 5.1 shows that the problem tackled by the modeler requires a connection between the reality of the modeled system and the physicomathematical abstraction. This implies that field measurements, information about the system, and the physical processes enter, in a direct or indirect way, into each step. Furthermore, if inconsistencies arise at one step, all the previous steps must be reconsidered, so that the relationships between different steps are very complex.

In Figure 5.1 and in the rest of this chapter, a clear distinction is made between two kinds of data about the physical system. The first kind consists of measurements of physical quantities. They are sometimes divided into hard and soft data. The former correspond to quantities that are directly related to the process under study (e.g., hydraulic conductivity of the subsoil), whereas the latter correspond to quantities whose relevance is only indirect (e.g., electrical resistivity, which can be related to hydraulic conductivity).

Measurements can also be divided into experimental data, which are collected in experiments under controlled conditions, similar to what is done in a laboratory, and monitoring data, which correspond to the measurement of physical quantities under "natural" conditions—that is, when the conditions of the natural system are not altered and kept under the control of the experimenter.

Information is considered qualitative data, which cannot be represented by a numerical quantity, but consists of economical, social, and financial data that may be important (for example, for the assessment of natural hazards) but that are not strictly linked to the physical model. Finally, from a practical point of view, neither measurements nor information are of any value unless their location in space and time is known.

Unfortunately, available data, above all monitoring measurements, are seldom collected with the aim of developing and applying a model of the physical system. For instance, piezometric heads or stratigraphic and lithological data are usually collected in existing wells, which are generally sparse and sometimes clustered in relatively small areas of a large aquifer, so that they do not provide the database necessary to know the actual field variability. Moreover, the model results are relevant at scale lengths that depend on the space and time distribution of available data (see Section 5.5); however, the dimension of the "measurement window" (Cushman, 1986), the space and time domain that influences the measure, is often different from the scale lengths and times at which the natural phenomena are simulated with the model.

Figure 5.1 gives a pictorial representation of the importance of the data for the development and application of a mathematical model. The data provide the connection between the physical and mathematical abstraction and the real world. If this connection is missing, the modeler's work can be interesting from a cultural point of view but has minor relevance to practical problems.

This point is stressed so early in the chapter because model users often forget this fundamental issue. On the other hand, modelers are often so deeply involved with the mathematical aspects of the model development and application that they fail in making the customer of the product aware of the necessity of efficient data collection. Model application is not a secondary product of the survey of a natural system, but it can be effective and useful only if the core of the project is the model itself. Professionals who offer development and application of mathematical models at low cost and without collecting appropriate field data usually ignore the real complexity of the problem..

## 5.2 MODELING GOALS

The first step of model development is definition of the goals of the model. Why is the model being developed? What is the phenomenon that we want to describe? At what accuracy do we want to approximate the actual processes? These are the basic questions that the modeler must answer at this stage.

In the scientific and academic world, a model can be used to understand the basic characteristics of a phenomenon. In the professional and engineering world, a model is often used to give answers to applications. The latter is the point of view of many perspective readers of this book and will be the point of reference for this chapter. Within this framework, practical problems drive the modeler or the final user to define the goals of the model; however, the definition must be consistent with the available information and data.

This is defning the goals of the model is not a trivial task; instead, failure of applied models is often due to the errors arising from ignoring or giving imprecise and even wrong answers to the questions asked. Figure 5.1 shows that the goals of the model stand at the top of the process of development and application and drive all the following steps.

## 5.3 MODEL STRUCTURE

The next step is the definition of the structure of the model (or *configuration space* in the terminology of Chapter 8), which includes several aspects:

1. The extension and the geometry of the domain and the time period to which the model is applied.
2. The space and time scales of the model. These correspond to the space and time scales at which the phenomena of interest are modeled. The choice of the model scales is fundamental because it controls the ability of the model to reproduce real features and the discretization of the domain for the numerical model. Moreover, confidence in the model forecast requires that the model scales are consistent with the space and time distribution of the field measurements.
3. The conceptual model, which is a schematic representation of the essential features of the physical system. For instance, this corresponds to the hydrogeological scheme for groundwater flow models.
4. The parameterization of heterogeneity and anisotropy of the physical system.
5. The boundary conditions. These require determination of the characteristics of the interactions between the modeled domain and the rest of the world.
6. The kind of sources and sinks.

Once again, it is apparent that all the decisions and the assumptions at this stage must be coherent with the available data and with the goals of the model. Furthermore, this stage is facilitated from the use of GIS tools, which permit the handling of distributed data in an easy way and aid model development.

For instance, if the goal of the model is to describe the regional flow in an aquifer, it is impractical to work on scale lengths shorter than a few kilometers. On the other hand, if one wants to estimate the recharge area of a design well field in an area of a few square kilometers, a model that describes groundwater flow at a scale of tens of meters has to be applied and requires a similar distribution of data.

This step includes the parameterization not only of the heterogeneity and the anisotropy of the physical system, but also the statement of boundary conditions and the description of sinks and sources. It is necessary to choose a number (possibly small) of parameters through which the modeler can introduce the boundary conditions, the heterogeneity structure, the source terms, and so forth in the equations (see Section 5.6). Determining the numerical values of the chosen parameters is carried out in the calibration phase, described in Section 5.5.

# 5.4 PHYSICOMATHEMATICAL EQUATIONS, DISCRETE EQUATIONS, AND COMPUTER CODES

## 5.4.1 Physical Equations and Discrete Mathematics of the Model

After the definition of the goals and structure of the model, one can proceed to study the physics of the problem. In particular, the modeler seeks the proper equations, which must take into account the results of the previous steps. For instance, the use of Darcy's law, combined with the approximation of two-dimensional horizontal flow, is reasonable if the large-scale groundwater flow in a regional aquifer is the objective of the model. On the other hand, a fully three-dimensional scheme can be important for studying local effects close to pumping wells or seepage faces or for studying the effect of a complex hydrogeological structure on the flow directions, which affect both contaminant transport and the determination of recharge areas.

The physicomathematical equations usually consist of a set of partial differential equations, whose solutions cannot be computed analytically but for simple geometries, boundary conditions, and homogeneous media. Therefore, the development and application of a model requires the use of numerical techniques (e.g., finite differences, finite elements, and boundary elements). Some of these techniques are described in Chapters 2 and 8. An example of finite differences techniques is given in Section 5.4.2.

Applied mathematicians have developed a number of numerical methods to find approximate solutions to partial differential equations. This makes sense if the discretization grid can be refined at will and if the numerical approximation is consistent; in that case the approximate numerical solution tends to the true analytical solution when the grid spacing tends to zero.

On the other hand, the modeler of a physical system must take into account several constraints. Decreasing the grid spacing can improve the ability of the model to reproduce fine-scale features (e.g., the local effects of pumping wells). However, this requires a detailed knowledge of the physical system; this could be a difficult task if the natural system is heterogeneous or anisotropic, which is often the case for geological formations or for turbulent flows. Therefore, the modeler of the physical system must work with numerical techniques that provide "good solutions" even for grids with large spacing. Good solutions are solutions that satisfy the physical principles at the base of the model (i.e., the conservation principles). Large spacing means that the grid elements or cells could be larger than the optimal value coming from the purely mathematical theory, but nevertheless consistent with the distribution of the available measurements. Once again, the significance of the field data is evident in this apparently theoretical step.

Previous remarks can be rephrased as follows: The modeler should prefer numerical techniques that discretize the integral balance equation over a cell or an element, in such a way as to guarantee that the conservation principles are satisfied at the model scales. Conversely, the modeler should avoid those numerical schemes that require fine grids and excessive regularity of the functions that represent physical quantities.

## 5.4.2 Example: Steady Water Flow in a Confined Aquifer

The concepts introduced in Section 5.4.1 are now specified for a particular discrete model, which considers water flow in a confined aquifer with the following hypotheses (see, e.g., Bear, 1979; de Marsily, 1986; Walton, 1992):

1. The aquifer is isotropic.
2. The flow is two dimensional in the horizontal direction and the piezometric head is constant along the vertical throughout the whole thickness of the aquifer.
3. The flow is stationary.
4. Darcy's law is valid.

The continuity equation is as follows:

$$\frac{\partial}{\partial x}\left(T(x,y)\frac{\partial h}{\partial x}(x,y)\right) + \frac{\partial}{\partial y}\left(T(x,y)\frac{\partial h}{\partial y}(x,y)\right) = f(x,y) \qquad (5.1)$$

where $T(x, y)$ is the hydraulic transmissivity (length²/time), $h(x, y)$ is the piezometric head (length), and $f(x, y)$ is the source term corresponding to the extracted flow rate per unit surface [(length/time)/length]

Equation (5.1) is used as the basic reference for developing a finite differences discrete model, for the regular finite differences network of Figure 5.2: The domain is subdivided in square cells whose centers are the nodes. A physically based model can be obtained with an equation that rests on the mass conservation principle applied to each cell of the network. With reference to the generic cell shown in the enlargement of Figure 5.2 and recalling hypotheses 2 and 3, one can state that the sum of the flow rates entering through the four sides of the cell is equal to the rate of water extraction from the cell. This can be expressed in mathematical form as

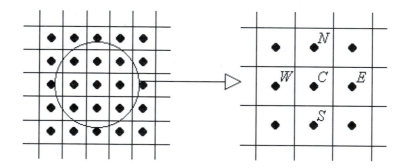

**Figure 5.2** Finite differences regular grid; dots and squares correspond to, respectively, nodes and cells of the grid, respectively. A cell, centered at node $C$, is shown in the enlargement; the four adjacent nodes ($N$, $E$, $S$, $W$) are also represented.

$$\Phi_{NC} + \Phi_{EC} + \Phi_{SC} + \Phi_{WC} = F_C \qquad (5.2)$$

where each term of the left-hand side corresponds to the flow rate through each of the four sides of the cell with center at node $C$; the right-hand side of equation (5.2) corresponds to the sum of the extracted flow rates of the cell.

The flow rates can be related to the discrete hydraulic gradients. If $A$ and $B$ are two adjacent nodes (e.g., nodes $N$ and $C$ of Figure 5.2), a discrete version of the Darcy's law can be introduced:

$$\Phi_{AB} = -T_{AB}(h_B - h_A) \qquad (5.3)$$

where $(h_B - h_A)$ is the difference between the piezometric heads at the two adjacent nodes and $T_{AB}$ is called the internode or interblock transmissivity. The internode transmissivity is the physical parameter that is used to parameterize the heterogeneity of the physical system and must satisfy conditions of symmetry

$$T_{AB} = T_{BA}$$

and of physical congruence.

$$T_{AB} \geq 0$$

Inserting equation (5.3) into equation (5.2), one obtains

$$T_{NC}(h_N - h_C) + T_{EC}(h_E - h_C) + T_{SC}(h_S - h_C) + T_{WC}(h_W - h_C) = F_C \qquad (5.4)$$

Equation (5.4) is a well-known equation and is the basis of one of the most widespread groundwater modeling programs (MODFLOW); the details of this derivation have been reviewed in this section only for clarifying the concepts introduced in Section 5.4.1 with a simple example.

### 5.4.3 The Computer Code

The development of the discrete scheme often leads to a (possibly large) system of (possibly nonlinear) equations, whose solution can be computed with the development of computer codes that implement appropriate algorithms for the solution of these systems of equations. The choice of the algorithm and of the computer code can be crucial, in particular when large systems of nonlinear equations are involved. In fact, these conditions often make it difficult to find a solution with the required accuracy unless the proper algorithms and computer codes are used.

Before using a computer code, its validation is mandatory; this can be done by comparing the results of a computer code with reference values, which are obtained with different techniques and are, or are assumed to be, correct. A list of possible comparisons follows:

1. Comparison with analytical solutions of the problem. In this case the numerical results are compared with analytical solutions of the equations; this kind of validation permits not only the test of the computer code, but also of the discrete equations, which

are assumed to be an accurate approximation of the continuous equations. In this case it is advisable to perform several tests with different values of the grid spacing in order to evaluate the rate of convergence of the numerical solution to the analytical solution.

2. Comparison with analytical solutions of the discrete problem. In this case the numerical results are compared with analytical solutions of the discrete equations. This test attempts to assess the validity of the code.

3. Comparison with results of other codes. If other computer codes related to the same problem are available and are considered to be validated already, the numerical results can be compared with the reference values obtained with these existing codes. The modeler must remember that different codes could also implement different algorithms or even different discrete equations.

4. Comparison with laboratory or field measurements. The validation involves the whole model rather than the computer code only.

If the comparisons are satisfactory—that is, if the differences between numerical results and the reference values are smaller than an acceptable discrepancy—the computer code is considered to be validated and can be used in all the following phases. However, the validity of the computer code is acceptable only for conditions similar to those used in the validation itself. For instance, if the code is validated for a homogeneous medium, this does not imply that it is validated also for heterogeneous media.

On the other hand, if the comparisons are not satisfactory, one must consider all the possible causes of errors, which include programming errors, errors in input files, and the wrong choice of an algorithm (for instance, an algorithm of solution of the discrete equations). If the comparisons are made with analytical solutions of the equations (item 1 of the preceding list), then errors in the development of the discrete equations should be taken into account. If the comparisons are performed with results of other computer codes (item 3 of the preceding list), then the modeler should carefully consider the possible differences between the physicomathematical equations and the discrete equations forming the basis of the code to be validated and the codes used for computation of reference values.

## 5.5 CALIBRATION OF THE MODEL

The objective of model calibration is to determine the numerical values of the model parameters. These values cannot be obtained in a straightforward way from direct field measurements. In fact, as already mentioned, the model scales are often different from the measurement window. For instance, the values of the internode transmissivities, which are introduced in a regional groundwater flow model based on the approximations described in Section 5.4.2, have to permit the simulation of water flow at scale lengths much greater than those involved in pumping tests.

This phase is sometimes referred to as parameter identification. Since model parameters must be optimal for simulating the actual behavior of the natural system, they are often determined with inverse modeling. Thorough and general descriptions of inverse problems can be found in the books by Tarantola (1987), Menke (1989), and Parker (1994). Some reviews of papers devoted to the inverse problem of groundwater hydrology have been published by Yeh (1986), Carrera (1988), and McLaughlin and Townley (1996). Specific descriptions of the stochastic approach to the inverse problem are given by Sun (1994), Ginn and Cushman (1990), and Zimmerman et al. (1998). In this section a few basic results are recalled.

A physicomathematical discrete model can be described as a set of equations, which translate into mathematical formulas the link between a set of parameter values, listed in the array **p**, and a set of physical parameters, listed in the array **d**, which describes the state of the physical system. Therefore, the discrete equations can be written as follows:

$$\mathbf{f}(\mathbf{p}, \mathbf{d}) = 0 \qquad (5.5)$$

If the values of the model parameters (i.e., the elements of the array **p**), are assigned, the model output corresponds to the array **d**, obtained by solving equation (5.5). With reference to equation (5.4), the array **d** consists of the values of the piezometric head at each node and the array **p** consists of the values of internode transmissivities. In this case, equation 5.5) can be written as

$$\mathbf{f}(\mathbf{p}, \mathbf{d}) = \mathbf{A}(\mathbf{d})\mathbf{p} - \mathbf{F} = 0 \qquad (5.6)$$

where the elements of the matrix $\mathbf{A}(\mathbf{d})$ are given by differences between piezometric heads of adjacent nodes, $(h_B - h_A)$. The calibration requires the solution of equation (5.5) for **p**. This is called the inverse problem.

The existence of the solution to the inverse problem in general cannot be guaranteed from a mathematical point of view without assigning very strict conditions. The existence of the solution to the inverse problem is justified by assuming that the hypotheses and the approximations introduced in the previous steps of the model development are correct. This means that model calibration actually involves all previous steps of the model development. In fact, if the calibration leads to satisfactory results, the confidence in the model outcome increases. On the contrary, if the model calibration shows some inconsistencies, all the previous steps must be revised. Problems can arise (1) if the expected goals of the model are not properly defined or if they are excessive for the present stage of development of the model, (2) if the choice of the discrete equations or the parameterization is not correct, (3) if the validation of the computer code is not complete, and (4) if observations and model forecasts used for the comparison refer to different time and space scales. This point will be discussed again in Section 5.5.3.

The uniqueness and stability of the solution to the inverse problem depend on the characteristics of the **A** matrix and are based on classical results of numerical analysis.

Before discussing the methods of solution of the inverse problem, it is worthwhile to stress the role that GIS can play in this phase. In fact, the representation and the comparison of different quantities, computed with the model or obtained from field measurements, are easy with GIS utilities. Furthermore, analysis of the results of model calibration, including comparison between different methods of solution, quantification of uncertainties (Section 5.5.4), and use of prior information even in the form of soft data (Section 5.5.5), take great advantage from the use of GIS tools.

The solution to the inverse problem has been often obtained, even in recent years, with a subjective trial-and-error procedure, which can be described as follows. The model outputs computed for different tentative values of model parameters are compared with the field data; parameter values, for which the differences between model output and field data are the lowest, are considered to be optimal.

Several different methods have been proposed to solve the inverse problem with automated techniques, which do not depend on the preconceptions of the modeler. Following the classification by Neuman (1973), these methods can be divided into two categories: direct and indirect methods.

### 5.5.1 Direct Methods Based on Equation Error

Direct methods consider (5.5) or (5.6) as equations for **p**. An example is given by the comparison model method for the solution of equation (5.6) proposed by Scarascia and Ponzini (1972) and then developed by Ponzini and Lozej (1982), Ponzini and Crosta (1988), and Ponzini et al. (1989).

More generally, direct methods are based on the minimization of the equation error defined as

$$\varepsilon_{eq}(p) \ = \ \|f(p,d)\| \tag{5.7}$$

For the model introduced in Section 5.4.2, if the norm of equation (5.7) is the sum of square errors, equation (5.7) becomes

$$\varepsilon_{eq}(\mathbf{p}) = (\mathbf{A}(\mathbf{d})\mathbf{p} - \mathbf{F}) \cdot (\mathbf{A}(\mathbf{d})\mathbf{p} - \mathbf{F}) \tag{5.8}$$

so this method corresponds to the linear least squares technique.

For the same model, the number of equations is generally lower than the number of unknowns (see, e.g., Giudici et al., 1995a, 1995b); therefore, the system (5.6) is underdetermined and it is not possible to find a unique solution. The problem can become overdetermined either by reducing the number of unknowns or by increasing the number of equations. The first approach can be applied, for instance, if the transmissivity can be assumed constant over a wide region of an aquifer; this approach is often referred to as zonation. The second approach is applicable if several sets of data are available, for instance, relative to different flow conditions, corresponding to variations of the source terms or of the boundary conditions. This approach is applied by Sagar et al. (1975), who use data from stationary conditions; by Ginn et al. (1990), who in turn use data from transient conditions; by Parravicini et al. (1995); and Giudici et al. (1995a), who developed the differential system method. The approach was also applied by Snodgrass and Kitanidis (1998) in the context of stochastic models.

Direct methods are very efficient from the computational point of view, but they suffer from some drawbacks. For example, in order to compute the function

$$\varepsilon_{eq}(\mathbf{p})$$

the data must be known at each node of the discretization grid. These data usually cannot be obtained without interpolation of data located at sparsely located, and often only a few, measurement points. This can be accomplished easily using the utilities available in GIS programs.

Furthermore, these methods might be affected by strong instability. For instance, the elements of the matrix $\mathbf{A(d)}$ are given by differences between piezometric heads at adjacent nodes $(h_B - h_A)$; if the absolute values of these differences are small, a limited error on the piezometric heads can lead to a large relative error in the computation of the corresponding element of the matrix $\mathbf{A(d)}$. In particular, if the absolute value of the difference $(h_B - h_A)$ is small, the sign of this difference could change and lead to results without physical significance (Giudici, 1994; Ponzini and Lozej, 1982).

## 5.5.2 Indirect Methods Based on Output Error

Alternative approaches have been proposed to overcome the problems mentioned at the end of Section 5.5.1. They are based on either taking into account the errors in field data in the solution of the inverse problem or using the available prior information (both soft and hard) on the physical parameters. Indirect methods aim to minimize the difference between measured data and model output.

Let $d_i^{(obs)}$ denote the value measured at a point of the domain; let $d_i^{(mod)}(\mathbf{p})$ denote the value of the physical quantity computed with the model by solving (5.5) for $\mathbf{d}$. Then the indirect methods of parameter identification are based on the minimization of the output error, defined as

$$\varepsilon_{out}(\mathbf{p}) = \sum_i \left( d_i^{(mod)}(\mathbf{p}) - d_i^{(obs)} \right)^2$$

where the summation includes all the measured quantities. Since $d_i^{(mod)}(\mathbf{p})$ is usually a nonlinear function of the parameter values, these methods are referred to as nonlinear least squares.

This approach can be derived by the method of maximum likelihood, which is based on the statistical analysis of the errors in the field data. In particular, the method of maximum likelihood reduces to the least squares technique when some hypotheses are satisfied (Gaussian distribution of errors, uncorrelated errors, etc.), as shown, for example, by Menke (1989). This explains in which sense indirect methods satisfy taking into account the errors in field data in the solution of the inverse problem.

However, even this approach suffers from the problems of non uniqueness and instability (Chavent, 1991), which depend on the convexity of the output error, a property that is usually difficult to test. In order to overcome the stability problem, the regularization approach is widely used. This method was originally proposed by Tychonov (1963) and was rigorously introduced for the inverse problem of groundwater hydrology by Kravaris and Seinfeld (1985). With this method, the identified parameter set corresponds to the array p, which minimizes the function

$$\varepsilon_{out}(\mathbf{p}) + \lambda \|\mathbf{p}\|$$

where $\lambda$ is a real positive number. The regularizing term, $\lambda \|\mathbf{p}\|$, gives higher weight to arrays with a large norm and has the effect of giving preference to smooth solutions, which do not exhibit variations at high wavenumbers (Giudici, 1994). This fact can also be considered as the introduction of a bias in the estimate given by the regularization method.

The regularization method can be modified so that it accounts for prior information in the values of the model parameters, and so using the available prior information (both soft and hard) on the physical parameters as stated previously. In particular, if $\mathbf{p}^*$ is the prior estimate of the model parameters, if $\mathbf{C_d}$ and $\mathbf{C_F}$ are the covariance matrices of the data and of the model parameters, respectively, the solution to the inverse problem corresponds to the vector $\mathbf{p}$, which minimizes the following function:

$$(d^{(obs)} - d^{(mod)}(p)) \, C_d^{-1} \, (d^{(obs)} - d^{(mod)}(p)) + (p^* - p) \, C_p^{-1} \, (p^* - p) \qquad (5.10)$$

This approach was described by Neuman and Yakowiz (1979) and considered within the framework of the maximum likelihood method in the sequence of papers by Carrera and Neuman (1986a, b, c).

Therefore, indirect methods provide a solution that can handle errors on the data and prior estimates of the model parameters in a natural way; they require, however, prior information on the covariance matrices of the data and the model parameters, which is not a trivial task (see Section 5.5.3). However, the modeler can take advantage of the utilities available in GIS programs, which handle distributed data with both deterministic and stochastic algorithms and produce not only raster grids, but also vector objects, surfaces, and so forth.

Indirect methods find the solution with iterative algorithms for function minimization. This implies that the following problems can arise:

1. The result can depend on the initial guess with which the iterative algorithm is started.
2. The algorithm can reach a local minimum, rather than the absolute minimum; probabilistic algorithms (e.g., simulated annealing), can overcome this problem.
3. The objective function given by (5.9) or (5.10) is sometimes flat over a large range of parameter values, so that the determination of the minimum could be ill conditioned (i.e., small errors in the measured data could lead to great differences in the estimated values of the parameters).
4. Convergence can be quite slow if the number of parameters to be identified is large.
5. Computation times are greater than for direct methods.

### 5.5.3 Model Calibration, Structure, and Scale Dependence of Parameters

The discussion of the solution to the inverse problem shows that the effectiveness of model calibration depends on all the previous steps of the model development. This can be explained and linked to the problem of scale dependence of model parameters in a more explicit and clear way, making reference to the model of Section 5.6.2.

The phenomenological Darcy's law is tested in laboratory experiments with samples whose length ranges from decimeters to meters. Models of regional groundwater flow usually involve scale lengths that vary between tens of meters and kilometers. Therefore, the discrete form of Darcy's law (5.3) cannot be validated by experimental studies at a scale at which it is applied, creating a problem—the validation of a phenomenological law at the model scale—for the modeler. This problem can be rephrased as follows: Under what hypothesis there is there a single value of internode transmissivity such that (5.3) holds?

The answer to this question requires determination of the relationship between the fine-scale transmissivity field and the internode transmissivity, which is significant at the scale length used for the model. The fine-scale field controls the flow, but the internode transmissivity reproduces the behavior at the model scale (see Section 5.5). Wen and Gomez-Hernandez (1996) and Renard and de Marsily (1997) give thorough reviews of the upscaling of physical parameters, whereas Ginn and Cushman (1992) discuss this problem in connection with the inverse problem. Generic models have been proposed to connect the lithological and geological structure with the values of the hydrodynamic parameters of sediments at different scales (Anderson, 1997; Koltermann and Gorelick, 1996; de Marsily et al., 1998).

Several phenomena (e.g., contaminant transport in the soil), are controlled by phenomena that take place at different scales. Finally, the values of the model parameters depend on the model scale; in addition, the results of the model are significant at the model scale. However, model output is often compared with data that are significant at scales different from the model scale. For instance, the piezometric head predicted with a regional groundwater flow model with cells whose side length is on the order of hundreds of meters cannot be compared in a straightforward way with the value of piezometric head measured in a piezometer at a point. In fact, the latter measurement might be influenced by pumping wells close to a piezometer or by local heterogeneity of the hydrogeological structure, whereas these effects are seldom taken into account for a regional groundwater flow model. Comparison between observed and predicted quantities is meaningful only if the dimensions of the "experimental window" are comparable with the scale of the model (Beckie, 1996; Cushman, 1986).

### 5.5.4 Uncertainty and the Calibrated Parameters

Calibrated values are always affected by some uncertainty, even in the best cases, when the inverse problem has a unique solution, because the measurement errors and the approximations introduced in the model prevent it from obtaining the "exact" values of model parameters. This uncertainty of model calibration leads to uncertainties in model predictions. Quantifying these uncertainties is of paramount importance when the results of a model are evaluated, especially if the model is used as a forecasting tool. Assessment of uncertainties can be obtained with stochastic modeling, with Monte Carlo techniques, or with a sensitivity analysis.

Zimmerman et al. (1998) compared seven different techniques of stochastic inversion, which were used to predict best estimates and uncertainties of travel time of nonreactive contaminants from an hypothetical radioactive waste disposal site. They found that the cumulative distribution functions of advective groundwater travel time of a conservative tracer were very different for the seven inverse methods. Moreover, the proper selection of the semivariogram of log transmissivity was found to have a significant impact on the accuracy and precision of the travel time predictions.

Here we recall the definition of the sensitivity coefficient. If $\varphi(\mathbf{p})$ represents the predicted value of a physical quantity obtained with the model using the set of parameter values given by $\mathbf{p}$, then the sensitivity of the model output on a given parameter, $p_m$, is given by

$$\frac{\partial \varphi}{\partial p_m} \varphi(\mathbf{p}) \tag{5.11}$$

Sensitivity can be defined by other expressions, which permit comparison of the sensitivity of different parameters (see, e.g., Hill, 1998). Recall that high sensitivity with respect to a model parameter means that a small error in that parameter can still lead to large errors in the predicted quantities. However, from the point of view of model calibration, this means that the value of the model parameter can be obtained in a more confident way. What is an acceptable value of the sensitivity coefficient for model application? There is no definite answer to this question. Once again, it mainly depends on the goals of the model.

### 5.5.5 Additional Data in Calibration: Joint Inversion and Soft Data

Problems with nonuniqueness, instability, and uncertainties of model calibration can be limited by the use of additional data, corresponding to different conditions of the physical system (Section 5.5.1). For instance, Carrera and Neuman (1986c), Ginn et al. (1990), and Vàzquez Gonzàlez et al. (1997) have shown that the simultaneous use of transient and stationary data improves the stability of the inversion; this result is general and independent from the inversion technique.

Several authors have proposed joint inversion of different kinds of data (e.g., hydrologic and geophysical data), in order to reduce parameter uncertainty. In the field of groundwater hydrology, this permits the use of geophysical data from seismic, geoelectric, or electromagnetic surveys to constrain results of the inversion of the purely hydrological data (Capilla et al., 1999). Geophysical data have been used for a long time in hydrogeological applications; their use in joint inversion has received new impetus from digital acquisition and processing systems that permit fast and fine-resolution tomographic surveys from surface and in boreholes (Hubbard et al., 1999; Hyndman et al., 1994; Rubin et al., 1992). Some key issues of joint inversion are discussed briefly in this section.

Both the geophysical (electrical resistivity, seismic velocity, etc.) and hydrodynamic parameters depend on sediment characteristics (porosity, grain size distribution, etc.), but the existence of a one-to-one mapping between geophysical and hydrodynamic parameters is unlikely. For instance, electrical resistivity is strongly influenced by the presence of dissolved salts in pore water, which can significantly alter the value of electrical resistivity of the same facies. Similarly, geophysical data can represent scales different from the model scale. For instance, one should recall that surveys conducted from the surface usually suffer from a decrease in resolving power with increasing exploration depth.

Joint inversion could also consider different physical quantities for the same process. For instance, calibration in groundwater hydrology is sometimes achieved by means of joint inversion of piezometric head and tracer data. The space and time evolution of conservative tracers is largely controlled by convective flow; furthermore, tracer trajectories are sensitive to fine-scale heterogeneities of hydrodynamic parameters and therefore can be important for this reason. These remarks support the use of concentration data in estimation of hydrodynamic parameters; however, they also show that the modeler must be careful in using concentration data and must pay great attention to the scales of the phenomena, especially to the comparisons between observation and model scales and to the goals of the model.

## 5.6 MODEL VALIDATION

A calibrated model is often believed to be a tool ready to forecast system behavior. Unfortunately, this is too simplistic! The approximations involved in the steps described in the preceding sections do not guarantee that the model can predict the evolution of the physical system with the accuracy required for the goals of the model. This is particularly true when model calibration is performed with measurements corresponding to specific situations of the natural system and characterized by a given accuracy. Despite what is commonly believed in practice, this does not mean that the model is also valid for simulating different situations and states of the system.

Confidence in model results requires careful validation. Some natural systems show slow dynamics, and the goal of the model could be the study of the evolution of the system over several decades or even centuries. A case in point is assessment of the long-term safety of underground disposal of radioactive wastes.

The model validation phase assesses the model's ability to predict the behavior of the system under conditions different from those used in the calibration phase. This is usually done by comparing the model forecasts and the observations for a time period following that for which the model was originally calibrated. This practice is often referred to as postaudit and is described by Konikow (1986), Anderson and Woessner (1991), Hassanizadeh and Carrera (1992), Konikow and Bredehoft (1992).

Is it possible to state in an unequivocal way whether the model is validated? No, it is not possible to have an objective and rigorous answer to this question, as shown by the cited authors. However, a few suggestions can be given.

First, the use of data corresponding to different flow situations in model calibration should give values of model parameters able to predict the behavior under a wide range of flow conditions.

Second, some of the available data, not used in the calibration phase, can be used to compare model forecast and observation following the calibration (see, e.g., Giudici et al., 2000). This permits a model validation even in the initial stage of model development.

Third, a model cannot be developed, used, and abandoned. It has to be updated continuously, using the new available data. It is important to improve continuously the structure of the model, revise the goals of the model, and even change the equations.

The model can eventually be used as a tool for predicting the natural evolution of a system and to help decision makers in their work. It can also be used as a tool to drive the collection of new and better data for monitoring purposes. Model outputs can assist in the planning of experiments (e.g., when model uncertainties could invalidate its forecast or when the postaudit analysis leads to unsatisfactory results). All these stages are easily carried out with the use of GIS utilities.

## 5.7 THE EVOLVING PROCESS OF MODEL DEVELOPMENT

Figure 5.3 gives a schematic representation of the development and of application of a model; this plot aims to give a qualitative indication of the increase in complexity, difficulties, and requirements, with the progress of the model development.

The complexity of the structure of the physical system and of the physical processes under consideration is assumed to show a linear increase with the stage of model development. For instance, the hydrogeological scheme and the equations used to model groundwater flow can evolve from simple radial one-dimensional flow in homogeneous media to more complex hydrogeological structures and flow conditions (e.g., two-dimensional flow in confined aquifers, quasi-three-dimensional flow in multilayered aquifer systems, fully three-dimensional models, including saturated-unsaturated zone and density effects, and so on). Physical processes and the mathematical equations with which they are described are more and more complex; however, these equations are often based on few basic phenomenological laws (different forms of Darcy's law for groundwater flow) and continuity equations, which translate the physical principles of conservation (of mass, energy, etc.) into mathematical language.

More complex equations induce an increase in the complexity of the numerical algorithms applied to solve them; the increase of numerical complexity is nonlinear, because the number of unknowns of the nonlinear equations might increase with the stage of model development, which can create computational problems. The model's ability to approximate the behavior of the real world improves with model development; however, after an initial effort that can produce a moderate improvement in the capability of the model to mimic the real behavior, the capability greatly increases when the model itself reaches an advanced stage. Further improvements require major changes in the model.

The importance of data for the development and application of models has been emphasized throughout this chapter. The model development is strictly linked to the availability of data that permit its application. In particular, necessary data do not follow a linear variation, but rather a "parabolic" trend, as functions of the model development level. In fact, when the model is very simple, it is sufficient to make few measurements in order to have a first satisfactory match between observations and model forecast. For advanced models, several processes are usually taken into account, more parameters are necessary, and strict experimental constraints must be introduced to set the model on a firm basis.

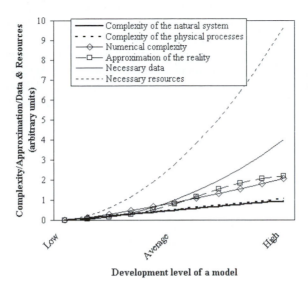

**Figure 5.3** Schematic dependence of several aspects of the model on the level of model development: the complexity of the structure of the natural system and of the physical processes, the numerical complexity, the approximation of the reality, and the necessary data and resources are considered. The units for the y-axis are arbitrary: 1 corresponds to the complexity of the physical processes for a high level of development of the model.

The necessary resources, which include human, technical, and financial resources, depend on all the aforementioned factors. The numerical complexity, which requires powerful software and hardware tools, is one of the most resource-demanding processes; however, the most resource-demanding aspect is the collection, processing, and interpretation of data.

## 5.8 CONCLUSIONS

The development and application of a model is a complex work that cannot proceed sequentially to the next step unless results of all previous steps listed in Figure 5.1 are fully satisfactory. In case they are not, all the preceding steps must be reevaluated and reanalyzed.

A model is a tool to simulate the behavior of the real world at given space and time scales; it is impossible to achieve the goals of the model without taking into account data on the physical system at any step of the model development.

The approximations introduced at any step of the model development involve different disciplines; therefore, the modeling team must include scientists and professionals with different expertise and able to work in a multidisciplinary framework.

The success of the modeling effort strongly depends on the quality and the quantity of the collected data. All the available data on the physical system must be considered and analyzed. Unfortunately, models are often based on few data collected with goals different from those of developing a model. This is the worst scenario under which to develop a model because one is constrained to use data that are not conformal to the need. This point has been addressed throughout this chapter and is repeated here to convince the reader that there are no short cuts to developing a useful mathematical model.

Is there a best model for a given real problem? No; model development is always an open-ended process that requires continuous adjustments. Since there is no best model for a given situation, it is a good practice to use several alternative models, possibly developed and applied by different modeling teams. A model is a dynamic tool. Its results must be verified and upgraded continuously if the final user wants more accurate and reliable answers to practical problems.

## 5.9 REFERENCES

Anderson, M. P. 1997. Characterization of Geological Heterogeneity. In Dagan, G., and Neuman, S. P. (eds.). *Subsurface Flow and Transport: A Stochastic Approach*, pp. 23–43. Cambridge, England: Cambridge University Press.

Anderson, M. P. and Woessner, W. W. 1991. *Applied Groundwater Modeling—Simulation of Flow and Advective Transport.* San Diego: Academic Press.

Bear, J. 1979. *Hydraulics of Groundwater.* New York: McGraw-Hill.

Beckie, R. 1996. Measurement Scale, Network Sampling Scale, and Groundwater Model Parameters. *Water Resources Research,* 32, 65–76.

Capilla, J. E., Rodrigo, J., and Gomez-Hernandez, J. J. 1999. Simulation of Non-Gaussian Transmissivity Fields Honoring Piezometric Data and Integrating Soft and Secondary Information. *Mathematical Geology,* 31, 907–927.

Carrera, J. 1988. State of Art of the Inverse Problem Applied to the Flow and Solute Transport Equations. In Custodio, E. et al. (eds.). *Groundwater Flow and Quality Modeling, pp.* 549–583. Dordrecht, The Netherlands: Reidel.

Carrera, J. and Neuman, S. P. 1986a. Estimation of Aquifer Parameters under Transient and Steady-State Conditions: 1. Maximum Likelihood Method Incorporating Prior Information. *Water Resources Research*, 22, 199–210.

Carrera, J., and Neuman, S. P. 1986b. Estimation of Aquifer Parameters under Transient and Steady-State Conditions: 2. Uniqueness, Stability and Solution Algorithms. *Water Resources Research*, 22, 211–227.

Carrera, J., and Neuman, S. P. 1986c. Estimation of Aquifer Parameters under Transient and Steady-State Conditions: 3. Application to Synthetic and Field Data. *Water Resources Research.* 22, 228–242.

Chavent, G. 1991. On the Theory and Practice of Non-Linear Least-Squares. *Advances in Water Resources*, 14, 55–63.

Cushman, J. H. 1986. On Measurement, Scale, and Scaling. *Water Resources Research*, 22 129–134.

Ginn, T. R. and Cushman, J. H. 1990. Inverse Methods for Subsurface Flow: A Critical Review of Stochastic Techniques. *Stochastic Hydrology and Hydraulics,* 4: 1-26.

Ginn, T. R., and Cushman, J. H. 1992. A Continuous-Time Inverse Operator for Groundwater and Contaminant Transport Modeling: Model Identifiability. *Water Resources Research*, 28, 539–549.

Ginn, T. R., Cushman, J. H., and Houch, M. H. 1990. A Continuous-Time Inverse Operator for Groundwater and Contaminant Transport Modeling: Deterministic Case. *Water Resources Research*, 26, 241–252.

Giudici, M. 1994. *Identifiability of Physical Parameters for Transport Phenomena in Geophysics,* Technical Report, University of Milan.

Giudici, M., Morossi, G., Parravicini, G., and Ponzini, G. 1995a. A New Method for the Identification of Distributed Transmissivities. *Water Resources Research*, 31, 1969–1988.

Giudici, M., Morossi, G., Parravicini, G., and Ponzini, G. 1995b. Some Considerations about Uniqueness in the Identification of Distributed Transmissivities of a Confined Aquifer. In Gambolati, G., and Verri, G. (eds.), *Advanced Methods for Groundwater Pollution Control,* pp. 149–160, New York: Springer-Verlag.

Giudici, M., Foglia, L., Parravicini, G., Ponzini, G., and Sincich, B. 2000. A Quasi Three Dimensional Model of Water Flow in the Subsurface of Milano (Italy): The Stationary Flow. *Hydrology and Earth System Sciences*, 4, 113–124.

Hassanizadeh, S. M., and Carrera, J. 1992. Validation of Geohydrological Models–Editorial. *Advances in Water Resources*, 15, 1-3.

Hill, M. C. 1998. *Methods and Guidelines for Effective Model Calibration.* Water resources investigations report 98-4005. Denver: United States Geological Survey.

Hubbard, S. S., Rubin, Y., and Majer, E. 1999. Spatial Correlation Structure Estimation Using Geophysical and Hydrogeological Data. *Water Resources Research*, 35, 1809–1825.

Hyndman, D. W., Harris, J. M., and Gorelick, S. M. 1994. Coupled Seismic and Tracer Test Inversion for Qquifer Property Characterization. *Water Resources Research*, 30, 1965–1977.

Koltermann, C. E., and Gorelick, S. M. 1996. Heterogeneity in Sedimentary Deposits: A Review of Structure-Imitating, Process-Imitating and Descriptive Approaches. *Water Resources Research*, 32, 2617–2658.

Konikow, L. F. 1986. Predictive Accuracy of a Ground-Water Model – Lessons From a Postaudit." *Groundwater*, 24, 173–184.

Konikow, L. F., and Bredehoft, J. D. 1992. Ground-Water Models Cannot Be Validated. *Advances in Water Resources*, 15, 75–83.

Kravaris, C., and Seinfeld, J. H. 1985. Identification of Parameters in Distributed Systems by Regularisation. *SIAM Journal on Control and Optimization*, 23, 217–241.

Marsily, G. de. 1986. *Quantitative Hydrogeology – Groundwater Hydrology for Engineers.* Orlando: Academic Press.

Marsily, G. de, Delay, F., Teles, V., and Schafmeister, M. T. 1998. Some Current Methods to Represent the Heterogeneity of Natural Media in Hydrogeology. *Hydrogeology Journal*, 6, 115–130.

McDonald, M.G., and Harbaugh, A. W. 1988. *A Modular Three-Dimensional Finite-Difference Ground-Water Flow Model.* Denver: United States Geological Survey, Techniques of Water Resources Investigations 06-A1.

McLaughlin, D., and Townley, L. R. 1996 A Reassessment of the Groundwater Inverse Problem. *Water Resources Research*, 32, 1131–1161.

Menke, W. 1989. *Geophysical Data Analysis: Discrete Inverse Theory.* San Diego: Academic Press.

Neuman, S. P. 1973. Calibration of Distributed Parameter Groundwater Flow Models Viewed as Multiple Objective Decision Process under Uncertainty. *Water Resources Research*, 9, 1006–1021.

Neuman, S. P., and Yakowitz, S. 1979. A Statistical Approach to the Inverse Problem of Aquifer Hydrology, 1. Theory. *Water Resources Research*, 15, 845–860.

Parker, R. L. 1994. *Geophysical Inverse Theory.* Princeton, NJ: Princeton University Press.

Parravicini, G., Giudici, M., Morossi, G., and Ponzini, G. 1995. Minimal a priori Assignment in a Direct Method for Determining Phenomenological Coefficients Uniquely. *Inverse Problems*, 11, 611–629.

Ponzini, G., and Crosta, G. 1988. The Comparison Model Method: A New Arithmetic Approach to the Discrete Inverse Problem of Groundwater Hydrology, 1, One-Dimensional Flow. *Transport in Porous Media*, 3, 415–436.

Ponzini, G., Crosta, G., and Giudici, M. 1989. Identification of Thermal Conductivities by Temperature Gradient Profiles: One-Dimensional Steady State Flow. *Geophysics*, 54, 643–653.

Ponzini, G., and Lozej, A. 1982. Identification of Aquifer Transmissivities: The Comparison Model Method. *Water Resources Research*, 18, 597–622.

Renard, P., and Marsily, G. de.1997. Calculating Equivalent Permeability: A Review. *Advances in Water Resources*, 20, 253–278.

Rubin, Y., Mavko, G., and Harris, J. 1992. Mapping Permeability in Heterogeneous Aquifers Using Hydrologic and Seismic Data. *Water Resources Research*, 28, 1809–1816.

Sagar, B., Yakowitz, S., and Duckstein, L. 1975. A Direct Method for the Identification of the Parameters of Dynamic Nonhomogeneous Aquifers. *Water Resources Research*, 11, 563–570.

Scarascia, S., and Ponzini, G. 1972. An Approximate Solution for the Inverse Problem in Hydraulics. *L'Energia Elettrica*, 49, 518–531.

Snodgrass, M. F., and Kitanidis, P. K. 1998 Transmissivity Identification Through Multi-Directional Aquifer Simulation. *Stochastic Hydrology and Hydraulics*, 12, 299–316.

Sun, N. Z. 1994. *Inverse Problems in Groundwater Modeling.* Dordrecht, The Netherlands: Kluwer Academic Publishers.

Tarantola, A. 1987. *Inverse Problem Theory: Methods for Fitting and Model Parameter Estimates.* New York: Elsevier.

Tychonov, P. N. 1963. Solution of Incorrectly Formulated Problems and the Regularisation Method. *Soviet Mathematical. Doklady*, 4, 1045–1064.

Vázquez González, R., Giudici, M., Parravicini, G., and Ponzini, G. 1997. The Differential System Method for the Identification of Transmissivity and Storativity. *Transport in Porous Media*, 26, 339–371.

Walton, W.C. 1992. *Groundwater Modeling Utilities.* Chelsea, MI: Lewis Publishers, Inc.

Wen, X. H., and Gomez-Hernandez, J. J. 1996. Upscaling Hydraulic Conductivity in Heterogeneous Media: An Overview. *Journal of Hydrology*, 183, ix–xxxii.

Yeh, W.-G. W. 1986. Review of Parameter Identification Procedures in Groundwater Hydrology: The Inverse Problem. *Water Resources Research*, 22, 95–108.

Zimmerman, D. A., Marsily, G. de, Gotway, C. A., Marietta, M. G., Axness, C. L., Beau-heim, R. L., Bras, R. L., Carrera, J., Dagan, G., Davies, P. B., Gallegos, D. P., Galli, A., Gómez-Hernández, J., Grindrod, P., Gutjahr, A. L., Kitanidis, P. K., Lavenue, A. M., McLaughlin, D., Neuman, S. P., RamaRao, B. S., Ravenne, C., Rubin, Y. 1998. A Comparison of Seven Geostatistically Based Inverse Approaches to Estimate Transmissivities for Modeling Advective Transport by Groundwater Flow. *Water Resources Research*, 34, 1373–1413.

## 5.10 ACKNOWLEDGEMENTS

The author gratefully thanks G. Ponzini (University of Milano), R. Bersezio (idem), and B. Testa (CNR—National Research Council) for discussions and for their review of this chapter. The chapter has been greatly improved following the suggestions and constructive criticism of two anonymous reviewers.

## 5.11 ABOUT THE CHAPTER AUTHOR

 Mauro Giudici is a professor in the Department of Earth Sciences at the University of Milano (Italy). He conducts research in geophysics, namely in forward and inverse modeling of groundwater flow and transport.

# 6

# Modeling Dynamic Systems and Four-Dimensional Geographic Information Systems

## 6.0 INTRODUCTION

Dynamic environmental processes vary simultaneously in space and in time. While the concept of space is easily understood, time may be thought of as a duration frame of reference for changes associated with objects in space. To capture in a four-dimensional (4D) model the essence of a space-time process, major limitations with the information processing and retrieval, and with data management, including context, content, updating, and storage, need to be resolved. Problems with space-time geographic information systems (GIS) modeling arise from the inability to conceptualize ramifications of the process, and analyze and visualize the full implications of the 4D information as it evolves in both time and space. These shortcomings can be overcome in part by modeling a process in time at a point and then interpolating or simulating spatial distributions for the time intervals of interest. However, while spatial aggregation is compatible with changes of scale, temporal averaging almost always leads to the loss of essential information. Many deterministic and stochastic space-time models operate outside the GIS framework, because too many simplifying assumptions are required to process a system of differential equations fully within the context of GIS. Part of the problem is that to cast a process within a GIS, an extensive exploratory data analysis is usually needed to decide on the appropriate scale and resolution and to determine whether the assumptions and simplifications are within the sensitivity bounds of the model. Promising approaches include current research efforts to develop 4D data storage, analysis, and visualization systems, and the as yet largely untapped potential of remote imaging and program induction approaches based on input data values. Case studies illustrate GIS modeling of seasonal non-point-source pollution potential, and hourly versus weekly soil water redistribution assessment following precipitation events during a dormant season.

# 6.1 SPACE-TIME MODELING IN GIS CONTEXT

The scientific community appears divided in its their approach to dynamic space-time modeling issues. Many GIS practitioners seem concerned with the appropriate data structure, visualization, and analysis tools for space-time applications but lack explicit space-time dynamic models to use in the GIS context. On the other hand, stochastic and deterministic modelers have developed many sophisticated space-time approaches that, other than perhaps with data access and display of final products, have little to do with GIS. A brief glance at the respective literature underscores the dichotomy of view—the two groups do not even seem to talk to each other.

Part of the problem may be in the semantics of the technical language, which are often foreign to each group, and part is undoubtedly due to differences in perspective. For example, as Koza (1992) points out, a computer program is seen as a formula, a plan, a control strategy, a computational procedure, a model, a decision tree, a game-playing strategy, a robotic action plan, a transfer function, a mathematical expression, a sequence of operations, or perhaps merely a composition of functions. Similarly, data inputs are referred to as sensor values, state variables, independent variables, attributes, information to be processed, input signals, input values, known variables, or arguments of a function. Practitioners speak of output as dependent variables, control variables, a category, a decision, an action, a move, an effect, a result, an output signal, an output value, a class, an unknown variable, or values returned by a function. It is thus small wonder that much confusion exists.

Although Newton-Smith (1980) expressed a view that no satisfactory understanding of space and time can be achieved at a purely semantic level, Table 6.1 lists some common analogies to help with the discussion that follows.

**Table 6.1.** Modified analogies between space and time semantics. Adapted from Lagran (1993).

| Description | Space | Time |
| --- | --- | --- |
| Entity | Process | System |
| Representation | Map | State |
| Components | Symbols | Events |
| Units | Cells, vectors | Days, hours, etc |
| Specific units | Objects | Stages |
| Separation | Boundaries | Changes |
| Position | Coordinates | Date |
| Size | Length, area, depth | Duration |
| Contiguity | Adjacent objects, cells | Past, present, future |
| Clustering | Aggregation | Averaging |
| Nearest neighbors | Infinite number | Only two |

### 6.1.1 Process Dynamics

Our purpose is to discuss the current status of modeling dynamic environmental processes in time, within a four-dimensional (4D) GIS framework, and to offer some direction for future research. A question that should also be explored in this chapter is why model at all, and specifically why model in GIS, especially when it comes to representation of dynamic systems as they develop in time.

We do not wish to enter into a philosophical debate as to what space and time are. Volumes have been written on the meaning and significance of space and time without a really satisfactory answer; see, for example, Newton-Smith, (1980), Morris, (1984), and Price (1996). Kelmelis (1991) provides an excellent overview and summary of the past and present writings on the subject.

What is important from the standpoint of modeling is that while we cannot see or feel time as such, we can readily perceive its effects (Newton-Smith, 1980). Thus time can be better understood when we consider it in terms of changes associated with objects in space. Some of these changes may involve transformations, others a movement of objects in relation to one another (Peuquet, 1994). We know from experience and it follows from Heisenberg's uncertainty principle (e.g., Marshall and Pounder, 1957) that an object can occupy only one position in space at a given time. When it moves or is moved to an alternate position, the move signifies change. Intuitively we realize that change does not occur instantaneously. Thus in a sense time becomes a convenient duration frame of reference for that change. It relates a number of time units to changes of position of the 3D objects in the Euclidean space $\Re^N$. When an object itself changes without actually changing position, we speak of transformation or aging and again rely on time to provide a duration frame of reference. Of necessity and by tradition, time frame of reference is based on the rhythms of our solar system (i.e., years, days, hours, etc.). Some attempts have been made, however; to relate it to a universal constant $c$—the speed of light (~$300 \times 10^6$ m/s). In such a 4D space-time system all axes would be given in the same units—light seconds. In this system best described by a light cone (Kelmelis, 1991), time is depicted as being at 45° to the

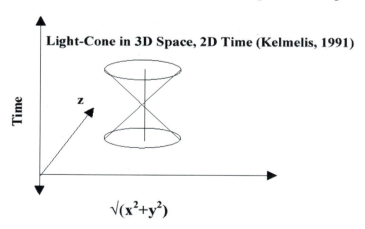

space axes (Figure 6.1). However, most commonly different units are used for time and space. Under these circumstances the light cone becomes very flat, and although time is represented as another (fourth) spatial dimension, orthogonal space curves can only show the state of the process at a given moment, while time curves describe the state of the system at a point (Hazelton et al., 1990).

All real-life objects and processes have a life span of some final duration. Additional temporal considerations may also involve changes in position, shape, and value of the object attributes. In view of an all-encompassing GIS perspective, the problem is how to identify, track in time, and represent different states of the same object, or different stages of the same process (Lagran, 1993). Space-time mapping of average conditions as they develop and persist over time is of some interest, particularly as a background or baseline information. However, the primary purpose of space-time GIS modeling in the future, apart from providing a comprehensive description of a process in time, could be a prediction of major abrupt changes and unusual or catastrophic events at some preselected level of certainty.

## 6.1.2 Perception of Change

All dynamic environmental systems describe spatial processes that change, or are subject to change, while the notion of a 4D GIS implies a model of a standard 2D or 3D process in time. Considering time as a metric dimension may be unfortunate since it does not have the same intrinsic nature and units as space and is not dimensionally compatible with either mass, length, area, or volume normally associated with objects as we perceive them in space. It may therefore be preferable, and dimensionally more nearly correct in what follows, to think of time in terms of relative changes of position or shape associated with dynamic systems in space. For example, Kelmelis (1991) concludes that space-time interaction may be viewed as a trajectory of a spatial feature through time. A trajectory implies motion in a 3D space and continuous change.

Changes occur at a variety of scales. When expressed in terms of time and space units they may range from seconds to millennia and points to regions. In ecology a frequently used device to provide a conceptual ordering of such phenomena is a graphical representation of the space-time diagram (Johnson, 1996). In such a 3D diagram peaks and ridges correspond to major events, nested within a hierarchical process in space and time. Events occurring within the same space-time neighborhood are likely to be correlated and take place on the same scale of resolution. Using logarithmic scales on the respective axes makes it possible to represent many orders of magnitude spanned by the same process as it progresses through space and time.

Attempting to model change at an inappropriate scale would be futile. Changes may be repetitive at a variety of scales, leading to a notion of process time series, or they may also appear to be seemingly random. Although this random appearance may be due to a simple measurement error, it often results from a poor choice of scale either in space or in time. In view of these difficulties, a space-time diagram approach, coupled with an in-depth exploratory data analysis, is a highly recommended first step prior to actual modeling.

### 6.1.3 Fourth Dimension

Multidimensionality is quite usual in mathematics or computer science. For example, parallel processing requires links and a communication mechanism among the multiple processors that may share a common memory. To find the best geometry to support a particular set of computations, multiple processors may have to be linked together. In general, *topology* defines geometry of properties that remain invariant under mapping; as the term is used in GIS, it addresses connectivity of structures and spaces in 4D. One such topological arrangement is the *hypercube,* where each processor is connected to two others for a 2D square, three others for a 3D cube, and four or more others for higher-dimension hypercubes. Specific hypercube algorithms are described and readily dealt with in computer literature (Ranka and Sahni, 1990). Therefore, processing of a 4D system should present no computational problems. However, accessing the voluminous data, their indexing, sequencing, and appropriate arrangement will continue to be a problem. Clearly, when modeling the space-time dynamic processes in four dimensions, it is particularly important to use an appropriately constructed database and to define a priori the context, resolution, and scale at which changes in the process are likely to take place. Thus the value of an in-depth exploratory data analysis, focused on choosing the correct space and time perspectives for the model, cannot be overemphasized.

To deal with temporal dynamics Peuquet (1994) has proposed a *triad* framework (Figure 6.2), concluding that it has good potential for representation of space-time data and systems in GIS context. A triad is a conceptual space-time formalization below within which data models can be designed and customized for applications. To do so, existing temporal state parameters in a system model (such as land use change) may need to be modified to simulate multiple realizations to compare with observed values in the context of the triad framework.

Figure 6.2 Pequot's (1994) Triad Framework

The triad framework is particularly well suited to deal with nondeterministic process models and can accommodate complex distributions of multiple events, locations, and participants. It may be used to uncover space-time patterns and associations through modeling. Future research needs to envisage additional extension of space-time similarity logic and integration of standard spatial and temporal analysis tools into the triad framework. However, a hierarchical ordering of the graphical space-time diagram in ecology provides a similar process framework, including the additional graphical information regarding appropriate scale.

## 6.2 MODELS AND MODELING

In general, as noted previously, any dynamic environmental model is a simplified description of a far more complicated process or system taking place in time and space. Making certain assumptions about the process makes working simplifications. The validity of a model depends to a large extent on how well such assumptions are met in practice. An environmental model usually uses one or more sets of field data, observed at a limited number of locations. The key assumption, which is often not correct, is that such field data are representative of a system. For data to be truly representative, we need to have some idea how they are likely to be distributed in time and space with relation to one another, and how dependent are they on other attributes that are part of the system under study. This requires a lot of exploratory analysis. We also need to be careful how, and with what type of instruments, we acquire the field data, since the very action of making measurements, placing instruments in the ground, or introducing an observer into the system will result in a certain amount of error. Cushman (1986) points out that to minimize such errors, both the field data and the instrumentation used to acquire them need to be scaled in a similar manner. For example, when measuring soil permeability in the field, strictly speaking the values are valid only at a scale and geometry of the instrument that was used. It is a simplification not strictly correct to assign the same values to a much larger (or smaller) area of different geometry.

Koza (1992) suggests that many problems from different disciplines can be rephrased as simply requiring a discovery of a computer program to produce a desired output when presented with a particular input. As pointed out earlier, computer programs, outputs, and inputs are known by many different names. However, regardless of differences the process of discovering a computer program that produces a desired output when presented with particular inputs is known as the problem of program induction. The perspective in this chapter will be that of a dynamic environmental model that represents the past, present, or future reality based on a series of observations, or changes in observations made at a point, or on a folded surface (2D manifold). A folded surface, analogous to a point in 1D, has no physical thickness (Hazelton et al., 1990) and can be used to visualize changes as successive GIS overlays with time on the vertical axis.

### 6.2.1 Scale

In a physical sense, the notion of scale refers to process resolution, defined by an equivalent length or area that represents scaled field reality (e.g., 1 : 20,000 map scale). In raster GIS, cell sizes uniquely define the map resolution and therefore the scale of aggregation for a model. This implies that a physical scale at which a process is modeled is dictated by the spatial extent at which the input and output properties associated with the process can either be assumed to be definable or vary in some predictable manner. The concept of space-time scale in GIS attempts to combine   processes operating at different levels (slices) of space and time. As scale changes, dominance of different processes may either decrease or increase (DeCoursey, 1996).

Purely spatial analysis is well suited to geostatistical techniques of interpolation and simulation. Simulation means simply the use of models to study the nature of phenomena. In geostatistics simulation is restricted to the synthetic set of values that arise from a random function and have the same spatial variance as the original realization (Olea, 1991). Interpolation may be defined as a process of finding a value of a function $f(x)$ at $x_0$ based on information provided by surrounding values $f(x_i)$. This is similar to Koza's (1992) concept of program induction.

When modeling by interpolation (kriging), geostatistics assumes that a value of a property $z(x,)$ at location $x$ is a random variable $Z(x)$, one of many generated by a random process. This allows the user to assign to $z(x)$ an expectation $E$ and invoke an *intrinsic hypothesis*, namely that expected value of a property is constant, does not depend on position, and is said to be stationary in the mean (Journel, 1996; Webster and Oliver, 1990;). In a dynamic environmental system such as soil, profile properties will vary as a function of position, time, and degree of aggregation reflected by scale and can be related to spatial and temporal dependence of field data, support or aggregation size, and the extent of the area under study (Webster and Oliver, 1990, pp. 213–240).

### 6.2.2 Aggregation and Averaging

As shown in Table 6.1, aggregation refers to space, averaging to time. We may note that while aggregation, as viewed in a remotely sensed image, is a perfectly legitimate spatial representation of surface properties, time averaging of spatial measurements generally leads to loss of information that may be very pertinent to a process. The question is whether at a scale of resolution (i.e., a watershed or region), such loss of information is significant. Accumulation is one type of time averaging that corresponds directly to aggregation in space. When individual measurements of a property at a given time interval are too small to record, it is customary to accumulate changes over an interval long enough to give measurable differences (Wolf and Rogowski, 1991).

Literature frequently illustrates different ways environmental system properties are thought of and represented. For example, when modeling a 3D dynamic process such as water movement in soil, the aerial extent of watersheds and catchments is usually perceived on a scale of kilometers (km), landscape topography is modeled in meters (m), and soil profile properties are generally considered at a scale of centimeters (cm) or less. What may be needed in environmental modeling is a system of compatible averaging and aggregation similar to the concept of fractals (that is, a function of both time and space). Invariance under displacement and change of scale are properties of homogeneous distributions on a line, plane, or in space. Most fractals are invariant under scaling (Maldebrot, 1983). However, no comparable alternative exists for time, and the closest notion may be the assumption of stationarity in geostatistics. However, for a small enough time interval, can two or more dynamic spatial processes be considered stationary in time? The very notion of dynamics is counterintuitive to stationarity.

Thus, in the case studies that follow, a space-time approach is illustrated in four dimensions in terms of changes that take place at specific time intervals assumed characteristic of the process. To distribute the process spatially we attempt to model it first on a folded surface at a number of points and depths, followed by aerial interpolation, or simulation using geostatistical methods. Time series of fundamental variables (e.g., precipitation) are used to alert us to possibilities of change in associated system variables (e.g., soil water content).

To summarize, modeling changes in GIS that take place in a dynamic environmental process may be approached in basically two ways, following an in-depth exploratory analysis of available data. First, choose a grid size and time interval appropriate to the process under study, utilizing exploratory data analysis (e.g., hierarchical ordering of the graphical space-time diagram). Second, characterize the spatial and temporal structure of the process change through standard geostatistical and time series methodology. Third, assign appropriate attribute values to each raster or vector cell through stochastic or deterministic interpolation or simulation. Fourth, manipulate resulting overlays with an appropriate algorithm on a cell-by-cell basis.

Alternatively, a process in time can be modeled only at locations where data are available, and the outcomes then distributed over an area according to some stochastic or deterministic scheme for each time interval of interest (Rogowski, 1999a). As long as the process remains linear, there should be no difference between the two approaches. Modeling of a process at a point and distributing modeled values over an area may appear preferable in specific instances, when dealing with limited data, a complex model, and a large GIS grid (Rogowski, 1996).

### 6.2.3 Inherent Problems

There are two fundamental problems when dealing with dynamic systems and modeling 3D processes in time (4D). One is that current models are simply not geared to handle the extensive data, except at a rather limited spatial scale; the other concerns our inability to see more then one frame of output at the time.

**Extensive Data.** For example, in hydrology raster data structures, based on digital elevation models (DEM), are normally used when modeling water budgets. Unfortunately, the number of standard 30-m DEM cells for a typical watershed or region can be very large (i.e., $10^4$ to $10^6$ or more), drastically reducing number of time periods that can be analyzed at one time (Maidment, 1996b). When attempting to model systems involving extensive use of time dynamics, vector data structures may be help to group similar objects into lines or polygons. But even then the product of $L$ spatial units, with $M$ variables each, for $N$ time periods can get large very quickly. Maidment (1996b) suggests that the product $LMN$ be $\leq 10^6$ for a reasonable computer run time. To get around these limitations, various methods of averaging and aggregation of both the spatial and temporal data have been used with limited success. Every time we average the space-time data some information is lost, unless the data can be described by a known distribution function.

**Visualizing Change.** To visualize change, we are limited to viewing it one GIS frame at the time. We can, of course, stack the time frame overlays one above the other, or even animate them to show change by a rapid succession of frames, but the fact remains, as pointed out earlier, that whenever time and space are represented in unrelated units, we can never view them in a 4D space, the way we represent dimensionally consistent results in the 3D space. Consequently, the question needs to be asked how best to represent environmental change in GIS, to enhance our knowledge of the processes, to draw conclusions, and to direct our future actions.

## 6.3 TIME DIMENSION AND THE GIS

The early work of Kelmelis (1991) has provided an overview of literature on the space-time topic and has developed a conceptual data model combining vector, partitioning, and feature approaches for use in the space-time modeling. The model envisaged a trajectory of a spatial feature through time, where time is an additional dimension, orthogonal to space. Much of the current focus of the GIS community appears to be on a database structure suitable for use in 4D models (Lagran, 1993; O'Connaill et al., 1993; Peuquet and Duan, 1994). An overriding problem is that 4D data sets are usually very large; additional problems include sparseness of sampling in some directions/locations and the use of sensors that are not always compatible with one another (Mason et al., 1994). Currently, in the ongoing Apoala project at the Pennsylvania State University, Peuquet and MacEachren (2000) are developing a novel integrated, multivariate, space-time geographic information, visualization, and analysis system to explore environmental data. Applications are characterized by large amounts of data that need to be stored, retrieved, and analyzed using parallel computing technology. The Apoala system architecture consists of three related modules: a data management module based on the triad concept discussed previously, an information extraction module, and a set of visualization tools. An approach suggested by MacEachren is that of extracting a knowledge base from a multivariate space-time data. Because large environmental data sets (e.g from Eos–Earth Observing System satellites) present a processing, accessibility, and retrieval challenge, developing a timely and comprehensive space-time data analysis and processing protocol is an utmost necessity (Miller, 1991). MacEachren et al. (2000) suggest a number of novel ways of visualizing and analyzing the 4D data by integrating knowledge discovery in databases (KDD) with the geographic visualization protocol (GVis). For example, a *geoview* representation is a 3D window to display space in the two or three dimensions and time. Coded, variable size, and color symbols (glyphs) are used to represent magnitude and location of events (e.g., precipitation). Similarly, glyphs are used within a 3D *scatter plot* representation to display spatial relationships among variables, where color again can be used to represent magnitude of additional attribute. *Parallel coordinate plots* and associated techniques of *assignment, brushing, focusing,* and *manipulation* allow visual exploration of relationships among the constituent variables and attributes of the process. *Color map* manipulation (replacing and swapping colors) is used to emphasize data patterns. Data sequencing renders display of one time slice after another, while categories extraction is handled with software that provides unsupervised classification using Bayesian approach.

Bayes's theorem describes conditional probability of sampling outcome based on relative composition of sample space (Olea, 1991). For example, the probability *(P)* of drawing a white ball *(A)* from one of several urns $(B_1, B_2, ... B_i ... B_n)$ when the relative proportions of red and white balls in each is known, is as follows:

$$P\left(B_j \mid A\right) = \frac{P\left(B_j\right)P\left(A \mid B_j\right)}{\displaystyle\sum_{i=1}^{n} P\left(B_i\right)P\left(A \mid B_i\right)}$$

Data representations that have been suggested in the literature are largely enhanced extensions of the traditional GIS formats. A time slice approach to collecting space-time data is common. It involves assigning all measurements to a given time. This is not always true, unless multiple automated and synchronized sampling locations are used. Apex points of a space data polygon may then be stored in 4D space at different orthogonal time intervals, thereby allowing time slices to represent the space continuously at different times (Hazelton et al., 1990). Such formats can readily be used with existing GIS programs and tools. O'Conaill et al. (1993) discuss optimal data structure relative to handling of space-time data sets, using parallel processing methods, and suggest an improved interpolation techniques to provide increased resolution.

Although Lagran's (1993) work contains an excellent discussion of past efforts and attempts to provide a conceptual, logical, and physical basis for a design of space-time GIS, much emphasis is still on data structure, indexing, and accessibility. Lagran (1993) recognizes the complexity of the space-time problems and proposes the use of cartographic time as a distillation of reality, similar to the way cartographic space is considered. The framework is a Newtonian concept of non-interacting space and time paradigm. Some major issues with space-time data structures (Lagran, 1993) appear to be problems of what to represent, how best to update the information over time, and the efficacy of storage and retrieval for future use. Lagran and Chrisman (1988) and Lagran (1989, 1993) sum up the situation as follows: (1) Indexing space-time data is a major problem; (2) structures for storing space-time data will be large and very complex; and (3) no single data structure is expected to be universally acceptable and useful.

## 6.4 MODELING SPACE-TIME SYSTEMS

Since GIS is particularly well suited to represent static, spatially distributed processes and events in two or three dimensions, multiple articles (Goodchild et al., 1996; NCGIA, 1996), workbooks (Eastman,1999; ESRI 1995), and monographs (Burrough, 1987) devoted to GIS have been written to address different aspects of spatial modeling and representation. It has long been recognized that most hydrologic and ecologic processes are driven by rainfall and are therefore time dependent (Maidment,1993). To describe their evolution in time and space, GIS needs to have time-dependent and spatially comprehensive data structures. Many deterministic models in applied environmental sciences include a time component in some form (Goodchild et al.,1996; Maidment, 1996a; Mitasova et al., 1995). Although recent literature has been increasingly involved with modeling of the space-time systems, modeling strictly within a 4D GIS framework is less common. One reason may be that basically GIS is not a dynamic system and has few procedures for indexing, accessing, and processing of spatial data sequentially in time and space simultaneously. There are additional practical constraints of computer time and memory needed to solve a system of differential equations describing changes as they occur, necessitating numerous simplifications and approximations. Approximations include the assumption of steady state, while simplifications attempt to describe individual events considered to be representative. Hydrological processes are routinely modeled over a number of years using daily or hourly time steps, a procedure that cannot be easily accommodated within a GIS (Maidment, 1993, 1996a).

Thus although bibliography on the 4D space-time modeling is very extensive, modeling within a 4D GIS framework is far less common, with some exceptions (Maidment, 1996a,b). In most cases the use of GIS is peripheral and limited mainly to accessing data and presentation of results in a form of time slices, often produced by a separate graphical software.

Current approaches to space-time modeling are either deterministic (Rogowski, 1996), stochastic (Wikle, 2000), or geostatistical (Stein et al., 1998). Some studies offer a combination of techniques; for example, Rogowski (1999a) models recharge flux at a point in time, distributing it over an area for selected time intervals of interest using geostatistical techniques, while Wikle and Cressie (1999a) use both the stochastic and geostatistical approaches to model regional distribution of atmospheric precipitation at a regional scale. A large number of geostatistical approaches, summarized by Kyriakidis and Journel (1999), extend 3D systems to incorporate time simply as an additional dimension in Euclidean space $\Re^N$, or at best as an anisotropic scaling, where anisotropy is characterized by having different features depending on direction for $n > 1$ (Olea, 1991). Since time is strictly not a part of Euclidean space $\Re^n$, it is conceptually difficult to conceive the validity of this approach. Nevertheless the results have been promising.

## 6.4.1 Deterministic Models

Geographical information systems comprise a set of tools used to store, retrieve, transform, manipulate, and display digital spatial data on a computer for select time intervals (Burrough, 1987; Moore et al.,1993). A basic operational unit in the GIS is an overlay, or a computer-generated map of a spatially distributed data set at a given instant of time. It is used to display or arrange information into layers, which are subsequently combined (intersected) according to some rule by the program modules, which carry out specific mathematical or statistical operations within an overlay on a pixel-by-pixel basis in raster GIS, and unit by spatial unit in triangular irregular network (TIN) or vector GIS. For example, a TIN, or an irregular vector, or contour-based representation of the surface requires considerably fewer points than a raster grid such as DEM (Maidment, 1993). However, from the standpoint of spatial analysis, raster data representation is preferred (Moore et al., 1993). If time dynamics are to be considered, Maidment (1996a) outlines a ten-step approach to hydrologic modeling in GIS. Starting with basics of site description and exploratory data analysis, soil water balance, flow, and constituent transport are computed, followed by accounting for withdrawals and presentation of results. These can be shown one time frame at a time or aggregated into seasonal or annual means, with progressive loss of detail.

It is usually preferable to use vector data structure when modeling in GIS. The

attribute table generally defines the properties of select spatial units. Geographic identifi-
ers are given as columns, and attribute values are listed in rows. Values of computed variables are defined in a time table, constructed by keeping time as an index of rows with a
new field for each spatial unit. Reading down the column, the time sequence of values for
a particular spatial unit is readily apparent and directly translatable to a GIS overlay representation for a given slice of time (Maidment, 1996a).

Process oriented deterministic models (e.g., Beven et al., 1994; Walton 1992) seldom
use GIS technology, relying instead on tabular and graphical representations of needed
inputs and outputs. However, Rogowski (1996) describes a GIS-based process model that
uses as input the field-measured parameters to delineate primary recharge contributing
areas on a large watershed. To generate spatial overlays of input using kriging in a GIS
framework, the procedure utilizes values of soil water content, bulk density, hydraulic
conductivity, and depth to water at 31 locations measured over time (Rogowski,1999b).

The number of locations is a compromise between sampling efficiency and requirements of geostatistical analysis. Mathematical operations on the overlays derived from
kriging are then carried out on a pixel-by-pixel basis to predict distributions of water flux
from below the root zone, travel time to the groundwater, and recharge flux pulse at the
water table for select times aggregated on a seasonal basis. One reason for modeling
longer (seasonal) time periods is that when individual recharge fluxes from successive
storms are plotted against their projected arrival times at the water table, a major storm
event generally sweeps the percolate from lesser events ahead of it, often arriving at the
water table as a single combined front (Rogowski, 1990). Another reason is that often pollutant amounts due to individual events are quite small and need to be accumulated over a
longer time period before they can be measured.

## 6.4.2 Geostatistical Models

Standard geostatistical methods have now been used in a number of environmental studies
to characterize variability in both space and time. Geostatistical space-time models are
used to describe trends in deposition of atmospheric pollutants, to characterize variability
of geophysical parameters, to model patterns of soil moisture or spread of disease, to characterize population dynamics, and to help design sampling networks. An extensive
detailed bibliography and summary of procedures and approaches is given in Kyriakidis
and Journel (1999).

Based on the assumed joint space and time dependence between observations, geostatistical models construct a probabilistic framework to take into account contributions
associated with physical processes as they occur in time. The space-time analysis may be
focused on representing system changes, evolution of a process in time, or modeling time
series associated with a process at different locations in space. The approach acknowl-
edges that basic differences exist between space and time (e.g., scales and units are different, the concept of isotropy has no meaning, measurements in time in general refer to the
past, and estimates obtained are exploratory). According to Kyriakidis and Journel (1999),
a space-time random variable (RV) $Z(u, t)$ can take on a number of outcomes at any location in space $(u \in \Re^n)$ or instant of time $(t \in T)$, according to a probability distribution.

The analysis in general assumes that $Z(u, t)$ is fully characterized by its cumulative distribution function. However, as pointed out earlier, by extension of Heisenberg's Uncertainty Principle, a 3D object can occupy only one position in space at a given time. It follows that all variables, random or otherwise, associated with such an object can have only one value at any location $u$ in the instant of time $t$, and the distribution of such values in time and space will denote change.

In geostatistical space-time models, decomposition into space-time trends is usually based on supplemental information rather than on information derived from data. The deterministic trend is generally written as a sum of products of a function used to model it and unknown fitted coefficients. In stochastic trend models, fitted coefficients are considered random, and the value of random trend is estimated by regression (Goovaerts, 1997, p. 148). An alternate approach requires constructing vectors of space random functions or vectors of time series. Conditional simulation approaches are customarily used to model spatial uncertainty (Deutsch and Journel, 1992, 1998), while temporal uncertainty is normally addressed via the time series. The present challenge is to extend conditional simulation approaches to space-time distributions.

To model the spread of plant disease and design optimal sampling scheme, Stein et al. (1998) utilized space-time variograms to ascertain the extent of dependence and simulated annealing to design an optimal sampling scheme. Although the space-time variogram indicates some differences between variability in space and time, the representation surface is rather difficult to interpret. The modeling and analysis of Stein et al. (1998) are carried out outside the GIS context, as are several other applied case studies. For example, Hohn and Gribko (1993) also compute the variogram in the space-time domain but use indicator kriging to predict spread of gypsy moth disease. Goovaerts and Sonnet (1993) describe a multivariate geostatistical space-time approach with data that are dense in space but few in time using a factorial kriging approach, while Goovaerts and Chiang (1993) analyze persistence in spatial patterns of field nitrogen over time. All geostatistical approaches appear to extend structural (variogram) analysis, interpolation (kriging), and conditional simulation approaches directly into a dynamic space-time domain utilizing time as a fourth dimension. Despite theoretical complexity and conceptual reservations, the approach appears to be working well, as illustrated by numerous successful case studies. Because individual approaches are largely case specific, they have to be carried out outside a standard GIS context. However, as Kyriakidis and Journel (1999) point out, there is a pressing need for an error-free representative database to allow validating what often now are only subjective model decompositions.

### 6.4.3 Stochastic Models

An extensive bibliography on space-time modeling is presented by Wikle (2000). When the temporal component is absent, geostatistical methods describe spatial variability adequately. To model space-time processes, geostatistics assumes that in general covariance structures are specified. For example, when standard kriging methodologies are extended into the space-time domain, time is treated as just another dimension. Similarly, multivariate or cokriging approaches consider individual time slices of spatial fields as variables, and for implementation purposes separation between time and space is assumed. When there is no space component, autoregressive error processes are generally used to represent temporal variability. When both the space and time components of variability are present, a process may have to be considered as a collection of spatially correlated time series in a continuous space (Wikle et al., 1998). A state-space formulation may include an additional error component with the solutions obtained using Kalman filters (Kalman, 1960).

True space-time dynamic models composed of a state process plus a measurement error consist of a spatial process that evolves in time and a component that accounts for spatial structure but does not change in time (Wikle and Cressie, 1999a). Although all these approaches have been very productive, it has been difficult to estimate the necessary parameters or include detailed descriptions of relationships between space and time variables. Generally the applications are limited, and much physical knowledge about the process cannot be incorporated into the model. It is believed that a Bayesian approach can overcome some of these shortcomings. It appears to have more flexibility in space-time interactions and can accommodate elements of physical knowledge as well as lack of stationarity, or linearity (Wikle and Cressie, 1999b; Wikle et al., 1998). Because of the complexity of the approach and case-specific formulations, it is doubtful whether it can be incorporated within a GIS framework.

A space-time hierarchical modeling approach appears promising in organizing variables into three levels that could be accommodated within a 4D GIS (Peuquet and MacEarchren, 2000). The three levels defined by Wikle (2000) are: (1) data; (2) state variables of interest; and (3) Statistical parameters and physical variables.

This leads to a formulation of the space time problem as

$$(process, parameters \mid data) \propto (data \mid process, parameters)[process \mid parameters][process]$$

where [ ] represents a probability distribution given ( | ) *data, process, or parameters*; $\propto$ means "proportional to"; $\in$ means "contained in"; and $\cup$ means "intersection". The objective is to find a posterior distribution of a space-time field $Y(s; t)$, given observations $Z(r; t)$ with $s$ and $r \in D$ a space domain, $t \in \mathfrak{I}$ an index of discrete times, and $(r; t)$ a member of the intersection of the time and space domains $(D \cup \mathfrak{I})$:

$$Z(r; t) = f_d(Y(s; t); \xi(r, s; t)) + \varepsilon(r; t) \tag{6.1}$$

$$Z(r; t) = k_t(r)Y_t) + \varepsilon(r; t) \tag{6.2}$$

$$Y(s; t) = \int_D w_s(u; t - \tau)Y(u; t - \tau) \, du + \eta(s; t) \tag{6.3}$$

where $f_d( )$ is a representation [equation (6.1)] of a relationship between a true process $Y(s; t)$ including [equation 6.3] spatial noise $\eta(s; t)$, and the observed values $Z(r; t)$; $\xi(r, s; t)$ designate parameters relating observations at $r$ to the process at $s$; and $\varepsilon(r; t)$ describes the measurement error [equation (6.1)]. In the predictive format [equation (6.2)], $k_t(r)$ is a vector to map observations from a location $r$ to prediction grid; $Y_t \equiv (Y(s_1; t)... Y(s_n; t)$ is a vector of $Y$ at $n$ prediction locations; and $w_s(u; t - \tau)$ in equation (6.3) is the space-time interaction function (Wikle, 2000).

In the associated case study Wikle (2000) applies the formulation to oceanic wind fields in the tropics represented by linearized set of partial differential equations considered within hierarchical space-time model for a relatively sparse data set. Results given for selected dates and times demonstrate how a model can be used. Simplifications include consideration of primarily linear equations and process interactions. Although the framework allows for coupling of different processes, such coupling is not explicitly considered.

In a related application, Wikle and Cressie (1999a, 1999b) utilize spatially a descriptive geostatistical approach combined with temporally dynamic time series to construct an SDTD model that exploits unidirectional passage of time. Implementation leads to the development of enhanced space-time form of Kalman filter. When applied to 30 years of monthly precipitation data, the spatially descriptive temporally dynamic (SDTD) model appeared to describe dynamic development of spatial processes adequately. The use of the enhanced Kalman filter allowed predictions to be made in time and space, accounting for missing data.

The elegant space-time models developed by Wikle (2000) and Wikle and Cressie (1999a, 1999b) appear to describe the evolution of space time processes over an area and in space adequately. They are, however, complex and to a large extent case specific. Generalizing and simplifying the associated methodology for use within a GIS would be difficult. As is the case with most other space-time approaches, results can be presented only one time frame at a time.

## 6.5 CASE STUDY 1: POTENTIAL FOR GROUNDWATER POLLUTION

This case study, cast in an Idrisi GIS framework (Eastman, 1999), illustrates an approach when temporal changes at a scale of a watershed are sufficiently slow to be considered on a seasonal basis. The approach demonstrates the potential distribution of contaminated recharge flux at the water table in space and time. It is assumed that recharge flux originates as seepage from below the root zone. Field observations, gathered over a period of one year, are designed to measure the quality of recharge at selected locations The model utilizes these measurements together with field estimates of soil parameters to predict the quality of recharge at the water table on a typical agricultural watershed (Rogowski, 1996, 1999a). The case study illustrates the application of the quantity-arrival-time distribution methodology of Nelson (1978a, 1978b). Probability kriging (Deutsch and Journel, 1992, 1998; Goovaerts, 1997) is used to project the spatial and temporal distribution of risk associated with exceeding threshold loading rates.

### 6.5.1 Model Development

**Recharge Flux.** The model (Rogowski, 1996), cast within a GIS framework, is based on the mass balance of soil water within a hydrologic catchment. Distribution and timing of recharge flux depend on the mass balance of soil water in the plant root zone, which controls the partitioning of precipitation, and can be written as

$$P = R_{chg} + R_{off} + ET + \Delta\theta$$

where $P$ is the precipitation and $R_{chg}$ is the deep percolation component of recharge flux, while $R_{off}$, $ET$, and $\Delta\theta$ represent losses due to surface runoff and evapotranspiration and changes in water storage.

To account for the spatial and temporal variability of soil water within a root zone, groundwater recharge flux is assumed to be a function of the spatially distributed profile storage capacity $\theta_s$, antecedent water content $\theta_{atc}$, and precipitation $P$, which change in time, and the profile physical properties, which remain constant or change very slowly. The average water content $\bar{\theta}$ at the bottom of the root zone, where there was little or no change in soil water, is used to model the $\bar{\theta}_{atc}$ distribution over a catchment.

Area-wide recharge, flux $\int_A q_j \, dA$, is computed on a pixel-by-pixel basis within a GIS as outflow from below the root zone over time:

$$\int_A q_j dA = -\int_{T_r}^{(T_r + T_{tot})} \{ K_0(j) / ([1 + \bar{\alpha}_j K_0(j) T / L_j]) \} c \qquad (6.4)$$

where $L_j$ is the spatial distribution of root zone depths, $\bar{\alpha}_j$ is the average, area-wide slope of the semilogarithmic soil hydraulic conductivity - $K(\theta)_j$ function, $T_r$ is the residence time of water in the root zone, and $T_{tot}$ is the event duration. The $K_0$ and $\theta_0$ values represent steady-state values of soil water content and hydraulic conductivity near field saturation ($\theta_{fs}$, $K_{fs}$) and the index ($j$) is used to denote location. Recharge flux from below the root zone will continue as long as there is water in excess of $\bar{\theta}_j$ in the root zone. The cumulative seasonal recharge flux ($\int_A q_{j-cum} \, dA$) can then be approximated as

$$\int_A q_{j-cum} dA = \sum_{T_r}^{T_r + T_{tot}} \left( \int_A q_j dA \right) \Delta t, \qquad \leq \left( \bar{\theta}_{Atc} - \bar{\theta}_j \right) \qquad (6.5)$$

where again $T_r$ is the residence time in the root zone, $T_{tot}$ represents total rain duration during a season, and $\Delta t$ refers to a time step used in the summation. Computations can be carried out on an event-by-event basis, or aggregated to represent any suitable time interval. One reason for modeling seasonal recharge is that a major storm can sweep the percolate from lesser events ahead of it, arriving at the water table as a single front; alternately, concentration of contaminants derived from a single event can be so low that some type of aggregation is required to measure them.

**Flux Pulse.** In coarse-textured, stony soils the effective porosity of the profile through which the flow takes place ($P_{eff}$) can be estimated as a fraction of the total pore space $P_{tot}$ (e.g., 0.7), or chosen from a uniform random distribution (Monte Carlo). Spatial distributions of field saturated hydraulic conductivity $K_{fs}$ provide initial estimates of recharge flux ($q_j \approx K_{fs}$) from below the root zone. When $q_j$ estimates are combined with $P_{eff}$, root ($L_r$), and vadose zone ($L_w$) depths, estimated residence time in the root zone $T_r(j)$ and travel time to the water table $T_w(j)$ can be written as:

$$T_r(j) = P_{eff} [L_r(j) / q_j] \qquad (6.6)$$

$$T_w(j) = P_{eff} [L_w(j) / q_j] \qquad (6.7)$$

To describe a time history of recharge flux departure from below the root zone and arrival at the water table, spatial distribution of the $\int_A q_{\text{j-pulse}}\, dA$ is reconstructed by computing the difference between the integrals of the first and last cumulative recharge flux fronts leaving the root zone and arriving at the water table. The first front will arrive at $T_{w\text{-}j}$, and the first and last fronts will be separated by a time interval $(T_r + T_{\text{tot}})_j$, and $T_{\text{tot}}$ represents duration of a precipitation event:

$$\int_{T_w}^{T} q_{j-cum}\, dT - \int_{T_w + T_r + T_{tot}}^{T} q_{j-cum}\, dT, \qquad q_{j-cum} \leq \overline{\theta}_{atc} - \overline{\theta}_j \qquad (6.8)$$

where $dT$ is the time increment and $T$ is the total time $[T \cong T_w + 2(T_r + T_{tot})]$. Spatial distributions of input parameters were interpolated by kriging from field measured data. The kriged values were assigned to each pixel, creating a GIS overlay. Such overlays were subsequently combined [equations (6.5) through (6.9)] pixel by pixel to give distributions of flux-pulse ($\int_A q_{\text{j-pulse}}\, dA$) below the root zone, and travel time ($T_{w\text{-}j}$) to the water table. Continuity of variables was tested using variogram analysis and cross-validation techniques. When combined with concentration of contaminant leaving the root zone the overlays describe spatial distribution of groundwater loading with a contaminant on a seasonal, annual, or event basis.

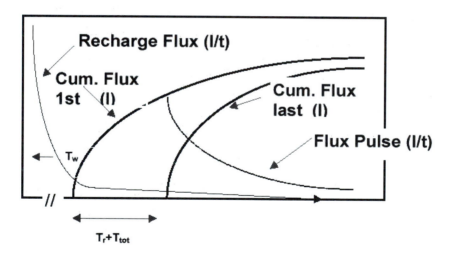

**Figure 6.3.** Schematic representation of quantity/arrival time distribution of Nelson (1978a, b) to describe recharge flux inflow history at the water table.

### 6.5.2 Model Implementation

**Catchment Description.** The model was applied to a 12,300-ha Mahantango Creek watershed in east-central Pennsylvania (Rogowski, 1999b). The watershed (Figure 6.4) is situated in the Valley and Ridge hypsographic province of the Appalachians. It is divided into upland hills, valleys, and mountain ridges dissected by streams. Elevation ranges from

150 m (MSL) in the valleys to 500 m at the ridge tops. The ridges and valleys run north-east-southwest, corresponding to the regional strike of the major rock formations. The area is underlain by interbedded sandstones, siltstones, and shales. Rock beds dip in the north-northwest direction in the northern half, and in the south-southeast direction in the southern half of the watershed. Land use is predominantly cropland in a rotation, contour strip management system of corn/small grain/meadow, with areas of permanent pasture, truck crops, woodlands, and orchards. Most ridges are forested, except for the areas that have been cleared for cultivation. The climate is humid with a udic moisture regime.

**Monitoring Protocol.** Moisture characteristic and hydraulic conductivity were evaluated on five representative farm areas ranging in size from 17 to 63 ha. All sampling locations were identified in UTM coordinates (Universal Transverse Mercator) using a commercial global positioning system (GPS). Monitoring included biweekly collection of root zone percolate from specially designed collection trenches at 31 base sites and biweekly measurement of soil water content at 0.6 m, evapotranspiration from 0.3 m deep columns of soil, and precipitation in wedge-shaped rain gages calibrated to the recording gage at the Met Site (Rogowski, 1999b).

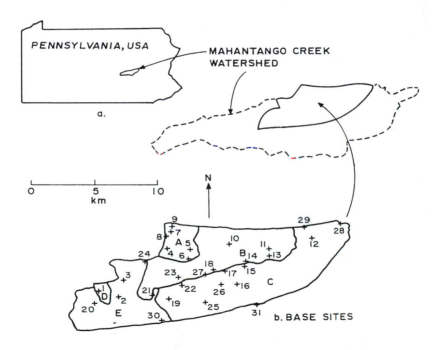

**Figure 6.4** Location of the Case Study I area, component sub-watersheds, and primary base sites.

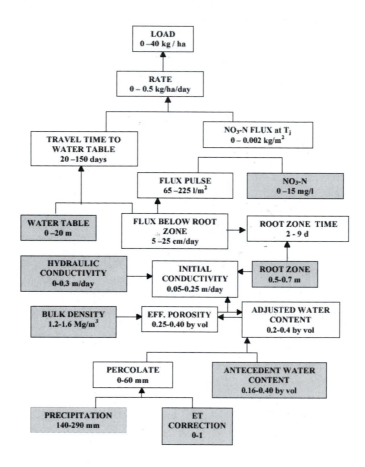

**Figure 6.5.** Flow diagram of the model representing individual overlays; lists the process/ variable, range of values, and algorithm; shaded overlays (KRIGE) represent kriged distributions.

**Nonpoint Source Pollution.** Nitrate ($NO_3$-N) concentration in root zone leachate was used as a potential source of nonpoint source pollution. Because concentrations of $NO_3$-N in leachate for individual storm events were very small, the conventions were captured by filtering leachate through the flow-through ion exchange resin (Riha, 1980; Rogowski, 1999b). Accumulated nitrate was washed off the resin with 2N KCl every 3 months, at the end of spring, summer, and fall, and analyzed using an automated cadmium reduction method.

Kriged overlays of $NO_3$-N concentration ($mg.L^{-1}$) in the root zone percolate were based on analysis of trench outflow. Nitrate flux from below the root zone was estimated by combining the flux pulse ($\int_A q_{j\text{-pulse}}\, dA$, 1 m$^{-2}$ d$^{-1}$) overlay below the root zone on a given day, with the average measured NO$_3$-N concentration (mg.L$^{-1}$) to give an overlay of spatially distributed nitrate flux (mg.m$^{-2}$) below the root zone. Divided by travel time to the water table $T_{w\text{-}j}$, converted to kilograms per hectare (kg/ha), a new overlay (RATE kg/ha/day) describes the potential rate of groundwater loading with nitrate. The value of $T_r +$ $T_{tot}$ will, in general, be short compared with $T_w$. Multiplying the RATE overlay by the number of days with recharge and integrating over the area gives an estimate of a seasonal load delivered to the groundwater.

   **Output.** Figure 6.5 shows a model flow chart, illustrating inputs, some of the inter-mediate results, and GIS output. Individual boxes represent GIS overlays for spring. Lines and arrows indicate internal relationships and the computational chain within the GIS. Shaded boxes (KRIGE) signify overlays where values have been kriged, otherwise an

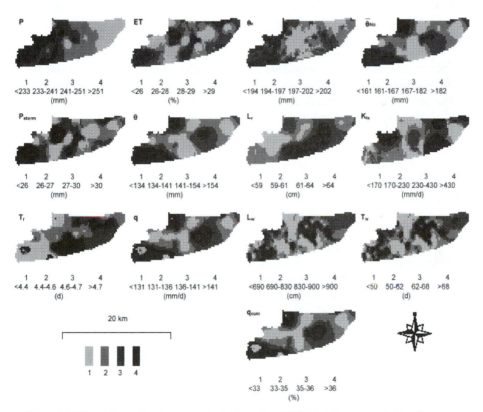

**Figure 6.6** GIS overlays of model parameters in the spring, 1990. First Row: precipitation ($P$), evapo-transpiration correction ($ET$), profile storage capacity ($\theta_s$), and antecedent water content ($\theta_{atc}$); Second Row: infiltrate corrected for $ET$ ($P_{storm}$), average seasonal water content ($\theta$), root zone depth ($L_r$), and field saturated hydraulic conductivity ($K_{fs}$); Third Row: residence time in the root zone ($T_r$), recharge flux ($\int_A q_j dA$) from below the root zone at one day ($q$), depth to water table ($L_w$), and travel time to water ($T_w$); Last Row: cumulative recharge flux ($\int_A q_{j\text{-}cum} dA$) as a function of Met Site precipitation ($q_{cum}$). The recharge flux was computed at $\alpha=2.1$ d$^{-1}$, all legends represent interquartile range: $<Q_1$ (1), $Q_1$ to $Q_2$ (2), $Q_2$ to $Q_3$ (3), and $>Q_3$ (4).

appropriate algorithm is indicated. Aggregating of precipitation into four major events for each season accounts for more than 90% of the precipitation. Table 6.2 shows standard statistics of the model variables. Distributions, except for $K_{fs}(j)$, appear normal or near normal, because most are based on interpolated values of input, and interpolation by kriging tends to smooth the differences in raw data. Figure 6.6 illustrates spatial distribution of model variables, where the legend, coded 1 to 4, represents the quartiles ($Q$).

Overlays in Figures 6.6 illustrate distributions of two sets of parameters during spring. Parameters $P$, $ET$, and $\bar{\theta}$ define antecedent conditions ($\theta_{atc}$), while parameters $L_r$, $L_w$, $K_{fs}$, $K_0(j)$, $\alpha_j$ and $P_{eff}$ enter directly into the computation of recharge flux ($\int_A q_j \, dA$) and travel times ($T_r$, $T_w$). Statistics in Table 6.2 and overlays in Figure 6.6 suggest that parameters that define the antecedent conditions do not vary sufficiently at a watershed scale to generate major seasonal differences in the $\bar{\theta}_{atc}$ overlays. This becomes particularly important when choosing how $K_0(j)$ values are to be determined. Estimated values of $K_0(j)$ were more often based on a less variable $K(\theta_j)$, from the Ahuja and Williams (1991) approximation:

$$K[(\theta_j)] = \exp[\ \bar{\beta}_1 + \ \bar{\beta}_2 \ln[(\theta_j)]] \tag{6.9}$$

than on the more variable $K_{fs}(j)$ measured with a Guelph permeameter (Rogowski, 1996), where in equation (6.10) $\bar{\beta}_1 = 1.31$ and $\bar{\beta}_2 = 0.98$ represent average values for all locations. The preceeding coefficients are specific to the Mahantango Creek soils; they are based on field measured data and are numerically different from those listed in Gregson et al. (1987) and Ahuja and Williams (1991).

Parameters that enter directly into the computation of travel time and recharge flux tend to influence model output strongly. The $L_w$ and $T_w$ overlay patterns in Figure 6.6 are very similar, while both the $q$ overlay and the $q_{cum}$ overlays bear a close resemblance to a kriged distribution of $K_0(j)$. When $K_0(j)$ values are set equal to the $K_{fs}(j)$ distribution, the overlay is very similar to $K_{fs}(j)$ distribution. The actual numerical values for $T_w(j)$ may differ, but the pattern will remain the same as long as $P_{eff}$ remains constant. If, however, $P_{eff}$ does not remain constant but varies spatially (as, for example, in the Monte Carlo simulation), the pattern of contributing areas will change.

Results suggest that soil water at the 0.6 m depth remains sufficiently stable in our area throughout the season, to model drainage flux from below a root zone under a unit gradient assumption. The model is very sensitive to the choice of parameter $\alpha_j$ [the slope of the $K(\theta_j)$] and the distributions of soil hydraulic conductivity $K_0(j)$ and effective porosity $P_{eff}$. Since at a catchment scale distributions of both $\alpha_j$ and $K_0(j)$ should remain reasonably constant or change only slowly in time, we would expect the largest differences in model predictions to be due to the changes in effective porosity $P_{eff}$ through which the flow takes place.

**Probability Kriging.** Probability kriging (PK) (Deutsch and Journel, 1998; Goovaerts, 1997) expresses spatial distributions of a variable as a probability of exceeding a chosen threshold. It is similar to cokriging, where two or more variables are estimated jointly, based on structural codependence. The indicator and rank order transformed data for each threshold are the two variables in PK system.

The procedure is as follows: A distribution of values is examined and a threshold and several cutoffs are chosen. The cutoffs divide the ranked and ordered data set into groups with approximately the same number of data points. For each cutoff, the numerical data values are transformed into a set of zeros and ones using an indicator transformation. Subsequently the ranked and ordered data set is subjected to a uniform transformation, whereby the $N$ ranked and ordered data values each receive a weight equal to $1/N$ times their rank.

**Table 6.2** Statistics[a] for model Input, Intermediate, and Output Variables,[b] during spring (S), summer (Su), and autumn (A).

## Kriged Inputs

| PARAM | UNIT | T | M | Cv% | MAX | MIN |
|---|---|---|---|---|---|---|
| $P_j$ | mm | S | 242 | 5 | 278 | 198 |
| | | Su | 262 | 5 | 288 | 230 |
| | | A | 187 | 5 | 203 | 141 |
| $ET(j)$ | % | S | 27.7 | 16 | 64.0 | 12.8 |
| | | Su | 8.9 | 35 | 17.3 | 3.8 |
| | | A[d] | 100.0 | — | — | — |
| $\overline{\theta}(j)^e$ | mm | S | 148 | 10 | 200 | 105 |
| | | Su | 131 | 5 | 146 | 120 |
| | | A | 154 | 7 | 187 | 127 |
| $K_{fs}(j)$ | mm/d | | 388 | 104 | 2620 | 70 |
| $\alpha_j$ | $d^{-1}$ | | 6.3 | 55 | 18.2 | 3.0 |
| $Bd_j$ | mg/m$^3$ | | 1.40 | 3 | 1.54 | 1.28 |
| $L_r(j)$ | m | | 0.62 | 5 | 0.70 | 0.55 |
| $L_w(j)$ | m | | 8.15 | 25 | 17.01 | 2.85 |
| $NO_3$-N $(j)$ | mg/L | S | 4.0 | 78 | 14.6 | 0.2 |
| | | Su | 5.8 | 28 | 10.1 | 3.0 |
| | | A | 3.4 | 79 | 14.8 | 0.5 |

## Computed Variables

| PARAM | UNIT | T | M | Cv% | MAX | MIN |
|---|---|---|---|---|---|---|
| $P_{storm}(j)^f$ | mm | S | 28.0 | 17 | 60.0 | 13.7 |
| | | Su | 8.4 | 35 | 16.7 | 3.4 |
| | | A | 49.2 | 6 | 53.5 | 37.1 |
| $\theta_s(j)$ | mm | | 205 | 3 | 225 | 181 |
| $\overline{\theta}_{atc}(j)$ | mm | S | 177 | 9 | 237 | 135 |
| | | Su | 140 | 5 | 158 | 126 |
| | | A | 203 | 6 | 235 | 172 |
| $K_0(j)$ | mm/d | S | 252 | 10 | 362 | 95 |
| | | Su | 203 | 5 | 229 | 183 |
| | | A | 270 | 31 | 860 | 93 |

**Outputs**

| PARAM | UNIT | T | M | Cv% | MAX | MIN |
|---|---|---|---|---|---|---|
| $\int_A q_j dA$ | cm/d | S | 13.5 | 6 | 16.0 | 7.1 |
| | | Su+ | 12.0 | 8 | 12.7 | 11.0 |
| | | A | 13.8 | 16 | 22.4 | 7.0 |
| $\int_A q_{cum}(j)dA^g$ | $1/m^2$ | S | 135 | 6 | 160 | 71 |
| | | Su | 120 | 8 | 127 | 110 |
| | | A | 138 | 16 | 224 | 70 |
| $T_r(j)$ | d | S | 4.6 | 25 | 8.3 | 3.8 |

| PARAM | UNIT | T | M | Cv% | MAX | MIN |
|---|---|---|---|---|---|---|
| | | Su | 5.2 | 25 | 5.7 | 4.6 |
| | | A | 4.6 | 28 | 8.7 | 2.8 |
| $T_w(j)$ | d | S | 60 | 15 | 117 | 21 |
| | | Su | 68 | 17 | 137 | 25 |
| | | A | 60 | 17 | 141 | 23 |
| $NO_3$-N flux | $g/m^2$ | S | 0.54 | 71 | 1819 | 21 |
| | | Su | 0.70 | 27 | 1240 | 340 |
| | | A | 0.49 | 81 | 2094 | 35 |
| RATE | g/ha/d | S | 96 | 77 | 379 | 3 |
| | | Su | 112 | 44 | 370 | 28 |
| | | A | 92 | 95 | 454 | 4 |
| LOAD | kg/ha | S | 9.3 | 77 | 36.7 | 0.2 |
| | | Su | 10.3 | 44 | 34.4 | 2.6 |
| | | A | 8.1 | 95.4 | 40.0 | 0.3 |

[a] Statistics: $M$ = mean; Cv% = coefficient of variation; Mode; MAX = maximum; MIN = minimum.
[b] Input: Kriged: $P$ = precipitation; $ET$ = evapotranspiration correction; $\theta$ = average water content at the bottom of the root zone; $K_{fs}$ = hydraulic conductivity ($K$) at field saturation; $\alpha$ = slope of the $K(\theta)$; BD = bulk density; $L_r$ = depth of the root zone; and $L_w$ = depth to the water table; $NO_3$-N $(j)$ = $NO_3$ - N concentration in the outflow from below the root zone. Intermediate: $P_{storm}$ = infiltrate corrected for $ET$; $\theta_s$ = profile storage capacity; $\theta_{atc}$ = average antecedent water content in the root zone; and $K_0$ = value of $K$ used in computation of recharge flux. Output: $\int_A q\, dA$ = flux below the root zone; $\int_A q_{cum} dA$ = cumulative flux at the water table; $T_r$ = residence time in the root zone, and $T_w$ = travel time to the water table.
[c] Spring (S) = 3/22 –6/22/90; summer (Su) = 6/23–9/22/ 90; autumn (A) = 9/23–12/19/90.
[d] No $ET$, 100% seepage below 0.3 m depth.
[e] Values of $\theta_j$, $\theta_s$ and $\theta_{atc}$ assume average profile depth of 0.62 m.
[f] Average estimated value per event at 4 events/season, normalized to Met Station reading. [g] Cumulative flux at the water table after one day, as percent of seasonal precipitation at the Met Site.

Following the computation of variograms and cross-variograms of the uniform and indicator data sets for each cutoff, the probability kriging estimator $F[\ ]$ is constructed to compute the probability of exceeding a given threshold value on a location-by-location basis. The estimator $F[\ ]$ is a conditional estimate drawn from a distribution of a random variable $Z(x)$, given $N$ neighboring observations $z(x)$:

$$F[z(x)|N] = P[Z(x) < z(x)|N]$$

$$= \sum_j a_j(z;x) \bullet i(z;x) + (1/N)\sum_j b_j(z;x) \bullet r(x_j) \qquad (6.10)$$

here $a_j$ and $b_j$ are the kriging weights, $i(\mathbf{z};x_j)$ refers to the indicator transformation and $r(x_j)$ to the ranked distribution.

The potential rate of groundwater loading with nitrate can be expressed in actual or probabilistic terms, and contributing source areas can be identified. In this case study the LOAD threshold was set at 5 kg/ha for each time period. Figure 6.7 shows the potential distribution of the total load (LOAD) for each time period, boundaries (dashed) of the component subwatersheds (A, B, C, and D, Figure 6.4), and the probability contours of exceeding the threshold during each season. Results suggest that subwatersheds A and C were potentially the largest contributors to $NO_3$-N load. In the spring and fall, the contributions were primarily from subwatershed C; in the summer major contributions were from subwatershed A followed by C and to an extent B. These differences may be due to differences in the amount of land farmed, farming practices, or nitrogen management.

Probabilistic assessment of nonpoint source pollution suggests areas where better

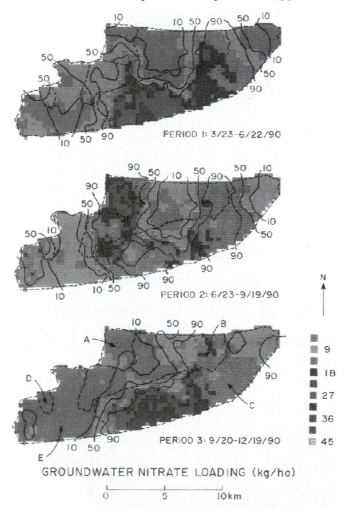

Figure 6.7 Distribution of total nitrate load (LOAD) on the case study area and isopleths of the probability of exceeding a threshold of 5 kg/ha $NO_3$-N in recharge.

management may be needed. For example, the 90% isopleth in Figure 6.7 delineates areas that could exhibit $NO_3$-N loads as large as 14 kg/ha, but also includes some locations where the loads are much lower. Should a threshold of 5 kg/ha be enforced, these locations would be the primary candidates for better management.   The areas falling within 50 to 90% may need attention, while those between 10 and 50% and below 10% would not.

### 6.5.3 Discussion

The case study describes a model used within a GIS framework to delineate areas potentially contributing $NO_3$-N to the groundwater recharge on a seasonal basis on an agricultural watershed. The model was assembled in a GIS format and applied to a 123 km$^2$ watershed in central Pennsylvania. Input data were based on field values measured in biweekly time intervals. Values for individual locations were interpolated by kriging and assigned to a GIS raster grid. Although the data in this example were aggregated on a seasonal basis, the mechanism can handle individual events. Univariate statistics appeared normal or near normal. Differences in $NO_3$-N concentration in the three seasons carried over to loading rates at the water table. There were five subwatersheds in the study area. Two of these were potentially the largest contributors to $NO_3$-N loads that varied with season. Potential risk of exceeding a given $NO_3$-N threshold in recharge was computed as a probability of exceeding a loading rate of 5 kg/ha. Primary concerns with the model involved realistic choices of input parameters. Some, such as water content at the 0.6 m depth in our area, remain sufficiently stable throughout the season to model hydraulic conductivity and recharge flux from below a root zone at steady state, with most variability due to temporal inputs of $NO_3$-N.

## 6.6 CASE STUDY 2: CHANGES IN SOIL WATER DISTRIBUTION

The second case study illustrates capturing short-term space-time changes in soil water resulting from intermittent precipitation events during a dormant season and their subsequent effect on stream flow and groundwater recharge. Data are analyzed within a GIS (Arc Info) framework utilizing geostatistics and overlay manipulation to delineate areas where changes in soil moisture are taking place. Changes are then related to the fluctuations of water level in shallow wells and stream flow. During a dormant season, lasting in central Pennsylvania from mid-November to mid-March, water content changes in the root zone are generally assumed to be very limited. Soil is often frozen, and some precipitation falls as snow. Despite the apparent weather limitations, shallow wells (piezometer tubes) placed in natural drainage locations and monitoring gages on streams draining the area register recharge to the water table and moderate runoff events fed by interflow. The objective of the study was to evaluate soil water distribution and mass balance at a farm scale in space and time and to relate them to a priori estimates of soil properties. The latter included detailed soil core identifications, a complete database of associated soil mapping unit properties, as well as digital terrain, soils, and land use maps, and infrared imagery (Rogowski, 1996, 1997, 1999b; Rogowski and Wolf, 1994).

### 6.6.1 Site Description

Figure 6.8 shows the location of instrumentation, topography, and land use at the case study site. A 48-ha farm scale catchment is located on the Mahantango Creek watershed in Pennsylvania. The study was designed to measure soil water at two depths (0.3 and 0.6 m) with a Troxler moisture (neutron) depth probe on a weekly basis in a dense network of 50 georeferenced sites, and to monitor water table response with four piezometer tubes strategically located in principal drainage areas. Stream outflow changes were recorded as event hydrographs at the catchment scale. Soil temperature at three depths (surface, 0.3 m, and 0.6 m), precipitation, and associated weather variables (air temperature, wind speed, relative humidity, snow depth) were recorded continuously by the weather station (Met Site) located at the site. Using automated Troxler Sentry probes, soil water was recorded hourly at two depths (0.3 and 0.6 m) and four locations near the Met Site. The land is farmed in alternating strips of corn and alfalfa with grassed or wooded waterways.

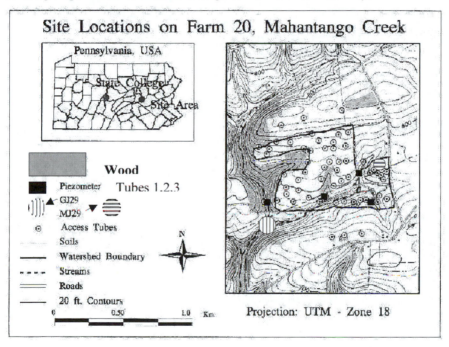

**Figure 6.8** A map of the Case Study 2 area on the Mahantango Creek watershed in Pennsylvania, underlying topography, soils, and location of the Met Site precipitation gage (MJ-29), streamflow recording gage (GJ-29), four non-recording piezometers, and 50 sample and soil moisture access

### 6.6.2 Results

Figure 6.9 shows distribution in time of the dormant season precipitation, piezometer tube response at four locations, and stream flow hydrograph. Negative piezometer depths (Tube 3) signify ponding conditions and surface runoff. Figure 6.10 illustrates time distribution of temperature and soil water at 0.3 and 0.6 m depths.

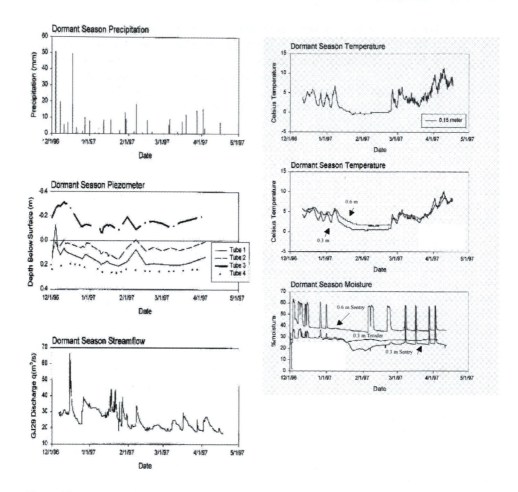

**Figure 6.9** (left) Distribution of dormant season precipitation at the Met Site (MJ-29), four weekly piezometer water levels referenced to soil surface, and stream flow at GJ-29.

**Figure 6.10** (right) Distribution of dormant season temperature at 0.15, 0.3, and 0.6 m, and distribution of soil moisture readings at 1-hr intervals.

Changes in soil water content at 0.3 and 0.6 m that in Figure 6.10 appear to take a form of pulses that correspond either to rapid percolation, or interflow following a rain event. Rapid percolation and recharge occur when both depths register a pulse accompanied by rise in water levels and increased stream flow in Figure 6.9. A pulse at only the 0.6 depth and a rise in water table and stream level following precipitation signify interflow. No pulse but a rise in ponded water and stream flow following rain signify surface runoff. Soil surface temperature in Figure 6.10 followed ambient and diurnal fluctuations. Temperature generally rose following a rain event. The ground never froze at 0.3 and 0.6 m; likewise, soil surface in December through the beginning of January was above freezing and then froze in January but began to thaw after about Feburary 20. Piezometer Tubes 1,

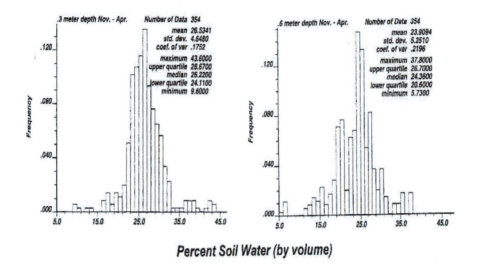

**Percent Soil Water (by volume)**

**Figure 6.11.** Histograms of the November to April distribution of soil water measured with a depth moisture probe at 0.3 and 0.6 m.

3, and 4 as well as the stream (Figure 6.9) responded rapidly to rain in December, accompanied by rapid percolation of event pulses and recharge (Figure 6.10). Water levels generally declined in January except for some surface runoff events in mid-January accompanied by a rise in ponded water (piezometer Tube 3), and increased stream flow. Beginning of February saw the first major interflow event, while the second one barely registered in the piezometer tubes. Rains in March and April contributed to a steady rise of water table at all locations, accompanied by high pulses of percolate recharge and modest hydrographs. Distribution of soil water, temperature, water table, and stream levels in time considered in the context of mass balance accounting looked promising. However, changes in soil water content were confined to select areas, which subsequently appeared to drain rapidly to the stream. At other locations frozen conditions and only minor changes in soil moisture prevailed. Figure 6.11 illustrates the distribution of soil water based on weekly depth moisture probe readings November to April. When compared with the maxima for the pulses in Figure 6.10 (about 60%), the interquartile range ($Q_1-Q_3$) appears narrow at both depths (24 to 29% by volume at 0.3 m, and 21 to 27% at the 0.6 m). The structure of both distributions is shown by raw monthly variograms in Figure 6.12. A substantial nugget at both depths suggests a considerable measurement error at a scale smaller than the scale of resolution, and continuity range to within 300 to 400m. The sill would be expected to be somewhat lower for the 0.3 m depth with the distribution less variable than at 0.6m.

Table 6.3 illustrates distribution of soil water relative to in situ landscape and soil properties, land use, and topography. Water contents were high on more level areas with slopes < 8%, concave-convex landforms, wood- or grass- covered waterways, and soils (Fragiudults) with a water-impeding fragipan layer. They appeared to be lower on concave-concave slopes of 15 to 25%, fallow fields where corn has been harvested, and coarse, stony soils (Dystrochrepts).

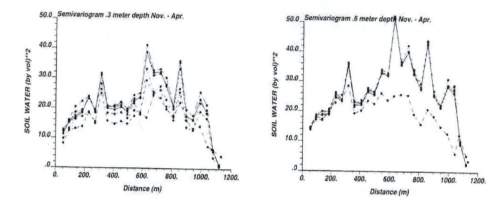

**Figure 6.12** Raw semivariograms of monthly soil water content measured with a depth moisture probe at 0,3 and 0.6 m.

**Table 6.3** Soil and landscape properties associated with high and low water contents

| Water Content | Slope | Land form | Land Use | Soil[c] |
|---|---|---|---|---|
| High | ≤ 8% | Concave-convex[a] | Woods/waterways | Fragiudults |
| Low | 15-25% | Concave-concave[b] | Corn ground | Dystrochrepts |

[a] Concave slope, convex contour

[b] Concave slope, concave contour

[c] Soil Survey Staff. (1975).

**Table 6.4.** Proportion of the area with an overlay ratio >1.0 indicating the amount of area with decreased water content in the current month (denominator) over the previous month (numerator). Values in brackets indicate the proportion of the area with increased water content.

| | Overlay Ratio Change | | | | |
|---|---|---|---|---|---|
| Depth | Nov/Dec | Dec/Jan | Jan/Feb | Feb/Mar | Mar/Apr |
| 0.3 m | 55%(45) | 13%(87) | 65%(35) | 14%(86) | 28%(72) |
| 0.6 m | 18%(82) | 0%(100) | 7%(93) | 3%(97) | 9%(91) |

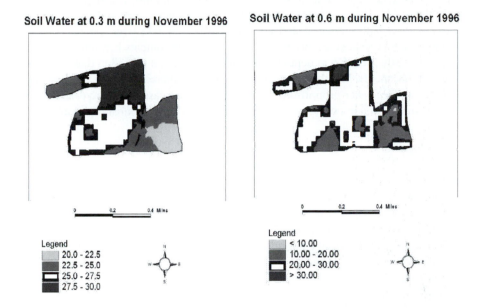

**Soil Water at 0.3 m during November 1996**          **Soil Water at 0.6 m during November 1996**

Legend
20.0 - 22.5
22.5 - 25.0
25.0 - 27.5
27.5 - 30.0

Legend
< 10.00
10.00 - 20.00
20.00 - 30.00
> 30.00

**Figure 6.13** Spatial distribution of soil water at 0.3 and 0.6 m in November. (a left, b right)

Figure 6.13 illustrates a typical spatial distribution of soil water in November at 0.3 m and at 0.6 m depths. The range at the 0.3 m was narrow, necessitating a corresponding change in legend. Lower water contents were associated with stony coarse Dystrochrept soils, and higher with steep locations and impeding horizons. Distributions for other months did not appear to be noticeably different. To evaluate what space-time changes in soil water there are from month to month, monthly ratios of spatial overlays are compare in Figure 6.14. Areas with overlay ratios greater than 1 indicate locations with decreased

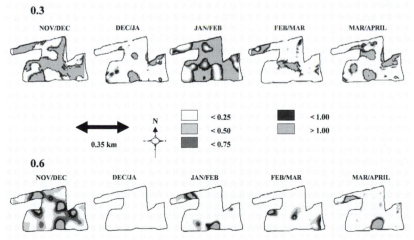

0.3
NOV/DEC       DEC/JA       JAN/FEB       FEB/MAR       MAR/APRIL

0.35 km

< 0.25
< 0.50
< 0.75
< 1.00
> 1.00

0.6
NOV/DEC       DEC/JA       JAN/FEB       FEB/MAR       MAR/APRIL

**Figure 6.14** Comparison of overlay ratios of soil water content in time on a month-to-month basis.

water content over the previous month, whereas areas with ratios less than 1 indicate locations with increased water content over the previous month. Table 6.4 shows the proportion of the area with increased and decreased water content over the previous month. Considered in that context, Figure 6.14 shows that water contents beginning in December increased steadily at the 0.6 m depth over most of the area. At the 0.3 m depth the response was mixed. In January, March, and April water contents increased over those for the previous month; in December proportions of the area with increased and decreased water content were similar; and in February water content decreased over most of the area.

### 6.6.3 Analysis

The case study showed that in general soil did not freeze at the 0.3 to 0.6 m depth during a dormant season, allowing for water movement in a form of percolation and interflow. Although water contents at the 0.3 to 0.6 m depth ranged from 10 to 44% by volume when measured with the depth moisture probe on a weekly basis, the interquartile range ($Q_3 - Q_1$) was narrow. In contrast, hourly measurements of soil water suggested rapid movement of water pulses through soil, either as percolate following a precipitation event or as interflow. Such short-term water pulses could be as high as 60% by volume locally. This suggests movement of percolate primarily through locations with water-filled large voids, or macropores. In a related study Rogowski (1997) found that in some areas on the Mahantango Creek watershed, where infiltration rates are high, more than 70% of flow is likely to occur through the macropores, which comprise about 11% of the area and include the case study site. The observed pulses of soil water following a precipitation event coincided with a rise in water table levels and increased streamflow.

On the month-to-month basis there was little visual change in water content at either depth. However, when ratios of successive monthly overlays were examined, specific locations appeared to be associated with either decreasing or increasing water content compared with the previous month, signifying a formation of sinks that drained rapidly to the water table, or else preferentially ponded areas. Alternate wetting and draining appeared to occur by layers in order of 0.3 m layer, 0.6 m layer, water table, stream. Soil properties and landscape configuration had predictable effects. Higher water contents were usually found on less steep, concave-convex areas with an impeding layer. The structure of the distributions appeared to vary little from month to month at either depth. However, both the range of continuity (~400 m) and the nugget effect were large. The large error noise component suggested by the nugget was also reflected by large fluctuations of the raw sill at distances larger than the range.

The implications for the space-time modeling of the soil water content during the dormant season are as follows: 1. As long as the soil is not frozen, considerable water movement following a precipitation event is likely to occur. 2. This movement will take a form of pulses either as percolate or interflow, which are reflected by rapid response at the water table and stream. 3. To record and model these series of events in time, a short time interval between measurements is usually required. 4. Long-term (month to month) changes in soil water contents and their preferential locations can be detected by comparing overlay ratios of the same areas in time.

## 6.7 CONCLUSIONS AND LESSONS LEARNED

Why do we need to model the 4D dynamic systems in GIS as they develop in time? The question becomes particularly pertinent in view of increasing availability of high-resolution space-based data. Continued development of both NASA's EOS and commercial EOS ranges from the LANDSAT series of imaging satellites recording reflectance at visible and infrared wavelengths (e.g., MODIS, a moderate-resolution imaging spectrometer; HIRIS, a high-resolution imaging spectrometer, and TIMS, the thermal infrared multispectral scanner) to microwave synthetic aperture radar (SAR) scanner and other more sophisticated instruments under development. These instruments, when coupled with detailed spaceborne radar elevation surveys and enhanced high-resolution GPS, may provide voluminous high-resolution data at a scale of 1 m$^2$ with relatively short revisit times. The challenge will be to relate such detailed but essentially 2D surface observations to 3D processes ongoing in the hydrosphere, atmosphere, and geosphere. The foundation of current models, both stochastic and deterministic, is the need to predict the spatial distribution of variables and temporal evolution of processes. It is driven by a lack of knowledge about the spatial and temporal workings of a process and the scarcity of data. This lack of information is replaced by interpolations or predictions that are subject to a number of simplifying assumptions and approximations. Should exhaustive space-time knowledge about environmental processes become available, there would be little need for such interpolations or predictions, and the principal task will be to project such knowledge into the future.

### 6.7.1 Future Directions

The direction of 4D space-time GIS modeling, apart from providing a comprehensive description of a process in time, is likely to be in forecasting major changes and unusual or catastrophic events at some preselected thresholds of certainty. Processing of a 4D system should present no computational problems. However, access to the voluminous multidimensional, multiparameter, continuously updated databases, and their appropriate indexing, sequencing, and arrangement will be difficult. Ideally, a progress milestone would be a 4D model in a GIS framework that runs in a diagnostic mode on a continuous basis, querying the past and making updated projections about the future. A simplified system already exists; it is known as weather forecasting.

Access is of particular importance if we adopt a perspective that such data may constitute input to future models, optimally cast in a framework of genetic algorithms. Rephrasing the problems to be solved as requiring a discovery of a computer program to produce a desired output when presented with a particular input (Koza, 1992) is known as program induction, an approach that promises data-rich GIS applications. Figure 6.15 shows how one such a model could be implemented within a GIS framework. With a heavy reliance on space-derived data imaging, such a water balance model would need to be supplemented by a methodology that can relate surface observations to processes operating within a 3D root zone profile and vadose zone to account for energy transfers, sources, and sinks not directly identifiable at the surface. The challenge would be to use the overabundance of data to characterize the process and project behavior in space and time.

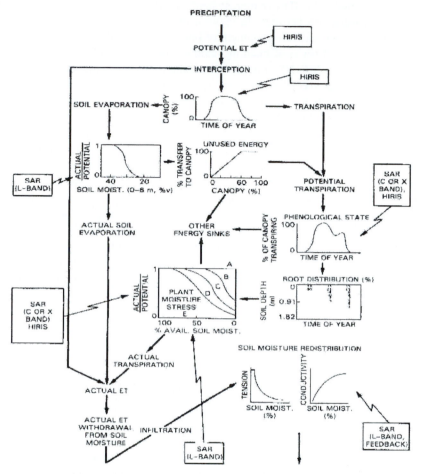

**Figure 6.15** A schematic diagram of water balance modeling (Saxton et al., 1974) showing where microwave imaging with SAR and other EOS instruments could be used within a GIS framework.

# 6.8 INFORMATION SOURCES

**Apoala Project**

http://www.geovista.psu.edu/apoala/index.htm

**Mahantango Creek Watershed Study Database.**

http://orser13.erri.psu.edu/web/orser4.htm

**GIS/EM3 Third International Conference**

http://www.ncgia.ucsb.edu/conf/SANTA_FE_CD-ROM/main.html

## 6.9 REFERENCES

Ahuja, L. R. and Williams, R. D. 1991. Scaling Water Characteristic and Hydraulic Conductivity Based on Gregson-Hector-McGowan Approach. *Soil Science Society of American Journal*, 55, 308–319

Beven, K., Quinn, P., Romanowicz, R., Freer, J., Fisher, J., and Lamb, R. 1994. *Topmodel and Gridatb, Users guide to Version 94.01.* CRES Technical Report TR110/94, Lancaster, England: Lancaster Univeristy.

Burrough, P. A. 1987. *Principles of Geographical Information Systems for Land Resources Assessment.* Monographs on soil and resources survey, No. 12. Oxford, England: Clarendon Press.

Cushman, J. H. 1986. On Measurement, Scale, and Scaling. *Water Resources Research,* 22, 129–134.

DeCoursey, D. G. 1996. *Hydrological, Climatological and Ecological Systems Scaling: A Review of Selected Literature and Comments.* Interim Progress Report. Fort Collins, CO: USDA-ARS-NPA.

Deutsch, C.V., and Journel, A. G. 1998. *GSLIB, Geostatistical Software Library and User's Guide.* New York: Oxford University Press, pp. 340–369.

Eastman, J. R. 1999. *IDRISI 32. Guide to GIS and Image Processing.* Worchester, MA: Clark University.

ESRI. 1995. *Understanding GIS.* New York: Wiley.

Goodchild M. F., Steyaert, L. T., Parks, B. O., Johnson, C., Maidment, D., Crane, M., and Glendinning, S. (eds.), 1996. *GIS and Environmental Modeling: Progress and Research Issues.* pp. 451–454. Fort Collins, CO: GIS World Books.

Goovaerts, P. 1997. *Geostatistics for Natural Resources Evaluation.* New York: Oxford University Press.

Goovaerts, P., and Chiang, C. N. 1993. Temporal Persistence of Spatial Patterns for Mineralizable Nitrogen and Selected Soil Properties. *Soil Science Society of American Journal,* 57, 372–381.

Goovaerts, P., and Sonnet, P. 1993. Study of Spatial and Temporal Variations of Hydro-Geo-chemical Variables Usinf Factorial Kriging Analysis. In A. Soares (eds.) *Geostatistics Tróia '92,* 745–756. Boston, MA: Kluver Academic Publishers.

Gregson, K., Hector, D. J., and McGowan, M. 1987. A One-Parameter Model for the Soil Water Characteristic. *Journal of Soil Science*, 38, 483–486.

Hazelton, N. W. J., Leahy, F. J., and Williamson, I. P. 1990. On the Design of Temporally Referenced, 3D Geographical Information Systems: Development of Four Dimensional GIS. In *Proceedings GIS/LIS'90*, Vol. 1, pp. 357–372. Bethesda, MD: ASPRS/ACSM/URISA/AAG.

Hohn, M. E., and Gribko, L. M. 1993. Forecasting Gypsy Moth Defoliation with Indicator Kriging. In A. Sores (ed.) *Geostatistics Tróia '92*, pp. 601–612. Boston, MA: Kluver Academic Publishers.

Johnson, A. R. 1996. Spatiotemporal Hierarchies in Ecological Theory and Modeling. In M. F. Goodchild, L. T. Steyaert, B. O. Parks, C. Johnston, D. Maidment, M. Crane, and S. Glendinning (eds.) *GIS and Environmental Modeling: Progress and Research Issues*, pp. 451–456. Fort Collins: GIS World Books.

Journel, A. G. 1996. Modeling Uncertainty and Spatial Dependence: Stochastic Imaging. *International Journal of Geographic Information Systems*, 10, no. 5, 517–522.

Kalman, R. E. 1960. A New Approach to Linear Filtering and Prediction Problems. *Journal Basic Engineering (ASME)*, 82D, 35–45.

Kelmelis, J. 1991. *Time and Space in Geographic Information: Towards a Four Dimensional Spatio-Temporal Data Model.* Unpublished Ph.D dissertation, Department of Geography, Pennsylvania State University, University Park, PA.

Koza, J. R. 1992. *Genetic Programming.* Cambridge, MA: The MIT Press.

Kyriakidis, P. C., and Journel, A. 1999. Geostatistical Space-Time Models: A Review. *Mathematical Geology*, 31, no. 6, 651–684.

Lagran, G. 1989. A Review of Temporal Database Research and Its Use in GIS Applications. *International Journal of Geographic Information Systems,* 17, no. 4, 291–299.

Lagran, G. 1993. *Time in Geographic Information Systems.* Washington, DC: Taylor & Francis.

Lagran, G., and Chrisman, N. R. 1988. A Framework for Temporal Geographic Information. *Cartographica,* 25, no. 3, 1–14.

Maidment, D. R. 1993. GIS and Hydrologic Modeling. In M. F. Goodchild, B. O. Parks and L. T. Steyaert, (eds.), *Environmental Modeling with GIS.* pp. 147–167. New York: Oxford University Press.

Maidment, D. R. 1996a. Environmental Modeling within GIS. In M. F. Goodchild, L. T. Steyaert, B. O. Parks, C. Johnston, D. Maidment, M.Crane, and S. Glendinning (eds.) *GIS and Environmental Modeling: Progress and Research Issues,* pp. 315–323. Fort Collins, CO: GIS World Books.

Maidment, D. R. 1996b. GIS and Hydrologic Modeling—an assessment of progress. In *Proceedings, Third International Conference/Workshop on Integrating GIS and Environmental Modeling,* Santa Fe, NM, January 21–26, 1996. Santa Barbara, CA: National Center for Geographic Information and Analysis. CD-ROM. (http://www.ncgia.ucsb.edu/conf/SANTA_FE_CD-ROM/main.html)

Maldebrot B. B. 1983. *Fractal Geometry of Nature.* New York: W. H. Freeman.

Marshall, J. S., and Pounder, E. R. 1957. *Physics.* New York: Macmillan.

Mason, D. C., O'Conaill, M. A. and Bell, S. B. M. 1994. Handling Four-Dimensional Georeferenced Data in Environmental GIS. *International Journal of Geographic Information Systems*, 8, no. 2, 91–215.

Miller H. J. 1991. Modeling Accessibility Using Space-Time Prism Concepts within Geographical Information Systems. *International Journal of Geographic Information Systems*, 5, no. 3, 287–301.

Mitasova, H., Mitas, L., Brown, W. M., Gerdes, D. P., Kosinovsky, I., and Baker, T. 1995. Modeling Spatially and Temporally Distributed Phenomena: New Methods and Tools. *International Journal of Geographic Information Systems,* 9, no. 4, 433–446.

Moore, D., Turner, A. K., Wilson, J. P., Jenson, S. K., and Band, L. E. 1993. GIS and land-surface-subsurface modeling. In M. F. Goodchild, B. O. Parks, and L. T. Steyaert (eds.), *Environmental Modeling,* pp. 196-230. New York: Oxford University Press.

Morris, R. 1984. *Time Arrows: Scientific Attitudes toward Time.* New York: Simon & Shuster.

NCGIA. 1996. Integrating GIS and Environmental Modeling. The Third International Conference, Santa Fe, NM, January 21–25. CD-ROM. http://www.ncgia.ucsb.edu/conf/SANTA_FE_CD-ROM/main.html

Nelson, R.W. 1978a. Evaluating the Environmental Consequences of Groundwater Contamination. 1. An Overview of Contaminant Arrival Distributions as General Evaluation Requirements. *Water Resources Research,* 14, no. 3, 409–415.

Nelson, R.W. 1978b. Evaluating the Environmental Consequences of Groundwater Contamination. 2. Obtaining Location/Arrival Time and Location/Outflow Quantity Distributions for Steady Flow Systems. *Water Resources Research,* 14, no. 3, 416–428.

Newton-Smith, W. H. 1980. *The Structure of Time.* London, England: Routledge.

O'Conaill, M. A., Mason, D. C., and Bell, S. B. M. 1993. Spatiotemporal GIS Techniques for Environmental Modeling. 103-112. In P. M. Mather (ed.) *Geographical Information Handling–Research and Applications.* New York: Wiley.

Olea, R. A. 1991. *Geostatistical Glossary and Multilingual Dictionary.* New York: Oxford University Press.

Peuquet, D. J. 1994. It's About Time: A Conceptual Framework for the Representation of Temporal Dynamics in Geographic Information Systems. *Annals of the Association of American Geographers,* 84, no. 3, 441–461.

Peuquet, D. J. and Duan, N. 1994. An Event Based Spatiotemporal Data Model (ESTDM) for Temporal Analysis of Geographical Data. *International Journal of Geographic Information Systems,* 9, no. 1, 7–24.

Peuquet, D. J., and MacEachren, A. M. 2000. *Apoala Project.* Pennsylvania State University. http://www.geovista.psu.edu/apoala/index.htm

Price, H., 1996. Time's Arrow and Archimides' Point. New York: Oxford Univ. Press.

Ranka, S. and Sahni, S. 1990. *Hypercube Algorithms.* New York: Springer Verlag.

Riha, S. J. 1980. *Simulation of Water and Nitrogen Movement and Nitrogen Transformation in Forest Soils.* Ph.D. Dissertation, Washington State University, Pullman, WA.

Rogowski, A.S. 1990. Estimation of the Groundwater Pollution Potential on an Agricultural Watershed. *Agricultural Water Management,* 18, 209–230.

Rogowski A. S. 1996. Quantifying the Model of Uncertainty and Risk Using Sequential Indicator Simulation. In W. D. Nettleton, A.G. Hornsby, R.B. Brown, and T. L. Coleman (eds) *Data reliability and risk assessment in soil interpretations.* pp. 143 –164. SSSA Special Publication 47. Madison, WI: Soil Science Society of America Inc.

Rogowski, A. S. 1997. Catchment Infiltration II: Contributing Area. *Transactions in GIS,* 1, no. 4, 267–284.

Rogowski, A. S. 1999a. Incorporating Soil Variability into a Spatially Distributed Model of Percolate Accounting. In H. T. Mowrer and R. G. Congalton (eds) *Quantifying Spatial Uncertainty in Natural Resources: Theory and Applications for GIS and Remote Sensing*, pp. 183–202. Chelsea, MI: Ann Arbor Press.

Rogowski, A. S. 1999b. Mahantango Creek (Sector #5) *Watershed Study Data Base.* Pennsylvania State University, Environmental Resources Research Institute Report ER9904. University Park, PA. (http://orser13.erri.psu.edu/web/orser4.htm)

Rogowski, A. S., and Wolf, J. K. 1994. Incorporating Variability into Soil Map Unit Delineations. *Soil Science Society of American Journal,* 58,163–174.

Saxton, K. E., Johnson, H. P., and Shaw, R. H. 1974. Modeling Evapotranspiration and Soil Moisture. *Transactions of the American Soceity of Agricultural Engineering.* 17, no. 4, 673–677.

Stein, A., Gronigen, J. W. Van, Jeger, M. J., and Hoosbeek, M. R. 1998. Space-Time Statistics for Environmental and Agricultural Related Phenomena. *Environmental and Ecological Statistics*, 5, 155-172.

Soil Survey Staff. 1975. *Soil Taxonomy.* Agriculture Handbook No.436. Washington, DC: US Government Printing Office.

Walton, W. C. 1992. *Groundwater Modeling Utilities.* Chelsea, MI: Lewis Publishers, Inc.

Webster, R. and Oliver, M. A. 1990. *Statistical Methods in Soil and Land Resource Survey.* New York: Oxford University Press.

Wikle, C. R. 2000. Hierarchical Space-Time Dynamic Models. In L. M. Berliner, D. Nychka, and T. Hoar (eds). *Studies in Statistics and Atmospheric Sciences.* pp. 45–82. New York: Springer–Verlag.

Wikle, C. R., Berliner, L. M., and Cressie, N. 1998. Hierarchical Bayesian Space-Time Models. *Environmental and Ecological Statistics*, 5, 117–154.

Wikle, C. R. and Cressie, N. 1999a. A Dimension Reduced Approach to Space-Time Kalman Filtering. *Biometrika*, 86, no. 4, 815–829.

Wikle, C. R. and Cressie, N. 1999b. Space-Time Statistical Modeling of Environmental Data. In H. T. Mowrer and R. G. Congalton (eds.) *Quantifying Spatial Uncertainty in Natural Resources: Theory and Applications for GIS and Remote Sensing*, pp. 213–235. Chelsea, MI: Ann Arbor Press.

Wolf, J. K. and Rogowski, A. S. 1991. Spatial Distribution of Soil Heat Flux and Growing Degree Days. *Soil Science Society of American Journal*, 55, 647–657.

## 6.10 ABOUT THE CHAPTER AUTHORS

Andrew S. Rogowski, a retired Soil Scientist, United States Department of Agriculture–Agricultural Research Service (SDA-ARS), is currently an Adjunct Professor of Soil Physics at the Pennsylvania State University and a self-employed Environmental Consultant. He holds M.S. and Ph.D degrees in Soil Physics from Iowa State University. His principal areas of research interest include risk assessment and modeling of spatial and temporal variability in three dimensions at different scales; applications of geostatistics to environmental problems; and assessment of uncertainty associated with measurement of soil attributes.

Jennifer L. Goyne, a GIS Analyst with Geo Decisions, a division of Gannett Fleming, Inc., holds a B.S. in Agricultural Engineering with a minor in Environmental Resource Management from Pennsylvania State University. Her interests include creating appropriate GIS databases and access software; extracting information for transforming GIS coverages into spatially and temporally correct positions within an application area; and providing technical assistance for the development of novel applications.

# 7

# *Modeling Human-Environmental Systems*

## 7.0 SUMMARY

This chapter focuses on the integration and development of environmental models that include human decision making. While many methodological and technical issues are common to all types of environmental models, our goal is to highlight the unique characteristics that need to be considered when modeling human-environmental dynamics and to identify future directions for human-environmental modeling. To achieve this goal, we have separated this chapter into several sections. First, we propose and define a conceptual framework for describing human-environmental models based on three critical dimensions: time, space, and human decision making. Second, using our framework, we summarize and compare whether and how different models (urban or rural systems, health, epidemiology, pollution, or hydrology) include space, time, and human decision making. This provides both an assessment of the models examined and a test of the framework. Third, we discuss the theoretical implications for linking human-environmental dynamics within the context of these three dimensions. Finally, we consider lessons learned and future directions for developing human-environmental models.

This chapter is not a guide for readers to learn how to model human-environmental systems. Rather, readers will find this chapter useful for understanding the basic issues that models of human-environmental dynamics must address; for developing the ability to assess the strengths and weaknesses of various human-environmental models; and for identifying future directions in modeling human-environmental systems. Ultimately, we hope to convince readers that modeling human-environmental dynamics is a useful and exciting activity that can complement biophysically based models and provide understanding of human-environmental systems.

## 7.1 INTRODUCTION

Models are often simplistic, but in many cases critical, abstract representations of the complex dynamics of human-environmental systems. They can be categorized in several ways. Models can focus on topics such as urban or rural systems, health, epidemiology, pollution, or hydrology (Landis, 1995; White and Engelen, 1993). Models can be used for many purposes, such as research, decision making (policy, planning, and management), and education (Costanza and Ruth, 1998; Ford, 1999; Grove, 1999). Models can also be categorized by the methods and techniques used, which might range from simple regressions to advanced dynamic simulations (Agarwal et al., 2000; Clarke and Gaydos, 1997; Clarke and Gaydos, 1998; Lambin, 1994).

Given the possible variety and ever-growing number of human-environmental models, we do not attempt to provide Noah's list of models. Rather, our goal in this chapter is to provide a framework for categorizing and summarizing models of human-environmental dynamics that is both inclusive of purpose, method, and topic while permitting fundamental comparisons of models along the dimensions of time, space, and human decision making. The framework we propose is not an end in and of itself. We anticipate that this framework will provide the basis for assessing current progress in modeling human-environmental dynamics and identifying and prioritizing directions for model development in terms of topic, purpose, and methods.

This chapter is separated into the following sections. First, we propose and define a conceptual framework for describing human-environmental models based on critical dimensions of time, space, and human decision making. Second, using our framework, we summarize and compare different examples of models (urban or rural systems, health, epidemiology, pollution, or hydrology) along these dimensions. This provides both an assessment of the models and a test of the framework. Third, we discuss the theoretical implications for linking human-environmental dynamics within the context of these three dimensions. Finally, we consider lessons learned and future directions for developing models of human-environmental dynamics.

## 7.2 KEY FEATURES OF HUMAN-ENVIRONMENT MODELS

### 7.2.1 Time, Space, and Human Decision Making: A Framework for Reviewing Human-Environmental Models

We propose a framework based on three critical dimensions for categorizing and summarizing models of human-environmental dynamics. Time and space are the first two dimensions and provide a common context in which all biophysical and human processes operate. In other words, models of biophysical and/or human processes operate in a temporal context, a spatial context, or both. When models incorporate human processes, a third dimension–what we refer to as the human decision making dimension–becomes important as well (Figure 7.1).

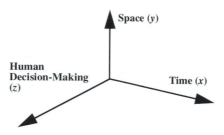

**Figure 7.1** The three dimensions of human-environmental models.

In reviewing and comparing human-environmental models along these dimensions, there are two distinct and important attributes that must be considered: *model scale* and *model complexity.* We begin with a discussion of scale since it is a concept that readers will probably find most familiar from earlier parts of this book.

**Model Scale.** Real-world processes operate at different scales (Allen and Hoekstra, 1992; Ehleringer and Field, 1993). When we discuss the temporal scale of models, we usually talk in terms of time step and duration. Time step is the smallest unit of analysis for change to occur for a specific process in a model. For example, in a model of forest dynamics, tree height may change daily. Duration refers to the length of time that the model is applied. Change in tree height might be modeled daily over the course of its life from seedling to mature tree by using a duration of 300 years. In this case, time step would equal one day and duration would equal 300 years.

When we discuss the spatial scale of models, we talk in terms of resolution and extent. *Resolution* refers to the smallest geographic unit of analysis for the model, such as the size of a cell in a raster system. Extent describes the total geographic area to which the model is applied. Consider a model of individual trees in a 50-ha forested area. In this case, an adequate resolution for individual trees might be 5 m and the model extent would equal 50 ha.

Most readers will find this discussion relatively straightforward and familiar. But how do we discuss human decision making in terms of scale? To date, the social sciences have not yet described human decision making in terms that are as concise and widely accepted for modeling as time step and duration, and resolution and extent. Like time and space, however, we propose that an analogous approach can be used to articulate scales of human decision making in terms of two components: agent and domain.

Agent refers to the human actor(s) in the model who are making decisions. The individual is the most familiar human decision making agent. But there are many human models that capture decision making processes at higher levels of social organization, such as household, neighborhood, county, state or province, or nation. These can all be considered agents in models and can be linked. For example, Figure 7.2 illustrates an example of a hierarchical approach to agents and domains for the study of urban ecosystems. While the agent captures the concept of who makes decisions, the domain describes the specific institutional and geographic context in which the agent acts. Representation of the domain can be articulated in a geographically explicit model through the use of boundary maps or GIS layers.

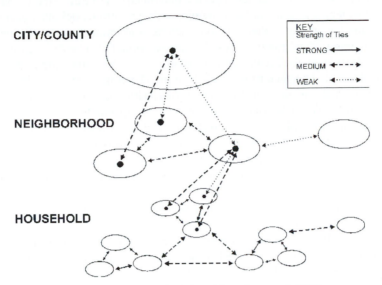

**Figure 7.2** Multiple agents and domains (adapted from Grove et al., 2000).

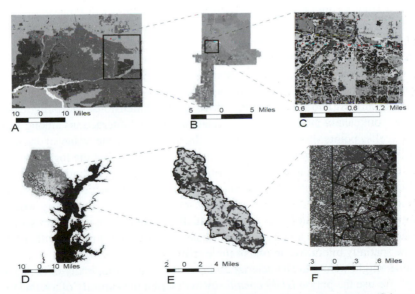

**Figure 7.3** Spatial representation of a hierarchal approach to modeling urban systems (Grimm et al., 2000). Figure shows three levels of spatial scale for the Central Arizona-Phoenix (upper) and Baltimore ecosystem (lower) studies.

In a model of farmer behavior (agent = individual) the farm is the domain within which farmers make decisions. In this example, we might also model other agents operating in the same region (e.g., other parcels), such as nonfarming households, public land managers, or conservation groups, whose boundaries would also be depicted by the same domain map. Institutionally, agents overlap spatially, since the farmer might receive extension advice about her livestock from an agent of the Department of Agriculture; have her cows inspected by the agent of an Department of Health; and receive financial subsidies from an agent of the Forest Service for planting trees in riparian buffer areas.

Using another example and a different scale of human decision making, consider a state forester (agent = state) who writes the forest management plan for the state forest (domain = state boundary) and prescribes how often trees (resolution) in different forest stands (extent) should be harvested (time step) for a specific period of time (duration) within state-owned property. In this case, the human decision making component to the model might include the behavior of the forester within the organizational context of the state-level natural resource agency.

**Model Complexity.** The second important and distinct attribute of human-environmental models is the approach(s) used to address the complexity of time, space, and human decision making found in real-world situations. We propose that the temporal, spatial, or human decision making (HDM) complexity of any model can be represented with an index, where a low score signifies only simple processing and a high score signifies more complex behaviors and interactions. Consider an index for *temporal complexity* of models: A model that is low in temporal complexity is a model that has one, or possibly a few, time steps and a short duration. A model with a middle to high score for temporal complexity is one that has many time steps and a longer duration. Models with a high score for temporal complexity are ones that have a large number of time steps, a long duration, and the capacity to handle time lags or feedback responses among variables, or have different time steps for different submodels.

An index of *spatial complexity* would represent the "spatial explicitness" of a model. There are two general types of spatially explicit models: spatially representative or spatially interactive. A model that is spatially representative can incorporate, produce, or display data in at least two and sometimes three spatial dimensions—northing, easting, and elevation—but cannot model topological relationships and interactions among geographic features (cells, points, lines, or polygons). In these cases, the value of each cell may change or remain the same from one point in time to another, but the logic that makes the change is not dependent on cells neighboring it. In contrast, a spatially interactive model is one that explicitly defines spatial relationships and their interactions (e.g., among neighboring units) over time. A model with a low score for spatial complexity would be one with little or no capacity to represent data spatially, a model with a medium score for spatial complexity would be able to represent data spatially, and a model with a high score would be spatially interactive in two or three spatial dimensions.

What might we use to characterize an index for model complexity of human decision making? We use the phrase *HDM complexity* to describe the capacity of a human-environmental model to handle decision making processes. In Table 7.1, we present a classification scheme for estimating HDM complexity using an index from 1 to 6. A model with a low score for human decision making complexity (1) is a model that does not include any

**Table 7.1** Six levels of Human decision making Complexity

| Level | |
|---|---|
| 1 | No human decision making -- only biophysical variables in the model |
| 2 | Human decision making assumed to be determinately related to population size, change, or density |
| 3 | Human decision making seen as a probability function depending on socio-economic and/or biophysical variables beyond population variables *without* feedback from the environment to the choice function. |
| 4 | Human decision making seen as a probability function depending on socio-economic and/or biophysical variables beyond population variables *with* feedback from the environment to the choice function. |
| 5 | One type of agent whose decisions are overtly modeled in regard to choices made about variables that affect other processes and outcomes. |
| 6 | Multiple types of agents whose decisions are overtly modeled in regard to choices made about variables that affect other processes and outcomes. The model may also be able to handle changes in the shape of domains as time steps are processed or interaction between decision making agents at multiple human decision making scales |

human decision making. In contrast, a model with a high score (5 or 6) is a model that includes one or more types of actors explicitly or can handle multiple agents interacting across domains, as shown in Figures 7.2 and 7.3. In essence, Figures 7.2 and 7.3 represent a hierarchical approach to social systems where lower-level agents interact
to generate higher-level behaviors and higher-level domains affect the behavior of lower-level agents (Grimm et al., 2000; Grove et al., 2000; Vogt et al., 2000;).

## 7.2.2 Application of the Framework

The three dimensions of human-environmental models—space, time, and human decision making—and two distinct attributes for each dimension—scale and complexity—provide the foundation for comparing and reviewing human-environmental models. Figure 7.4 presents the framework with the three dimensions represented and the models located in terms of their spatial, temporal, and HDM complexity along each axis.

Human-environmental models can be placed somewhere within the three dimensional space of Figure 7.4 to represent graphically their comparative focus, strengths, and abilities. Consider a time series modeling effort. Suppose a hydrologist is interested in modeling the quantity of water held in a city's reservoir over time and wants to use historic data on reservoir levels and other relevant information to forecast levels in the future.

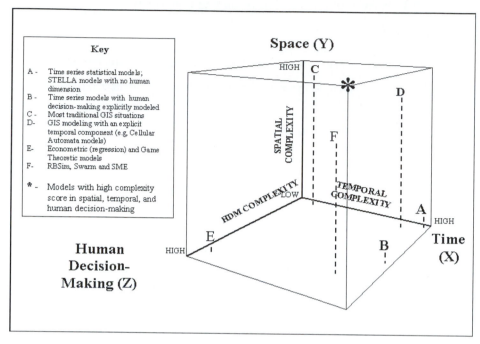

**Figure 7.4** A framework for reviewing human-environmental models.

He or she may decide that the appropriate analytic technique for this question is to develop a regression or autoregression model. Often, time series statistical analyses focus on variations over time (such as reservoir water levels) without considering spatial or human decision making complexity. Such modeling approaches might have a high score for temporal complexity but would have a low score for spatial or human decision making complexity (Figure 7.4, A). Continuing with our reservoir example, a time series model that explicitly included human decision making, such as household decisions over water consumption in response to changes in water costs or to drought-based conservation practices, would have a higher complexity score along the human decision making axis (Figure 7.4, B).

A modeling approach that would have a high score for temporal complexity is based on dynamic systems software (for example, STELLA, ModelMaker, Powersim). This type of software allows a modeler to represent systems as stocks, flows, and processes and to run the model over a series of time periods (Costanza and Gottlieb, 1998; Hannon and Ruth, 1997). STELLA does not have its own spatial modeling capabilities; and if a model based on STELLA does not include a human component, it would have a low score for both the spatial and human decision making complexity (Figure 7.4, A). STELLA models have the capacity to explicitly model human decision making; consequently, the complexity score for any STELLA model along the human decision making axis depends on the specific processes included in the model.

Geographic information systems (GIS) are the obvious spatial modeling technology and models that include GIS have high scores for spatial complexity. Many GIS applications in the 1970s, 1980s, and early 1990s have low temporal and human decision making scores (Figure 7.4, C) because most GIS applications developed during that period used data layers developed for only one or possibly two time points and concentrated on the heterogeneity of biophysical, landscape characteristics. Similar to statistical modeling, however, GIS has powerful abilities and can, in some applications, have high scores for temporal and/or human decision making complexity. For example, cellular automata (CA) models (Clarke et al., 1997; Clarke and Gaydos, 1998) are one special type of raster-based modeling that explicitly captures change over time in a spatial context. CA models have higher scores for temporal complexity than standard GIS systems (Figure 7.4, D). Similar to STELLA modeling, the relative location of a CA model along the human decision making axis depends on the specific social processes included in the model.

What about models with high scores for human decision making complexity? Econometric models are one type of modeling approach that would rank higher in this dimension because they often explicitly try to model human behavior (Figure 7.4, E). For instance, in political science or sociological research, surveys and regression analysis are often used to understand better how various factors influence individual behavior. Most of these types of regression models would have low scores for temporal and spatial complexity unless the modeler explicitly included spatial or temporal parameters (for an example of regression integrated with spatiotemporal modeling, see Veldkamp and Fresco, 1996).

Game theoretic modeling is another approach that models human decision making complexity very well (Figure 7.4, E). Game theorists explicitly try to understand why humans behave the way they do under certain bargaining or collective action situations (see, for example, Roth, 1985; Ostrom et al., 1994). This modeling approach can include some temporal complexity. For example, there has been substantial work by experimental economists who have developed predictive models of repeated human decision making over several time periods (Kagel and Roth, 1997; Smith et al., 1994).

Combinations of techniques to model all three dimensions have begun to emerge (Figure 7.4, F). For example, Swarm, is an agent based modeling framework that can handle temporal, spatial, and human decision making complexity (see, for example, Popper and Smuts, 1999; http://www.swarm.org). Swarm is a software package that allows for the development of multiagent simulation of complex systems and has the capacity to develop spatiotemporal models of human decision making. Recently, Swarm has been used to model agents in various economic decision making scenarios (Luna and Stefannson, 2000). These agent-based approaches can span space, time, and decision making dimensions and cross social and biophysical scales by modeling agents at different scales.

The Recreation Behavior Simulator, or RBSim, developed by Itami and Gimblett, provides another example of modeling in all three dimensions. RBSim combines GIS and autonomous human agents to simulate human decision making and movement over geographic space and time (Gimblett et al., 1998; see also http://www.dlsr.com.au/software/rbsim). In these models, programs are written to capture the logic or decision rules of various types of decision makers or agents.

Finally, Costanza and colleagues' Spatial Modeling Environment (SME) combines GIS capabilities with STELLA to create a system to model landscape change in a fashion similar to cellular automata (http://kabir.cbl.umces.edu/SMP/MVD/SME1.html). With the STELLA modeling embedded in SME, it is possible to model human decision making using the STELLA stock-flow-process environment. SME models with human decision making would also have high scores for all three dimensions of complexity.

We try to compare different types of models, but we should note that when two models are compared side by side, they might have identical scores of complexity for each dimension (Figure 7.4) even though they operate at different spatial, temporal, or human decision making scales. Alternatively, two models operating at identical scales could have different scores of complexity for each dimension. Thus, it is important to consider both sets of model attributes—scale and model complexity—when assessing whether human-environmental models are comparable.

The utility of this three-dimensional framework is that, first, it forces us to consider and clearly articulate the two important attributes of models: scale and complexity. Further, the framework encourages developers of human-environmental models to consider the appropriate scale(s) and levels of complexity necessary to address the problem or system they want to model. The following section considers how models of three different types of human-environmental systems relate to this framework.

## 7.3 EXAMPLES OF HUMAN-ENVIRONMENT MODELS

Researchers from a variety of disciplines have modeled a vast array of human-environmental systems. These various types of models differ in structure and design because of the nature of the systems being modeled, the methodological and disciplinary background of the modelers, and the different purposes of various models. Here we provide examples from three broad areas of human-environmental modeling in order to demonstrate the approaches that have been used to model human decision making; the effect of human populations on the environment; and the effect of the environment on human populations. This section discusses only a small, representative subset of different types of human-environmental models and is limited to models with (1) a substantial interface with the biophysical environment, and (2) some ability to represent spatial relationships. Elsewhere, several sources provide comprehensive reviews of various types of human-environmental models (Johnston and Barra, 2000; Southworth, 1995; Webster and Pauley, 1991; Wegener, 1994; Wilson, 1998)

The basic systems models discussed here deal with urban systems, land use/land cover change, and models related to environmental health. It should be noted that there is a considerable amount of content overlap among these three categories. The models described here can be distinguished in their focus, or more specifically, to what degree each model adds complexity or simplifies different model components. For example, both the California Urban Futures model (CUF) and the Agricultural Nonpoint Source Pollution model (AGNPS) deal with agricultural land use in some way. However, the AGNPS model has much more complexity for the ecological function of the agricultural system than the CUF model, and the CUF model has more complexity for location-economics, where the AGNPS has no content. This is natural since these two models were designed for entirely different purposes. These two examples illustrate how the purpose of a model determines how a particular system is modeled and which components are included in different models of similar systems.

## 7.3.1 Urban Systems Modeling

A significant problem for urban planners in the 1970s and 1980s was the process of rapid urbanization and suburbanization. Even prior to the advent of GIS technologies, planners used overlay techniques with mylar maps to examine the spatial characteristics of urban areas. As GIS tools and techniques were developed and distributed in more user-friendly software platforms, GIS became a natural tool to model processes of urban growth. Because of these conditions, GIS was rapidly adopted within urban planning as a visualization and modeling tool. A large number of models of urban change were developed in the late 1980s and early 1990s. These models were used for simulation and scenario testing, allowing policy makers to observe the predicted effects of various policy prescriptions as well as different population growth and development scenarios. Here we present three urban growth models that can be differentiated by the following characteristics: spatial unit of analysis, data visualization, model complexity, choice of exogenous factors, and calibration/validation tools.

The CUF model (Landis, 1995; Landis and Zhang, 1998) was originally developed to predict urban development in a 14 county area in Northern California but has since been applied to a variety of urban areas for urban development predictions under different scenarios. The CUF-1 model uses a bottom-up approach: Population growth is modeled for individual subareas (incorporated cities/counties and developable land units [DLUs]). Since the initial CUF model was designed, the model has been revised (CUF-2) to include multiple land uses and calibrates model output from more recent data. The CUF-2 model also uses a smaller spatial unit of analysis (100m * 100m cells) than the CUF-1 model, which used only several hundred DLU areas to model a 14 county area in Northern California.

One of the most widely applied urban systems models is the Metropolitan Integrated Land Use system (METROPILUS) (Putman, 1983; 1992). The METROPILUS model is actually the integration of a series of model components, each focused on a particular aspect of land use/land cover change processes. Development began in the 1970s and the latest incarnation includes a user-friendly graphical user interface linked to ArcView GIS software package. The main model components are a residential allocation component, an employment allocation component, and a land use change component. The residential and employment model components can be used to predict population changes, and the land use change component estimates changes in land cover based on the demands placed on the landscape by the residential and employment components. This compartmentalized or modular approach to modeling is a common approach in human-environmental modeling as well as ecosystems modeling.

An alternative approach applied in the late 1980s and 1990s was the use of cellular automata, where the state of each unit of analysis is a function of three factors: (1) the cell's prior state, (2) the neighboring area, and (3) a series of state transition rules (Deadman et al., 1993). Examples of cellular automata models applied to urban areas include White and Engelen (1993), Batty and Xie (1994a, 1994b, 1994c) and Clarke (Clarke et al., 1997; Clarke and Gaydos, 1998). In the case of White and Engelen and Batty and Xie, an urban area is represented by a raster data structure of cells where the state of each cell at a specific time point is the product of the state of that cell at the prior time point and the states of cells in a neighborhood surrounding that cell. These researchers have used this approach to model the affect of different policy prescriptions and scenarios, such as different population growth rates, zoning restrictions, and various economic development assumptions.

## 7.3.2 Rural Systems Modeling

Rural land use/land cover change or rural systems modeling is another major area of human-environmental systems modeling. This area of modeling overlaps with urban systems modeling in that the rural-urban interface is often an important factor in many urban and rural environments. However, rural systems modeling most often focuses on areas where agriculture and forest predominate and there is little direct effect from the encroachment of urban areas. These rural systems models can be differentiated from urban systems models in their focus and component complexity. In particular, rural systems models have more complexity in specifying the dynamics of agriculture and forestry land uses. A review of urban and rural systems models shows that most urban systems models focusing on urban expansion and land use change are concerned with the provision of services (e.g., transportation infrastructure) ( Johnson and Barra, 2000; Wegener, 1994). The exceptions are models of urban climates. In contrast, rural systems models have included more explicit linkages between land use decisions and landscape outcomes associated with environmental effects (e.g., carbon sequestration, groundwater contamination, biodiversity). For example, the CUF-2 model (Landis and Zhang, 1998) includes multiple land uses in modeling urban growth, but only one class is used to represent all agricultural land uses (crops, pasture, and forest land) while there are six classes of urban land uses. Furthermore, the CUF-2 model does not distinguish between crops or stages of forest growth.

Recently, rural systems models or land use/land cover change (LUCC) models have been the focus of researchers examining global change issues including deforestation/reforestation, biodiversity, carbon sequestration, and other land-atmosphere exchanges. (For a review, see Agarwal et al., 2000.) These models include spatially explicit models of land cover change, dynamic systems models, and models that predict emergent behavior in human systems.

One method of modeling LUCC is through the use of spatially explicit dynamic systems models. Costanza and colleagues have developed such a model for the Patuxent watershed by integrating a general ecosystem model with economic decision making (Voinov and Costanza, 1999; Voinov et al., 1999). This model has been implemented for the Patuxent watershed using the spatial modeling environment (SME), developed by the authors and colleagues. Originally focusing on hydrology and the surface and subsurface exchange of nutrients, the model uses the specification of economic development to incorporate land use change in the model rather than using land use change as an exogenous factor affecting the hydrological system. This dynamic approach provides powerful coupling techniques whereby feedbacks in each model component can be linked to other model components and provide a more realistic representation of human-environment interactions.

There are many models that predict the behavior of rural systems from an ecosystem perspective. These models generally treat land use decisions as exogenous drivers to the model rather than modeling the land use decision making process itself. Various models have been developed for a range of different ecosystem types and for applications such as predicting crop productivity and forest succession in publicly managed lands.

One such model is the LANDIS model (He and Mladenoff, 1999; He et al., 1999a, 1999b; Mladenoff and He, 1999), which was developed to simulate forest landscape change under different harvesting and disturbance regimes. The LANDIS model uses a spatially explicit raster structure to simulate spatial interactions, such as seed dispersal, and produces a species-level output assuming different management practices and disturbances. The creators of this model stress that the LANDIS model is most useful as a tool to project plausible landscape outcomes under certain conditions rather than as a spatially explicit prediction tool (Mladenoff and He, 1999). This is an important concept in human-environmental dynamics and ecosystems modeling. Data availability limits the ability to predict the behavior of systems even if models exist to properly simulate those systems.

The LANDIS model is most applicable to environments dominated by forest cover. A somewhat more complex system is an environment where there are a variety of land uses (e.g., forestry, agroforestry, crops, pasture) and numerous actors. The LANDIS model has a great deal of complexity in terms of forest dynamics and deals with a specific type of ecosystem (largely, forested). Landscapes with a high degree of human activity are typically characterized by a broad range of land uses (for example, forests, crops, pasture, residential). Models of these systems with a broader range of land uses often simplify different components in order to produce a model that performs acceptably for the research question at hand but lacks specificity in some areas. For example, Dale and colleagues created a model of forest cover dynamics for a study in Rondonia, Brazil (Dale et al., 1993, 1994), but this model lacks the species level complexity of the LANDIS model. Using a spatially explicit approach, Dale and others (1993, 1994) modeled the effects of land cover change drivers under different management scenarios on spatial patterns and composition of land cover. This model was unique in that it linked a dynamic model of land cover change to a spatial representation where each parcel was composed of multiple cells rather than using a coarser unit of analysis (e.g., parcel, municipal area, or region). This approach allowed land use activities to be modeled at the household level, the same level at which land use decisions were made in the particular study area, and predicted specific land cover outcomes at a parcel or regional level. This model used a robust spatiotemporal framework but treated land management scenarios as exogenous rather than modeling the occurrence of these land use activities.

Taking model integration one step further, some researchers have moved toward developing a modeling approach integrating land use, decision making, and ecological change in rural environments at fine spatial scales. One such effort is the Forest Land Oriented Resource Envisioning System (FLORES), a model constructed from a multidisciplinary team of researchers (Vanclay, 1998). During a three-week workshop in 1999, a team of researchers constructed a model of rural land use change for a test area in Indonesia by including the following major model components: crops—soils, trees—forest, household decision making, and biodiversity—fauna. This integrated model was developed using a dynamic modeling package and loosely coupled to a spatial information system.

A loosely coupled model is where model components exchange data through input/output of different model components rather than operating in an integrated modeling framework. Future plans are to produce a model that more closely integrates the existing model components in a spatial framework that would allow more explicit spatial interactions to occur in the model. What was unique about the FLORES model was the effort to balance the complexity of the agriculture, forest ecology, and human decision making components within a spatial framework. In regard to the space-time-HDM modeling framework, the FLORES model is well developed in terms of the temporal dynamics and human decision making but currently is not well developed along the spatial dimension.

### 7.3.3 Health, Epidemiology, Pollution, and Hydrology

A natural interface of human-environmental modeling is in health-related applications such as disease contagion, environmental impact assessments, and pollution modeling. This is a vast area of research and modeling. Here we present a very limited set of examples to demonstrate types of models that have been constructed for these applications. In particular, we present models from the areas of environmental impact modeling and spatial epidemiology.

The Agricultural Nonpoint Source Pollution model (http://www.sedlab.olemiss.edu/AGNPS98.html) has been widely used to predict nutrient and fertilizer passport in agricultural systems. It is a raster-based model that uses different exogenous land use decisions (e.g., fertilizer applications) and landscape characteristics (soil characteristics, topography, etc.) to predict soil erosion and nutrient transport in agricultural environments. The AGNPS model is an example of a spatially explicit model, including spatial interactions, that operates in a temporal framework. While the AGNPS model lacks a component to model land use decisions, the model has the ability to examine the effect of different land use strategies by using them as model drivers. There are a tremendous number of models related to environmental health issues in agricultural environments, and AGNPS is a widely used model that is representative of this area of modeling.

Agricultural and hydrological models are less commonly referred to as human-environmental models because the model often treats human decisions as exogenous to the system (e.g., AGNPS). A variety of researchers have been successful in developing loosely coupled models linking existing social and biophysical models. A loosely coupled model can be referred to as a model where the data are exchanged between model components through input/output, but the model behavior is not integrated between the model components. We would argue that a major research challenge facing the modeling community is in developing tightly coupled models that balance model complexity on both the social and biophysical sides.

Other modeling applications related to the health aspect of human-environmental dynamics include the spaBraziltial modeling of disease vectors and transmission (Smith and Harris, 1991). Much of this research involves spatial examination of a combination of features, and thus GIS provides an ideal framework for these applications. There is a wide literature related to the application of GIS to epidemiology, although most of this work is empirical rather than modeling based. However, some researchers are developing models as a tool for predicting risk and exposure to different disease vectors or to predict disease incidence. For example, Castro and Singer (2000) used a spatial model to examine the relationship between land cover and malaria incidence in the Rondonia, Brazil. Other examples include spatial epidemiological models of HIV incidence (Salzberg and Mac-Rae, 1993). Because their applications are more focused on exposure or threats to human population from some source, these models do not explicitly model human decision making. However, there is utility in this integration. One example would be a model of rural land use expansion as it relates to the encroachment on malaria-infested areas. A dynamic simulation could model how the decision making process determines the spatial expansion of settlements and how the associated land cover changes might affect the habitat for mosquitoes and, in turn, the exposure to malaria.

### 7.3.4 Summary

Figure 7.5 is an example of how some of the different models can be described in terms of spatial, temporal, and HDM complexity. As we have discussed previously, models that are positioned high on the space complexity axis are spatially explicit and allow for spatial

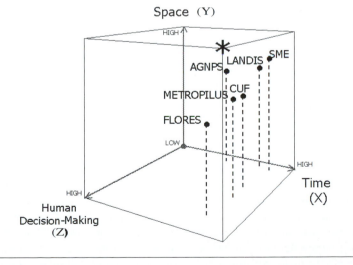

**Figure 7.5** Positions of models in relation to space, time, and decision making.

interactions. Models that are positioned high on the temporal complexity axis are dynamic, demonstrate feedbacks and equilibria in model states, and allow for varying time intervals. Models that are positioned high on the HDM complexity axis explicitly model agent decision making based on a set of heuristics defined in the model at multiple institutional scales. Some models were not designed to address all of these axes. Therefore, a low position on one axis does not mean the model is not as well constructed as another model, just that it is not as complex for that axis. In addition, model positions are approximate and adjusted to make the figure more readable. Within the context of a model's purpose, an important goal of future human-environmental models is to specify and develop models that have as high a position on all three axes as necessary (the asterik in figure 7.5)

## 7.4 MODELING COMPLEXITY AND HUMAN-ENVIRONMENTAL DYNAMICS

The physical and biological sciences have struggled to develop appropriate frameworks for environmental models at different spatial and temporal scales and levels of model complexity. The difficulty of this challenge increases greatly when theories of human decision making, scales, and complexity are included. As we noted earlier, scales and complexity of human decision making range from individuals to groups of increasingly large size until they encompass global networks. This section discusses specific theoretical issues for linking human-environmental dynamics within the context of space, time, and human decision making. Many of these issues have important implications for the complexity modelers might choose to include in their models.

### 7.4.1 General Issues

As modelers work to develop human-environmental models, it is essential that they identify the optimal scale(s) for their specific questions. In this context and because human-environmental dynamics are complex, it is important to recognize that certain human-environmental processes may be associated with specific scales in some cases, while processes may occur across multiple scales in other cases. Further, human-environmental processes to be addressed by the model might not operate at the same scale(s), and linkages may have to connect across scales (Redman et al., 2000).

### 7.4.2 Time

Human and environmental processes might work at different rates. Further, rates of change, such as the land cover change shown in Figure 7.6, are not necessarily linear over time. Thus, modelers need to consider whether there are time lags, nonlinear relationships, defining events, and positive and negative feedback loops that affect the responses among social and environmental processes (Costanza and Ruth, 1998; Gladwell, 2000; Grove, 1999). For instance, time lags might exist between changes in land use and transport of nitrogen in groundwater since groundwater flows might occur at a much slower rate than land use change from an agricultural to a residential land use. Similarly, forest stand characteristics (structure and composition) are more likely to reflect historic, selective harvesting preferences and practices than current ownership preferences and practices. This is due to the fact that changes in vegetation growth, species dynamics, and soil fertility change at

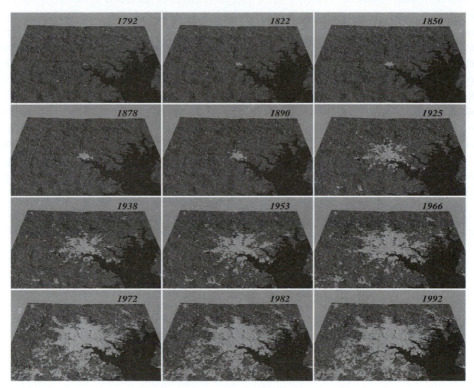

**Figure 7.6** Perspective view of urban growth in Baltimore, Maryland. over 200 years (1792–1992). Yellow polygons are built-up areas as determined from historical maps and satellite imagery, green areas are forests, and blue areas are water (Chesapeake Bay). *Source*: Penny Masuoka, UMBC, NASA Goddard Space Flight Center and William Acevedo, USGS, NASA Ames Research Center.

much slower rates than land ownership. In both cases, human and environmental legacies affect current human-environmental dynamics.

### 7.4.3 Space

Studies of how spatial characteristics affect ecological dynamics, particularly with GIS and computer modeling, have been an area of great interest (Forman and Godron, 1986; Gustafson, 1998; Naveh and Lieberman, 1994; Pickett and Cadenasso, 1995; Turner, 1989; Turner and Gardner, 1990). Examples of types of spatial metrics include measures of (1) landscape composition (for example, number of categories, proportions, diversity [evenness and richness]), (2) landscape configuration (such as size, shape, density, connectivity, fractal, and patch neighborhood), and (3) scale/structure (for instance, trendsurface, correlogram, and semivariogram) (Gustafson, 1998; McGarigal and Marks, 1995). However, comparable GIS and modeling efforts have not focused on how the spatial characteristics of certain phenomena, such as adjacency, shape, and matrix, affect human processes and the relationships among human-environmental processes (Grove, 1999). For instance, spatial adjacency (neighborhood analysis) might be included in models of human decision making to account for whether neighboring industrial areas affect residential locational decisions. The size and shape of an area (spatial metrics of a patch) might affect human processes. For example, as people commute from one commercial area to another, are individuals more likely to travel across a narrow or a wide residential area? Finally, the location of an area within a regional matrix (patch matrix analyses) might affect human decision making. For instance, how does access to a diversity of work, recreation, and other leisure amenities affect residential choices made by single adults and retired couples in urban areas? Each of these spatial examples can have implications for modeling human–environmental dynamics. These spatial interactions are complicated further by the fact that boundary conditions of areas might need to be considered as well, since the permeability of social and environmental areas might affect the flows of materials, nutrients, fauna, flora, persons, diseases, and ideas.

### 7.5.4 Human Decision Making

Coucelis (2001) has noted in Chapter 2 of this book that although human-environmental modeling might be primarily an applied field, it is not exempt from the need to be theoretically well grounded. It is too easy to develop models that look good but have an underlying ontology that is less plausible than a computer game. Since it is unlikely that we will ever have a "theory of everything" for human-environmental systems, we must develop our ability to assemble a wide variety of partial theories from the physical, biological, and social sciences. In other words, integrated approaches to human-environmental models are more than a matter of replacing integrated models with coupled models: rather, integrated approaches have to make sure that the assemblage of concepts, ontologies, approaches, theories, degrees of confidence, and spatiotemporal structures within a single framework respect the strengths and weaknesses of each part and yield a whole that is logically coherent (Coucelis, 2001).

   This issue of an overall, coherent approach to modeling human-environmental systems is critical. Coucelis (2001) cites Smyth (1998, p. 192) to observe that it is convenient to think of the modeled world as a microworld defined by an ontology consisting of contents, spatial structure, temporal structure, "physics" (rules of behavior), and rules of inference or logic. The notion of such a microworld is useful for reminding us that models are not the real thing and that they need to be internally consistent (Coucelis, 2001).

   We agree with Coucelis (2001) that human-environmental models often include a variety of disciplines and syntheses, which means that human-environmental models will assemble several kinds of "physics," some based on causal hypotheses (A appears to cause B), some on statistical regularities (A is statistically associated with B), others on empirical rules of thumb (when A, usually B), and others still on arbitrary rules of behavior specified by the modeler (if A is the case, then do B). Combining such a variety of partial "physics" into a complete model that is free of internal contradictions is a challenge for which few guidelines exist, and which becomes more difficult as the assemblage from different domains become more remote from each other. While it might be challenging to link models of rainfall and runoff to determine the likelihood that an area will flood, it is quite another to link models of industrialization and species extinctions in order to understand the relationship between urbanization and biodiversity (Coucelis, 2001).

We noted earlier that we are unlikely to have a "theory of everything" for human-environmental systems. However, many insights into the challenge of developing integrated approaches to human-environmental models that are based on different types of "physics" can be found in existing literature. Thus, we propose that the theoretical approaches to human decision making are most likely to be found among midlevel theories and not grand, unified theories of human-environmental systems (Burch and Grove, 1999; Ostrom, 1998; Parker et al., 1999; Pickett et al., 1999). This idea of midrange theory and its utility comes from Merton (1968, p.39), who notes that midrange theories "lie between the minor but necessary working hypotheses that evolve in abundance during day-to-day research and the all inclusive systematic efforts to develop unified theory that will explain all the observed uniformities of ... behavior, ... organization, and ... change. Mid-range theories are empirically grounded theories—involving sets of confirmed hypotheses—and not merely organized descriptive data or empirical generalizations which remain logically disparate and unconnected." Midrange theories connect observations, inferences, hypotheses, and empirically based research. While midrange theories may not be logically derived from a single, all-embracing theory, they may be consistent with one (Merton, 1968). Finally, rather than deriving a model of human-environmental dynamics from a single, all-embracing theory, a midrange theory approach provides the basis for progressively developing a more general model that is adequate for consolidating groups of midrange theories.

Many midrange theories that are appropriate for human-environmental modeling have been available for some time. For instance, social scientists have worked for a long time to include human-environmental interactions. Firey's (1990) comprehensive review of sociological work since 1926 demonstrates that substantial theoretical and empirical efforts have existed for some time and anticipated many of the concerns for integrating human-environmental dynamics. Firey's (1990, p.23) analysis of these works translates diverse terms into a unified lexicon. As he notes,

*When Mukerjee speaks of the "entire circle of man's life and well-being," he is placing into a single system such diverse factors as social organization, flora, fauna, fertility, climate, and topography. When Vance refers to the "cotton system" he has in mind a "complex whole" partaking of certain attributes of the physical environment–chemical, climatic, and genetic–and of certain attributes of the sociocultural order–structural, attitudinal, and organizational. When Odum speaks of "balance," Zimmerman of "real communities," Landis of "patterns," Kaufman of "stability," and Gibbs and Martin of "sustenance organization," there is implied some reference to a system whose components are not exclusively physical nor exclusively social or cultural. In these expressions there is a dual reference to two orders of phenomena, both of which have been articulated into a single conceptual construct. This concept of a resource system has two features, which particularly recommend it as a point of departure for further sociological research on natural resources. First, it is congruent with a great deal of important work that is being done in systematic sociological theory, centering on the concept of "social system." Second, it is noncommittal as to the mode of formulating causal relationships among the variables that enter into a system. In other words, either physical or sociocultural variables can be taken as independent, so that, with appropriate measures of both, a wide variety of hypotheses can be formulated and tested.*

We propose that many of these studies are relevant, informative, and important to rediscover for modeling the "physics" of different scales and levels of complexity of human-environmental dynamics. For this integration of midlevel theory in human-environmental models to occur, however, it is crucial to link appropriate midlevel physical, biological, and social theories (Pickett et al., 1999) to appropriate temporal and spatial scales and levels of complexity (Grove, 1999; Redman et al., 2000).

# 7.5 LESSONS LEARNED AND FUTURE DIRECTIONS

What then are the pressing needs related to future human-environmental modeling efforts? The preceding discussion proposed that much existing theoretical and empirical research on human-environmental systems is relevant and important to modeling efforts of human-environmental systems today. It also emphasized that human-environmental processes can be, and usually are, temporally and spatially complex as they interact with various scales of human decision making. Given the need to mine existing literature that might be relevant to specific human-environmental modeling questions, the following discussion focuses on three sets of activities or issues we think are particularly important in general to the development of future human-environmental models: (1) standard conventions for reporting scale across time, space, and human decision making, (2) closing the data gap, and (3) new forms of collaboration in the development of human-environmental models.

### 7.5.1 Conventions for Reporting Scale and Complexity

A significant hurdle we must overcome in the context of human-environmental modeling is the failure to articulate and document temporal, spatial, and human decision making scale(s). Many modeling techniques have the capacity to model across multiple scales of time, space, or human decision making. But in literature documenting applications of certain models, even though it is possible to articulate the temporal and spatial scale of a model application, we often find that many model summaries do not do this. Further, when models include a human decision making component, we are constrained by the lack of a well-specified language of scale that researchers can agree on and consistently report.

In our view, this failure to articulate and document the scale(s) of human-environmental models becomes problematic when we try to compare model results of similar systems, since it is well known that relationships among variables change depending on the scale of analysis (Root and Schneider, 1995; Turner et al., 1989). If we unknowingly compare results of two models operating at different scales, we might draw incomplete or incorrect conclusions that could lead to false theoretical understandings of the processes at work.

We cannot stress enough the theoretical implications of this issue, since scale probably drives many of the conflicts perceived to exist among different disciplines in the social sciences. For example, the arguments and differences existing among psychologists (Lynch, 1960; Sommer, 1969), sociologists (Bailey and Mulcahy, 1972; Catton, 1992, 1994; Field and Burch, 1988; Firey, 1945; Schnore, 1958; Young, 1974, 1992;), geographers (Agnew and Duncan, 1989), and political scientists (Masters, 1989) might be attributed more to the use of different scales and criteria (vis. Allen and Hoekstra, 1992) than questions of who is right or wrong. For instance, psychologists and sociologists argue about whether individual behaviors create social structures or whether social structures determine individual behaviors. Rather than seeing this as a mutually exclusive dichotomy, it may be more appropriate to conceive of such a question as a matter of scale and to ask about the relative relationship between individual behavior and social structure for a given question (Vogt et al., 2000). With this approach, questions are more resolvable by actually promoting discussions among modelers of human-environmental systems.

In the context of future modeling endeavors, we propose that *any* paper reporting model results should clearly report the scale(s) used. Temporal and spatial scale are relatively straightforward: Each of their scale components in Table 7.1 can be articulated clearly using generally accepted scientific measurements. This could be true as well for the scale components of human decision making if we can come to some agreement on a standard language and definitions for agent and domain. The terms we have proposed are our attempt to move us toward such a common set of terms. Regardless of the final set of terms, we propose that the clear articulation of temporal, spatial, and human decision making scale(s) is as essential to a paper as the abstract or the list of keywords.

A similar argument could be made for documenting a model's complexity, but more discussion is probably needed to build consensus for indices of temporal, spatial, and human decision making complexity. Earlier, we presented an index for estimating human decision making complexity for individual models (Table 7.1). Similar indices need to be developed for estimating the temporal and spatial complexity axes of Figure 7.4. Once we come to some agreement about these measures of scale and complexity, we will have moved forward significantly in our ability to compare the results of models using similar scales and/or complexity and to know when not to compare results because of differences in scale and/or complexity. Further, we might be able to evaluate whether various models using different scales and complexity can be linked if we understand the location of a particular model within the human-environmental modeling framework (Figure 7.4). This is crucial for developing multiscale or hierarchical approaches to modeling human-environmental systems.

### 7.5.2 Closing the Data Gap

An important challenge to modeling human-environmental systems is our lack of digitally available data. Several research organizations are starting to collect data for particular geographic regions, and these data are purposefully relevant to human-environmental modeling (see, for example, the Human Dimensions of Environmental Change research program sponsored by the U.S. National Science Foundation at http://www.nsf.gov/sbe/hdgc/hdgc-cntr.htm, and the U.S. NSF funded Long-Term Ecological Research groups at http://www.lternet.edu/ or the International Human Dimensions of Global Change research program at http://www.uni-bonn.de/ihdp/). While these groups collect time-series, spatially distributed, physical, biological, and social data for their areas of geographic focus, data for many other areas are either dispersed or absent. This data gap can hinder many human environmental modeling efforts that are empirically based.

With the tremendous advances made in the Internet (largely because of World Wide Web technologies), the sharing of data has become much easier. Recent endeavors such as the U.S. National Spatial Data Inventory (NSDI, located at www.fgdc.gov/nsdi/nsdi.html) and the Federal Geographic Data Committee (FGDC, located at www.fgdc.gov/fgdc/fgdc.html) have made significant progress in establishing standards for documenting information about spatial datasets (commonly called metadata) and developing a network of spatial data clearinghouses (see www.fgdc.gov/clearinghouse/clearinghouse.html). These are important advances for modelers because they provide global access to well-documented data sets (i.e., usable) that might be available for a particular area. Further, we are beginning to see Internet browser technologies that allow people to search for spatial data for a particular geographic location (e.g., MapInfo's "metadata browser").

Finally, there have been several attempts recently to overcome the quantitative data gaps for modeling purposes by integrating statements about qualitative changes in human behavior and environmental impacts (trajectories) based on regional case studies. For example, see Kuipers (1994), Kasperson and others (1995), Petschel-Held and others (1999), and Petschel-Held and Lüdeke (2000). In short, important progress is being made for solving the data gap problem. While much of the effort has focused on complete spatial datasets, additional attention will need to focus on time series data and data related to human decision making.

### 7.5.3 New Collaborative Forms for Development of Models

Models involving time, space, and human decision making can be incredibly complex and depend on knowledge from many disciplines. Until now, most models have developed in isolation. This is related to the fact that modelers have been funded through grants or focused funds from a particular organization with an interest in human-environmental modeling. Even in the context of large interdisciplinary research centers like the NSF networks cited previously, their efforts have been constrained by funds, staff, and expertise.

In contrast to traditional approaches to model development, recent advances in Internet and Web technologies have created new types of opportunities for collaboration in the development of human-environmental modeling. Already, "open source" programming efforts have been used to solve complex computing problems (see, for example, Kiernan, 1999; Learmonth, 1997; McHugh, 1998, and http://www.opensource.org). The principle of open source programming is based on a collaborative licensing agreement that enables people to download program source code freely and utilize it on the condition that they agree to provide their enhancements to the rest of the programming community. There have been several very successful, complex programming endeavors using the open source concept, the most prominent being the Linux computer operating system. There have also been some open source endeavors that have failed. The Linux model has shown that extremely complex problems can be tackled through collaboration over the Internet and that this kind of collaboration can produce extremely robust results. For instance, Linux is known to be a very stable software program, and it is largely because of what is referred to as "Linus's Law" (Linus Torvalds is the initial developer of Linux): "Given enough eyeballs, all [problems] are shallow" (Raymond, 1999). In other words, if we can get enough eyes with various skills and expertise working on a problem, every problem, regardless of complexity, can be solved because an individual or a team of individuals will come up with elegant solutions.

Yet how is an open source approach to computing connected to human-environmental modeling? We propose that a similar approach to the development of human-environmental models provides the basis for focusing enough eyeballs on important human-environmental problems (Schweik and Grove, in press). A similar argument has been made for open source endeavors in other areas of scientific research (Gezelter, 2000). Initiating such an open source modeling effort will require three components: (1) a Web site to support modeling collaboration (e.g., data and interactions among individuals, such as bulletin boards and FAQs); (2) establishment of one or more modeling kernels (these would be core components of models using various technologies) that are designed in a modular fashion and allow relatively easy enhancements from participants; and (3) development of mechanisms for sharing model enhancements that encourage participation and provide incentives that are comparable and as valued as publishing in peer-reviewed journals.

We recognize that the application of the open source programming concept to human-environmental modeling might appear daunting and even seem radical. However, the Linux example shows how extremely complex problems can be solved when enough people look at them. Given the complexities involved in modeling time, space, and human decision making, the open source programming concept might be a vital modeling approach for creative solutions to difficult human-environmental modeling problems.

## 7.6 CONCLUSION

Our goal in this chapter has been to contribute to the further development of human-environmental models. To achieve this goal, we proposed a conceptual framework for summarizing and comparing human-environmental models. We then reviewed several types of human-environmental models and related them, in a general way, to the framework. Based on these discussions, we identified some key issues that are inherent to modeling temporal, spatial, and human decision making scale and complexity. Finally, we discussed some new directions for modeling human-environmental dynamics. In the end, however, we have only mapped some possibilities and ideas for modeling human-environmental dynamics. We hope that others will improve on this initial effort for two reasons. First, we believe that modeling human-environmental dynamics is an interesting and exciting activity. Second, as the prevalence and significance of human-environmental interactions continue to grow, decision makers, researchers, and educators will find it increasingly important to have accurate, timely, and extensive information and understanding about the systems they inhabit and depend on for life.

## 7.7 REFERENCES

Agarwal, C., Green, G., Grove, M., Schweik, C., and Ostrom, E. 2000. Assessing Land Use Change Models and their Impact: Spatial, Temporal and Human Decision-Making Elements. General Technical Report (GTR). Northeastern Research Station, U.S. Department of Agriculture (USDA) Forest Service, Newtown Square, PA.

Allen, T. F. H., and Hoekstra, T. W., 1992. *Toward a Unified Ecology.* New York: Columbia University Press.

Bailey, K. D., and Mulcahy, P., 1972, Sociocultural versus Neoclassical Ecology: A Contribution to the Problem of Scope in Sociology. *The Sociological Quarterly,* 13 (Winter), 37–48.

Batty, M., and Xie, Y. 1994a. From Cells to Cities. *Environment and Planning B,* 21, S31–S34.

Batty, M., and Xie, Y. 1994b. Modeling Inside GIS: Part 2. Selecting and Calibrating Urban Models Using ARC-INFO. *International Journal of Geographic Information Systems,* 8, no. 5, 451–470.

Batty, M., and Xie, Y. 1994c. Modelling Inside GIS: Part 1. Model Structures, Exploratory Spatial Data Analysis and Aggregation. *International Journal of Geographic Information Systems,* 8, no. 3, 291–307.

Burch, W. R. Jr., and Grove, J. M. 1999. Ecosystem Management–Some Social, Conceptual, Scientific, and Operational Guidelines for Practitioners. In N. C. Johnson, A. J. Malk, W. T. Sexton, and R. Szaro (eds.), *Ecological Stewardship: A Common Reference for Ecosystem Management,* vol. 3, pp. 279–295. Oxford : Elsevier Science Ltd.

Castro, M. C. D., and Singer, B. 2000. *Agricultural Colonization, Environmental Changes, and Patterns of Malaria Transmission on the Tropical Rain Forest: The Case of Machadinho d'Oeste, Rondonia, Brasil.* Paper Presented at the Annual Meeting of the Population Association of America. Los Angeles, CA.

Catton, W. R., Jr. 1992. Separation versus Unification in Sociological Human Ecology. In L. Freese (ed.), *Advances in Human Ecology,* pp. 65–99, Vol. 1. Greenwich, CT: JAI Press.

Catton, W. R., Jr. 1994. Foundations of Human Ecology. *Sociological Perspectives,* 37, 75–95.

Clarke, K. C., and Gaydos, L. 1998. Loose Coupling: A Cellular Automaton Model and GIS: Long-Term Growth Prediction for San Francisco and Washington/Baltimore. *International Journal of Geographical Information Science,* 12, no. 7, 699–714.

Clarke, K. C., Hoppen, S., and Gaydos, L. 1997. A Self-Modifying Cellular Automaton Model of Historical Urbanization in the San Francisco Bay Area. *Environment and Planning B: Planning and Design,* v. 24, 247–261.

Costanza, R., and Gottlieb, S. 1998. Modeling Ecological and Economic Systems with STELLA: Part II. *Ecological Modeling,* 112, no. 2-3, 81–84.

Costanza, R., and Ruth, M. 1998. Using Dynamic Modeling to Scope Environmental Problems and Build Consensus. *Environmental Management*, 22, no.2 , 183–195.

Coucelis, H. 2001. Modeling Frameworks, Paradigms and Approaches. pp. 36–50. In K.C. Clarke, B. E. Parks, and M. P. Crane (eds.), *Geographic Information Systems and Environmental Modeling.* Upper Saddle River, NJ: Prentice Hall.

Dale, V., O'Neill, R. V., Pedlowski, M., and Southworth, F. 1993. Causes and Effects of Land-Use Change in Central Rondonia, Brazil. *Photogrammetric Engineering and Remote Sensing,* 59, no. 6, 997–1005.

Dale, V. H., O'Neill, R. V., Southworth, F., and Pedlowski, M. 1994. Modeling Effects of Land Management in the Brazilian Amazon Settlement of Rondônia. *Conservation Biology,* 8, no. 1, 196–206.

Deadman, P., Brown, R. D., and Gimblett, H. R. 1993. Modelling Rural Residential Settlement Patterns with Cellular Automata. *Journal of Environmental Management,* 37, 147–160.

Ehleringer, J. R., and Field, C. B. 1993. *Scaling Physiological Processes: Leaf to Globe.* San Diego: Academic Press.

Field, D. R., and Burch W. R. Jr., (eds). (1988). *Rural Sociology and the Environment.* Middleton, WI: Social Ecology Press.

Firey, W. 1945. Sentiment and Symbolism as Ecological Variables. *American Sociological Review,* 10 (April), 140–148.

Firey, W. 1990. Some Contributions of Sociology to the Study of Natural Resources. In R. G. Lee, D. R. Field, and W. R. Burch, Jr. (eds.), *Community and Forestry: Continuities in the Sociology of Natural Resources,* p. 256. Boulder, CO: Westview Press.

Ford, A. 1999. *Modeling the Environment: An Introduction to System Dynamics and Models of Environmental Systems.* Washington, DC: Island Press.

Forman, R. T. T., and M. Godron. 1986. *Landscape Ecology.* New York: Wiley.

Gezelter, D. 2000. http://www.openscience.org/talks/bnl/index.html

Gimblett, H. R., Richards, M. T., and Itami, R. M. 1998. A Complex Systems Approach to Simulating Human Behavior Using Synthetic Landscapes. *Complexity International,* 6. http://life.csu.edu.au/complex/ci/vol6/gimblett/gimblett.html

Gladwell, M. 2000. *The Tipping Point: How Little Things Can Make a Big Difference.* New York: Little, Brown and Company.

Grimm, N., Grove, J. M., Pickett, S. T. A., and Redman, C. 2000. Integrated Approaches to Long-Term Studies of Urban Ecological Systems. *Bioscience*, 50, no. 7, 1–11.

Grove, J. M. 1999. New Tools for Exploring Theory and Methods in Human Ecosystem and Landscape Research: Computer Modeling, Remote Sensing and Geographic Information Systems. In H. K. Cordell and J. C. Bergstrom (eds.), *Integrating Social Science and Ecosystem Management*, pp. 219-236. Champaign, IL: Sagamore Press.

Grove, J. M, Hinson, K., and Northrop, R. 2000. Education, Social Ecology, and Urban Ecosystems, with examples from Baltimore, Maryland. In A. R. Berkowitz, C. H. Nilon, and K. S. Hollweg (eds.) *Understanding Urban Ecosystems: A New Frontier for Science and Education*, p. 22. New York: Springer-Verlag.

Gustafson, E. J. 1998. Quantifying Landscape Spatial Pattern: What is the State of the Art? *Ecosystems,* 1, 143–156.

Hannon, B., and Ruth, M. 1997. *Modelling Dynamic Biological Systems.* New York: Springer-Verlag.

He, H. S., and Mladenoff, D. J. 1999. Spatially Explicit and Stochastic Simulation of Forest-Landscape Fire Distrubance and Succession. *Ecology,* 80, 81-99.

He, H. S., Mladenoff, D. J., and Crow, T. R. 1999a. Linking an Ecosystem Model and a Landscape Model to Study Forest Species Response to Climate Warming. *Ecological Modeling*, 112, 213–233.

He, H. S., Mladenoff, D. J., and Crow, T. R. 1999b. Object Oriented Design of LANDIS, a Spatially Explicit and Stochastic Forest Landscape Model. *Ecological Modeling,* 119, 1–19.

Johnston, R. A., and Barra, T. D. L. 2000. Comprehensive Regional Modeling for Long-Range Planning: Linking Integrated Urban Models and Geographic Information Systems. *Transportation Research A,* 34, no. 2, 125–136.

Kagel, J. H., and Roth, A. E. (eds.). 1997. *The Handbook of Experimental Economics.* Princeton, NJ: Princeton University Press.

Kasperson, J. X., Kasperson, R. E. and Turner, II B. L. (eds). 1995. *Regions at Risk. Comparisons of Threatened Environments*. New York: United Nations Publications.

Kiernan, V. 1999. The 'Open Source Movement' Turns Its Eye to Science *The Chronicle of Higher Education*, http://chronicle.com, Section: Information Technology, A51.

Kuipers, B. 1994 *Qualitative Reasoning: Modeling and Simulation with Incomplete Knowledge.* Cambridge, MA: MIT Press.

Lambin, E. 1994. *Modeling Deforestation Processes: A Review.* TREES Project, European Union.

Landis, J. D. 1995. Imagining Land Use Futures: Applying the California Urban Futures Model. *Journal of the American Planning Association,* 61, no. 4, 438–457.

Landis, J. D., and Zhang, M. 1998. The Second Generation of the California Urban Futures Model: Part 1: Model Logic and Theory. *Environment and Planning B,* 25, no. 5, 657–666.

Learmonth, M. 1997. *Giving It All Away,* http://www.metroactive.com/papers/metro/05.08.97/cover/linus-9719.html. 2/15/2000.

Luna, F., and Stefannson, B. 2000. *Economic Simulations in Swarm: Agent-Based Modelling and Object Oriented Programming.* Dordrecht, The Netherlands: Kluwer Academic Publishers.

Lynch, K. 1960. *The Image of the City.* Boston: MIT Press.

Masters, R. D. 1989. *The Nature of Politics.* New Haven, CT: Yale University Press.

McGarigal, K., and Marks, B. J. 1995. *FRAGSTATS: Spatial Pattern Analysis Program for Quantifying Landscape Structure.* USDA Forest Service General Technical Report, PNW-351.

McHugh, J. 1998. For the Love of Hacking. *Forbes.* August 10. http://www.forbes.com/forbes/98/0810/6203094a.htm

Merton, R. K. 1968. (enlarged edition). *Social Theory and Social Structure.* New York: Free Press.

Mladenoff, D. J., and He, H. S. 1999. Design and Behavior of LANDIS, an Object-Oriented Model of Forest Landscape Disturbance and Succession. In D. J. Mladenoff and W. L. Baker (eds.) *Spatial Modeling of Forest Landscape Change: Approaches and Applications.* Cambridge, England: Cambridge University Press.

Naveh, Z., and Lieberman, A. S. 1994. *Landscape Ecology: Theory and Application,* 2nd ed. New York: Springer-Verlag.

Opensource.org 2000. *Frequently Asked Questions About Open Source.* http://www.opensource.org/faq.html. 2/15/2000.

Ostrom, E. 1998. A Behavioral Approach to the Rational Choice Theory of Collective Action. Presidential Address, American Political Science Association, 1997. *American Political Science Review,* 92, no. 1, 1–22.

Ostrom, E., Gardner, R., and Walker, J. 1994. *Rules, Games, and Common-Pool Resources.* Ann Arbor: University of Michigan Press.

Parker, J. K., Sturtevant, V., Shannon, M., Grove, J. M., and Burch, W. R. Jr. 1999. Partnerships for Adaptive Management, Communication and Adoption of Innovation, Property Regimes, and Community Deliberation: The Contributions of Mid-Range Social Science Theory to Forest Ecosystem Management. In N. C. Johnson, A. J. Malk, W. T. Sexton, R. Szaro (eds.), *Ecological Stewardship: A Common Reference for Ecosystem Management.* Vol. 3, pp. 245-277. Oxford, England: Elsevier Science Ltd.

Pepper, J. W., and B. B. Smuts. 1999. The Evolution of Cooperation in an Ecological Context: An Agent-based Model. In T. Kohler, and Gumerman (eds.) *Dynamics in Human and Primate Societies.* Oxford, England: Oxford University Press.

Petschel-Held, G., Lüdeke, M. K. B., and Reusswig, F. 1999. Actors, Structures and Environments: A Comparative and Transdisciplinary View on Regional Case Studies of Global Environmental Change. In B. Lohnert, and H. Geist (eds.) *Coping with Changing Environments: Social Dimensions of Endangered Ecosystems in the Developing World.* Ashgate, Aldershot, Brookfield, Singapore, Sydney. - or as internet document: http://www.pik-potsdam.de/~gerhard/chap11.pdf

Petschel-Held, G., and Lüdeke, M. K. B. 2000. *Integration of Case Studies by Means of Artificial Intelligence.* Submitted to *Integrated Assessment.* http://www.pik-potsdam.de/~gerhard/pdfs/int_assess.pdf.

Pickett, S. T. A., Burch, W. R., Jr., and Grove, J. M. 1999. Interdisciplinary Research: Maintaining the Constructive Impulse in a Culture of Criticism. *Ecosystems,* 2, 302–307.

Pickett, S. T. A. and Cadenasso, M. L. 1995. Landscape Ecology: Spatial Heterogeneity in Ecological Systems. *Science,* 269 (July 21, 1995), 331–334.

Putman, S. 1992. *Integrated Urban Models 2: New Research and Application of Optimization and Dynamics,* London: Pion Press.

Putman, S. H. 1983. *Integrated Urban Models: Analysis of Transportation and Land Use.* London: Pion Press.

Raymond, E. 1999. *The Cathedral and the Bazaar,* March 16, 2000, http://www.tuxedo.org/~esr/writings/cathedral-bazaar/cathedral-bazaar.html.

Redman, C., Grove, J. M, and Kuby, L. 2000. *Toward a Unified Understanding of Human*

*Ecosystems: Integrating Social Sciences into Long-Term Ecological Research.* White Paper of the Social Science Committee of the LTER Network. http://www.lter-net.edu/research/pubs/informal/socsciwhtppr.htm.

Root, T. L., and Schneider, S. H. 1995. Ecology and Climate: Research Strategies and Implications. *Science.* 269, 334–341.

Roth, Alvin E. (ed.). 1985. *Game-Theoretic Models of Bargaining.* Cambridge, England: Cambridge University Press.

Salzberg, A. M., and MacRae, D. 1993. Policies for Curbing the HIV Epidemic in the US: Implications of a Simulation Model. *Socio-Economic Planning Sciences,* 27, no.3, 153–169.

Schnore, L. F. 1958. Social Morphology and Human Ecology. *The American Journal of Sociology,* 63 (May), 620–634.

Schweik, C. M., and Grove, M. 2000. Fostering Open-Research Via a Web System. *Public Administration and Management: An Interactive Journal* (http://www.pamij.com), 5(4).

Smith, G. C., and Harris, S. 1991. Rabies in Urban Foxes (*Vulpes vulpes*) in Britain–the Use of a Spatial Stochastic Simulation Model to Examine the Pattern of Spread and Evaluate the Efficacy of Different Control Regimes. *Philospophical Transactions of the Royal Society of London Series B-Biological Sciences,* 334, no. 1271, 459–479.

Smith, V., Hoffman, E., McCabe, K., and Shachat, K. 1994. Preferences, Property Rights and Anonymity in Bargaining Games. *Games and Economic Behavior.* 7, 346–380.

Smyth, C. S. 1998. A Representational Framework for Geographic Modeling. In M. J. Egenhofer and R. G. Golledge (eds.), *Spatial and Temporal Reasoning in Geographic Information Systems.* pp. 191–213. New York: Oxford University Press.

Sommer, R. 1969. *Personal Space: The Behavioral Basis of Design.* Englewood Cliffs, NJ: Prentice Hall.

Southworth, F. 1995. *A Technical Review of Urban Land Use–Transportation Models as Tools for Evaluating Vehicle Travel Reduction Strategies.* Oak Ridge, TN: Oak Ridge National Laboratory.

Turner, M. G. 1989. Landscape Ecology: The Effect of Pattern on Process. *Annual Review of Ecology and Systematics.* 20, 171–197.

Turner, M. G., and Gardner, R. H. (eds.). 1990. *Quantitative Methods in Landscape Ecology.* New York: Springer-Verlag.

Turner, M. G., O'Neill, R. V., Gardner, R., and Milne, B. T. 1989. Effects of Changing Spatial Scale on the Analysis of Landscape Pattern. *Landscape Ecology,* 3, no. 4, 153–162.

Vanclay, J. K. 1998. FLORES: For Exploring Land Use Options in Forested Landscapes. *Agroforestry Forum,* 9, no. 1, 47–52.

Veldkamp, A., and Fresco, C. O. 1996. CLUE: A Conceptual Model to Study the Conversion of Land Use and Its Effects. *Ecological Modelling,* 85, 253–270.

Vogt, K. A., Grove, J. M., Asbjornsen, H., Maxwell, K., Vogt D. J., Sigurdardottir, R., Dove, M. 2000. Linking Ecological and Social Scales for Natural Resource Management. In J. Liu, and W. W. Taylor (eds.), *Integrating Landscape Ecology into Natural Resource Management,* p. 28. New York: Cambridge University Press.

Voinov, A., and Costanza, R. 1999. Landscape Modeling of Surface Water Flow: 2. Patuxent Case Study. *Ecological Modeling*, 119, 211–230.

Voinov, A., Costanza, R., Wainger, L., Boumans, R., Villa, F., Maxwell, T., and Voinov, H. 1999. Patuxent Landscape Model: Integrated Ecological Economic Modeling of a Watershed. *Environmental Modeling & Software Journal,* 14, no. 5, 473–491.

Webster, F. V., and Pauley, N. J. 1991. Overview of an International Study to Compare Models and Evaluate Land-Use and Transport Policies. *Transport Reviews,* 11, no. 3, 197–222.


Wegener, M. 1994. Operational Urban Models. *Journal of the American Planning Association,* 60, 17–29.

White, R., and Engelen, G. 1993. Cellular Automata and Fractal Urban Form: A Cellular Modeling Approach to the Evolution of Urban Land-Use Patterns. *Environment and Planning A,* 25, 1175–1199.

Wilson, A. G. 1998. Land Use/Transport Interaction Models: Past and Future. *Journal of Transport Economics*, 32, no. 1, 3–26.

Young, G. L. 1974. Human Ecology as an Interdisciplinary Concept: A Critical Inquiry. *Advances in Ecological Research,* 8, 1–105.

Young, G. L. 1992. Between the Atom and the Void: Hierarchy in Human Ecology. In Freese, L. (ed.), *Advances in Human Ecology.* pp. 119–147, Volume 1. Greenwich, CT: JAI Press.

## 7.8 ACKNOWLEDGMENTS

Support for this study was provided by the U.S. Forest Service's Burlington Laboratory (4454) and Southern and Northern Global Change Programs (particularly Cooperative Agreement 23-99-0074); the National Science Foundation (NSF Grant #DEB-9714835); the Environmental Protection Agency (EPA Grant #R-825792-01-0); and Indiana University's Center for the Study of Institutions, Population, and Environmental Change (NSFSBR 9521918).

## 7.9 ABOUT THE CHAPTER AUTHORS

**Morgan Grove** has worked for the U.S. Forest Service's Northeastern Research Station since 1996 as a Research Forester/Social Ecologist in Burlington, Vermont; Durham, New Hampshire, and Baltimore, Maryland. He is a Principal Investigator in the National Science Foundation, Long Term Ecological Research Program's Baltimore Ecosystem Study and a developer of the Northeastern Research Station (NED) set of decision-support tools. His research is on human ecosystem and landscape studies of forested areas, participatory action research approaches, and the development of technology transfer tools.

**Charlie Schweik** is Assistant Research Professor with the Center for Public Policy and Administration and the Resource Economics Department at the University of Massachusetts, Amherst. His recent research focuses on the human dimensions of environmental change, specifically applying GIS and satellite imagery to study human incentives, actions, and environmental outcomes. He has worked for IBM and the Department of Energy.

**Tom Evans** is an Assistant Professor in the Department of Geography and research associate at the Center for the Study of Institutions, Population, and Environmental Change (CIPEC) at Indiana University. His research is related to the social and biophysical factors affecting land cover change, especially deforestation and reforestation. He has worked on land cover change worldwide. His current research is focused on land cover change modeling using dynamic systems models and agent- based modeling.

**Glen Green** is a Research Fellow at the Center for the Study of Institutions, Population, and Environmental Change (CIPEC) at Indiana University, Bloomington Indiana. His research has focused on the use of Remote sensing to map and monitor land cover change and to understand the human dimension of global environmental change across multiple environments.

# 8

# *Modeling Physical Systems*

## 8.0 SUMMARY

Human society and ecosystems depend on the quantity and quality of water, air, soil, and other natural resources. Our understanding of atmospheric, hydrologic, and landscape processes is therefore crucial for environmental sustainability and ecological stability. Besides being key components of life, water and air are also the main carriers of pollutants, and modeling of these physical systems is an inherent part of modern environmental research. In this chapter, the general principles used for building models of physical systems are introduced and illustrated by examples describing distribution of surface water and sediment. The role of computational modeling and simulations in providing better understanding of physical processes is discussed and key modeling approaches are explained.

Modeling of physical processes is based on representation of spatial phenomena as fields and their discretization, according to the approach used for the solution of governing differential equations. The equations are usually solved by numerical approaches, such as finite difference, finite element, and path sampling methods. The role of geographic information systems (GIS) remains primarily in the area of input data processing, analysis, visualization, and management of model results. Simpler models, especially raster-based models, can be developed and run within the GIS, using GIS commands, map algebra tools, or modeling languages. Several modeling systems include specialized GIS capabilities used for creating the conceptual site models and for customized visualization, analysis, and presentation of the results. Because of their role in the environment, models of physical systems are often the core modules in integrated models and decision support systems.

## 8.1 INTRODUCTION

One of the key challenges in environmental research is the description of interacting physical processes  with sufficient realism. It is clear that a rapid development of computer technology offers new opportunities to tackle extremely complex environmental problems. In fact, computational simulation and modeling is becoming a third way of doing scientific research complementing the traditional experiments and analytical theories. Computational approaches belong to "young" methodologies that were developed only over the past few decades, and their progress is closely tied to the advances in computer capabilities. As such, they have their own rules, challenges, successes, and limitations. The role of algorithms, data structures, computationally efficient methods, advanced visualization, and exploration of parallelism is crucial for new advances in environmental research and requires close collaboration between traditional research disciplines and computational science.

Originally, GIS applications were focused on static spatial data processing, analysis, and computer cartography. However, development of new geospatial data collection technologies  and computer capabilities together with acute environmental problems have pushed the GIS applications into more sophisticated levels. Advanced geoscientific applications involve multidimensional phenomena and dynamics (see Chapter 6; Burrough, 1998,  Mitas et al., 1997; Mitasova et al., 1995), supercomputing class simulations, as well as real-time processing of large sets of measured data (Cantin and Fortin, 2000). Nevertheless, the process-based modeling of geospatial phenomena involves substantially more uncertainty than modeling in physics or chemistry. One of the key reasons is the aforementioned complexity of studied phenomena. The practical solutions then have to rely on the best possible combination of physical models, empirical evidence, intuition, and available measured data. In physics, the accuracy is usually understood in a much stricter sense, because many fundamental laws are known over a broad range of scales in energy, distance, or time. For example, the Schrodinger equation describes the matter at the electronic level virtually exactly, which means within spectroscopic accuracy of 6 to 12 digits. This is seldom the case in complex geoscientific applications, where 50% differences between measurements and model predictions can be in many instances considered satisfactory. New generation distributed models supported by high-resolution data provide an opportunity to improve the predictive capabilities of models and provide reliable information for decision making.

This chapter introduces the basic types of models and their components and explains the building of physical process models described by differential equations. Possible methods of solution are mentioned, with a newer path sampling method described in more detail. Different approaches to integration of models with GIS are explained and illustrated by several practical examples.

## 8.2 PHYSICAL PROCESSES IN ENVIRONMENTAL MODELING

The physical processes relevant for environmental modeling can be categorized as follows:

1. Atmospheric, such as global and regional air circulation
2. Hydrologic and ocean studies, for processes involving water and related phenomena
3. Geomorphologic, geologic, and geophysical processes, describing surface and subsurface solid earth phenomena

These categories are conveniently based on disciplines; however, many important processes permeate more than one or even all of them. One example is the water cycle, which involves the movement of water between the atmosphere and earth surface/subsurface layers. Another case is the carbon cycle, which describes the evolution of carbon in the full variety of its forms: generation and transport of gases such as carbon dioxide and methane, life cycles of organic matter of plants and animals, organic matter in soil, etc. It is becoming obvious that water and carbon cycles will play a key role for sustainability of life on earth, and they are the focus of major efforts in environmental modeling (e.g., Krysanova and Wechsung, 2000).

### 8.2.1 Model Types and Components

Computational approaches to investigations of physical systems are based on simulation and modeling (Ceperley, 1999). By simulation we understand a computer representation of reality in which the simulated system is governed by a set of known physical laws expressed in mathematical language. In this case the model is already in place and its range of validity and accuracy is supposed to be well known and verified. The task of simulation is to solve the underlying equations for a particular realistic situation. The fact that the fundamental laws are known does not mean that simulations are straightforward or easy. The corresponding equations are often difficult to solve, and barriers in computational feasibility and efficiency often limit accuracy, resolution, or size of the modeled system.

On the other hand, by modeling we understand a process by which the scientist is trying to build a simplified version of reality for phenomenon for which the fundamental laws are either unknown, impractical to use, or simply do not exist. This typically involves systems that are very complex, with many constituents and variety of interactions between them and with limited amount of available experimental information. Typical examples include many ecological models and systems that involve anthropogenic activities. Modeling process often involves trial-and-error effort, and in some cases its predictive power, accuracy, and relation to reality might be a research problem on its own.

Physical systems are often described by a combination of physically based and empirical models. Empirical models are based on observations and statistical analysis of observed data, and their applicability is limited to the conditions for which they were developed. They can provide a rough picture of the phenomenon under study, but they often cannot explain how the system works. Because of their simplicity, they are widely used for practical applications and as components of more complex models for the subprocesses for which the physical model is unknown or too complicated. Physically based models take into account the nature of physical processes behind the observed phenomena and are usually based on the following:

1. Definition of model constituents and corresponding physical quantities such as concentration, density, velocity, etc. Typically, the physical quantities depend on position in space and on time. To emphasize this distributed and dynamic character, such quantities are characterized as physical fields. The fields can be scalar (single component) or multi component, such as vector or tensor fields. Sometimes the physical constituent has a discrete character, and then instead of fields, particle(s) or ensembles of particles are used. There exists a transformation between particles and fields, in the sense that density of particles in space defines a field and vice versa (i.e., field can be represented by particles with corresponding spatial distribution).

2. Description of configuration space for fields and/or particles and the corresponding range of its physical validity. This includes specification of relevant initial, external, or boundary conditions, as well as physical conditions and parameters.

3. Description of interactions between the constituents, such as the impact of one field on another or interaction between particles and fields (e.g., gravitation).

4. Governing equations derived from natural laws or evolution rules, which describe the behavior of the system in space and time. Typical examples are continuity, mass, energy, and momentum conservation, diffusion-advection, reaction kinetics, and similar types of equations. In addition, various constituent or state equations that relate the constituents to physical conditions and parameters need to be included (e.g., a state equation that relates pressure, volume, and temperature).

Models can be further categorized according to several criteria. Based on the treatment of spatial distribution, models can be divided essentially into two classes:

1. Models based on homogeneous or spatially averaged units. In hydrologic applications, the units represent watershed hierarchies, channels, and stream networks, lakes, wetlands, or hillslope segments (e.g., Srinivasan and Arnold, 1994; Helleweger and Maidment, 1999). The processes are then described by unit-to-unit transport rules or by ordinary differential equations for quasi-1D flow. This approach is very effective for systems that include human-made structures (urban hydrology, agricultural fields); however, adequate selection of units, their network topology, and hierarchies can be time consuming for larger areas and can require substantial expertise. This is true especially for complex, natural environments that cannot be easily described by simple geometrical features (e.g., a complex hillslope by a tilted plane, curved stream by a line segment).

2. Distributed models rely on various discretizations of fields, such as regular or irregular grids or meshes, derived from the numerical methods used for solving the governing equations. The most common approaches are based on finite differences, finite elements, and variety of spectral projection methods (Saghafian, 1996; Tucker et al., 1999; Vieux et al., 1996). Recently, path sampling strategies were employed for some of the water and soil transport problems because of their simplicity, robustness and scalability (Mitas and Mitasova, 1998).

## 8.2.2 Equation Examples

Many physical processes can be described by partial differential equations. A 2D continuity equation approximately describes steady state shallow water flow over a hillslope:

$$\nabla \cdot [h(\mathbf{r})v(\mathbf{r})] = i\,(\mathbf{r}) \tag{8.1}$$

where $\nabla\cdot$ is the divergence operator, $\mathbf{r} = (x, y)$ is a position, $h(\mathbf{r})$ is the water depth, $v(\mathbf{r})$ is the water flow velocity vector field, and $i(\mathbf{r})$ is the rainfall excess. If we assume uniform flow velocity, the explicit solution of this equation can be expressed as a function of upslope contributing area (Moore and Foster, 1990; Moore et al., 1993), which can be computed using standard GIS flow tracing functions (Figure 8.1b, c). A more realistic solution that takes into account variable flow velocity (due to topography and land cover) requires numerical solution of equation (8.1) (Figure 8.1d, e, f; Mitas and Mitasova, 1998). If we are interested in dynamical effects, the water flow equation will become

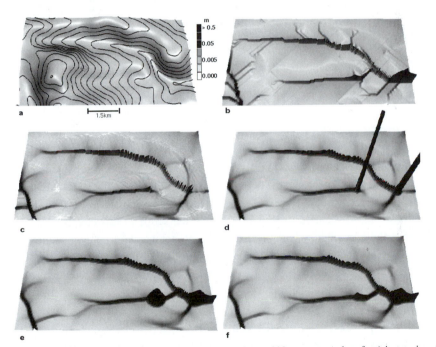

**Figure 8.1** Modeling steady state surface water depth $h$ (shown as 200x exaggerated surface) in terrain with depressions and uniform soil and cover conditions: (a) input 10 m resolution digital elevation model (DEM); (b) $h$ as a function of upslope contributing area computed by D8 algorithm: artificial flow pattern on hillslopes, geometrical flow through depressions (**r.watershed** in GRASS5); (c) $h$ as a function of upslope contributing area computed by vector-grid algorithm: depressions act as sinks, (Mitasova et al., 1996, **r.flow** in GRASS5); (d) 2D kinematic wave solution of continuity equation: water accumulates in depressions to infinity (SIMWE with diffusion→0); (e) 2D approximate diffusive wave solution: water fills depressions creating ponds with subsequent outflow (SIMWE); (f) 2D approximate diffusive wave solution with a channel with predefined gradient in one of the depressions: water flows rapidly through the depression. The SIMWE model is described by Mitas and Mitasova (1998).

$$\partial h(\mathbf{r}, t) / \partial t + \nabla \cdot [h(\mathbf{r}, t)v(\mathbf{r}, t)] = i(\mathbf{r}, t) \tag{8.2}$$

where the first term represents the change of water depth due to time $t$, and the second term represents the change in water depth due to space.

Surface water flow can detach soil and transport it over landscape, which can be described by an equation similar to equation (8.2) with additional terms:

$$\partial c(\mathbf{r}, t) / \partial t + \nabla \cdot [c(\mathbf{r}, t)v(\mathbf{r}, t)] - D\Delta c(\mathbf{r}, t) = \text{sources}(\mathbf{r}, t) - \text{sinks}(\mathbf{r}, t) \tag{8.3}$$

where $c(\mathbf{r}, t)$ is the sediment concentration, $\Delta$ is Laplace's operator, and the term $D\Delta c(\mathbf{r}, t)$ describes the diffusion that is a result of random influences. On the right side is the sources/sinks term, which describes inflows or outflows of solute mass $c(\mathbf{r}, t)$. This equation includes both deterministic and stochastic influences, which is typical for most natural processes. An even more general equation would include a rate term:

$$\partial c(\mathbf{r}, t)/\partial t + \nabla \cdot [c(\mathbf{r}, t) v(\mathbf{r}, t)] - D \Delta c(\mathbf{r}, t) + U(\mathbf{r}, t) c(\mathbf{r}, t) = \text{sources}(\mathbf{r}, t) - \text{sinks}(\mathbf{r}, t) \tag{8.4}$$

where $U(\mathbf{r}, t)$ is a potential that describes the local rate of proliferation or diminishing of $c(\mathbf{r}, t)$.

### 8.2.3 Methods of Solution

The differential equations describing the spatial processes are usually solved in a discretized form. Typical approaches include finite difference (see Chapter 5) and **finite element** methods (e.g., Burnett, 1987; Tucker et al., 1999; Vieux et al., 1996). The detailed description of these standard methods is beyond the scope of this chapter, but we briefly introduce a newer method, which has a promising potential for many geoscientific applications. The approach can be called path sampling (other names, such as random walks, are also common [Gardiner, 1985]) and relies on the duality: field <=> particle density. Suppose that we have an equation

$$L[c(\mathbf{r}, t)] = S(\mathbf{r}, t) \tag{8.5}$$

where $L$ is an operator, $c(\mathbf{r}, t)$ is the unknown quantity, and $S$ denotes sources-sinks. Suppose that $L$ is linear, such as in equation (8.4) (although weak nonlinearities can be treated, too). Symbolically, the solution can be written as

$$c(\mathbf{r}, t) = L^{-1}S(\mathbf{r}, t) \tag{8.6}$$

where $L^{-1}$ is the operator inverse to $L$. Of course, this implies that the inverse operator is known, which is seldom the case. Nevertheless, we are able to simulate what is the action of $L^{-1}$ on $S(r, t)$. This can be done as follows (Figure 8.2):

1. Sample the source term field by a set of points in the configuration space.

2. Apply the action of $L^{-1}$ on $S(r, t)$ by using appropriate expression for the Green's function (Stakgold, 1997); points representing $S(r, t)$ evolve and create paths.

3. Transform the solution represented by the path samples to continuous field $c(r, t)$ by evaluating the path densities.

    The path sampling method has been successfully used for a number of linear (or weakly nonlinear) transport and stationary problems (Ceperley and Mitas, 1996; Moglestue, 1993; Nuclear Energy Agency, 1995). It has several important advantages when compared with more traditional approaches. The method is very robust and can be easily extended into arbitrary dimension. It is also mesh free because it is based on evolving sampling points, which are independent and unrelated to any prescribed geometrical construction. The independence of the sampling points also enables an efficient and scalable by definition implementation on parallel architectures. Because the method does not rely on meshes, it is also rather straightforward to implement in a multiscale framework by an appropriate rescaling of the sample populations in particular regions (Mitasova and Mitas, in press). The method has been used in the SIMWE (SIMulation of Water Erosion) model (Mitas and Mitasova, 1998).

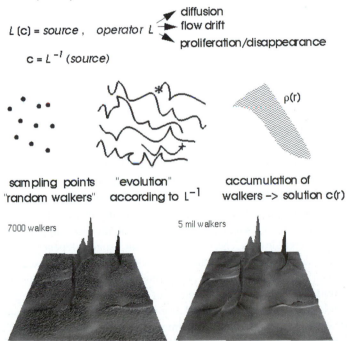

Figure 8.2 Path sampling method for solution of partial differential equations.

### 8.2.4 The Problem of Multiple Scales

Many of the natural processes involve more than a single scale and exhibit multiscale, multiprocess type of phenomena (Green et al., 2000; Steyaert, 1993). The problem of multiple scales now permeates a number of scientific disciplines, such as materials research, geosciences, and biology. Some multiscale problems can be partitioned into a system of nested models in the direction from fine to coarse scales. This requires development of an effective model on each scale level that incorporates simplified or "smoothed out" effects coming from finer, more accurate levels. On the other hand, in the direction from coarse to fine scales, one develops a set of effective embeddings that determine boundary and/or external conditions for the processes on finer scales. The high accuracy, resolution, and fine scale processes are then used only in "hot spots" of the studied system that require such a treatment.

It is interesting that in many cases such partitioning/nesting of models might be very difficult. This is true whenever the system exhibits fluctuations that appear over a large range of scales, which can reach up to the size of the whole system. Such examples exist in nature (e.g., an onset of magnetism in magnetic materials, self-organized criticality [Favis-Mortlock et al. 1998; Jensen, 1998]). In these cases one must carefully filter out the inherently multiscale effects first. Such models are being studied and have been constructed; however, their investigation is a research problem by itself. Another case for which a simple scale decomposition and nesting might not work are nonlinear and unstable systems. Whenever a small perturbation can rapidly overtake a large part or even the whole system (detonation phenomena, collapse of a large engineering or natural structure, avalanches, etc.), accurate description and decomposition is difficult and often crude approximation must be made to enable feasible studies.

## 8.3 MODELS AND GIS

The use of GIS for environmental modeling has substantially increased over the past five years, moving from research to routine applications (Goodchild et al., 1993, 1996, 1997; Parks et al., 2000). The implementation of new modeling functions, improvements in capabilities to import/export data in various formats, and introduction of object-oriented technology facilitate more efficient use of GIS functionality and support coupling of models with GIS at different levels. The type of coupling usually depends on the model complexity as well as on the target audience. Developers and researchers with full understanding of the underlying theory, algorithms, and source code often work with GIS-independent models, while using a wide range of software tools (including GIS) to support the model development and applications. Land owners, decision makers, and the public in general need a different modeling environment, with easy-to-use models, analysis, and visualization tools, supporting adequate understanding and interpretation of results. With the growing use of GIS by users with different backgrounds, the linkage between GIS and models has taken many forms and levels of complexity.

## 8.3.1 Full Integration of Models within GIS: Embedded Coupling

Models that are useful for a wide range of GIS applications are often developed and implemented within a GIS using its programming tools, such as Application Programming Interface (API), scripting language, or map algebra operations. The model is then run as a GIS function or command, the inputs and outputs are stored in a GIS database, and no data transfer is needed. Portability of the model is restricted, and the enhancements as well as the maintenance of the model are dependent on the GIS. This type of model development is further supported by customization and application development tools (ESRI, 1996), extensions to map algebra (Park and Wagner, 1997; Wesseling et al., 1996, and visual modeling tools (Murray et al., 2000).

Implementation of simple models within a raster GIS is relatively easy using the combination of map algebra operations and raster GIS functions. Site-based models, which are location dependent but do not incorporate spatial interaction, can be implemented as spatial models by applying the underlying equation(s) to each spatial unit (grid cell or polygon) using map algebra. Models that involve spatial interactions, such as water flow, can be computed using a flow tracing function (Mitasova et al., 1995; Moore et al., 1993) or the neighborhood syntax of the map algebra (Shapiro and Westervelt, 1992). Accuracy and realism of the resulting water flow pattern is then dependent on the underlying algorithm used for the flow tracing and on the quality of input data, as illustrated by figure 8.1b, c.

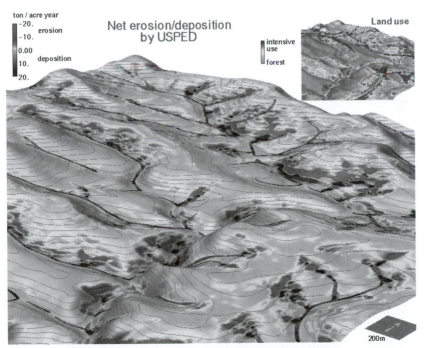

**Figure 8.3** Land use and net erosion/deposition pattern computed by unit stream power-based erosion deposition (USPED) (Mitasova and Mitas, 1999) using general GIS (GRASS5).

Map algebra operations have been successfully used for the development of simple dynamic models using cellular automata or finite difference methods for solving underlying partial differential equations. This type of dynamic models have been developed both within the general GIS as well as for more specialized systems and GIS extensions, such as the PCRaster with its high-level programming language supporting development of dynamic models (Burrough et al., 2000; Sluiter et al., 2000; Wesseling et al., 1996) or cellular automata extension of IDRISI (Park and Wagner, 1997).

Full integration of complex models involving solutions of coupled partial differential equations has been limited, in spite of several successful implementations (Doe et al., 1996; Saghafian, 1996; Vieux, 1995). Among the basic obstacles is the fact that the support for double precision floating point data, an essential requirement for effective modeling of nonlinear phenomena, has been less than adequate. For example, in the most widely used GIS, the floating point grids are supported; however, the tools for creating suitable nonlinear color tables or categories are not available and the number of significant digits can be cut off during processing without giving any warning, leading to the loss spatial variability for entire regions (Mitasova and Mitas, 1999). With the growing number of 3D dynamic models (see Chapter 6) better and/or cheaper support for temporal 3D data and finite element meshes would be needed. Even with all the necessary capabilities available within the GIS, the biggest disadvantage of full integration of complex models is that the models become too dependent on the development and fate of a particular GIS. Changes in the GIS data structures, functionality, interface, libraries, or programming tools, which are beyond the control of the model developers, may require time-consuming changes in the models or the models become incompatible with the latest version of the GIS software. Also, the fully integrated model is less portable and users have to install entire GIS even if they need the model only for a one-time application. Therefore, alternative methods of coupling have been more widely adopted, especially for complex models.

## 8.3.2 Integration under a Common Interface: Tight Coupling

The tight coupling level of integration has been widely used for existing models that were developed outside GIS. The GIS and model are linked through a common interface that guides the user through the steps needed for input data processing, running the models, and analyzing and presenting the results. The model has its own data structures, and the interface provides the tools to extract the input data from the GIS layers and create the associated databases in a format required by the model. The interface also allows the user to visualize the results using both the GIS display tools and specialized graphical and numerical outputs. This type of integration has proven to be effective for several hydrologic and non–point source pollution models (Khairy et al., 2000; Srinivasan and Arnold, 1994: SWAT;  Rewerts and Engel, 1991: ANSWERS; Srinivasan and Engel, 1991: AGNPS; Environmental Protection Agency, 2000: BASIN-2). Several models have been coupled with more than one GIS (e.g., SWAT, ANSWERS with both GRASS and ArcView). This approach have been used not only for coupling of a single model with GIS but also for development of modeling systems supporting simulation of numerous interacting processes (see Chapter 9).

### 8.3.3 Linkage through the Inport/Export of Data: Loose Coupling

Improvements in GIS import/export capabilities as well as the increased availability of digital geospatial data have made the loose coupling a routine procedure used with almost any stand-alone landscape process model. In this case, the model is developed and run independent of GIS while input data are processed and exported from GIS and results are imported to GIS for analysis and visualization. Using a common data format (e.g., regular grid or a vector data structure designed consistently with a given GIS) makes this coupling more efficient. However, changing the input data for simulation of different scenarios is not very convenient, and this approach is therefore suitable for applications where the need for modifications of input data is small. The model is to a large extent independent from GIS; it can therefore be used with different systems and changes in GIS will have minimal impact on the functionality of the model. There is a large number of models with this type of GIS coupling; for example, SIMWE erosion and deposition model (Mitas and Mitasova, 1998), and SWMS2D for hydrologic modeling of watersheds with drainage (Badiger and Cooke, 2000).

### 8.3.4 Incorporation of GIS Functionality into Modeling Systems

Large, professional modeling systems, most often aimed at engineering applications, use both loose coupling with a GIS and their own specialized GIS capabilities. A general GIS is used for storing, managing, and processing of basic topographic data and for generating the cartographic output. The modeling system includes support for GIS functions where tight coupling with the model is necessary, such as the design of a conceptual model for the given site, adjustment of finite difference/finite element grids and meshes, as well as modifications of the model parameters (conditions of simulations) based on the simulation results. Because many of the professional systems are dynamic and 3D, they also have their own visualization capabilities specific to a given discipline. State-of-the-art systems with this type of GIS coupling have been developed for groundwater and surface water modeling (Danish Hydrologic Institute, 2000: MIKE11,21; Environmental Modeling Research Laboratory, 2000: GMS/SMS/WMS; Heinzer et al., 2000; Holland and Goran, 2000), urban hydrology (Danish Hydrologic Institute, 2000: MOUSE; Johnston and Srivastava, 1999: HydroPEDDS), or oil spill simulation (Cantin and Fortin, 2000). Several geoscientific models are coupled with systems that are not considered a GIS; however, they have spatial data processing and visualization capabilities and a high-level programming language that allows the users to write the models efficiently—for example, the landscape evolution model CHILD (Tucker et al., 1999) is coupled with Matlab.

### 8.3.5 GIS and the Web-based Models

With the explosive growth of the Internet, physical process models useful for a wider range of users, such as farmers, land owners, city planners, or public land managers, are being implemented as Web-based applications. The successful applications include not only the modeling tools but also the databases with input data and model parameters so that the user does not have to deal with the time-consuming tasks of finding, processing, and submitting the input data for the model runs. Usually only selection of the location and land use management scenario is needed from the given set of options. Spatial data are stored in a GIS on the server and the digital maps, representing the inputs and model results, are served using Internet map serving technology, such as ArcView IMS, MapObjects IMS (ESRI, 1999/2000), or GRASSlinks (Neteler, 2000). The Web-based models of physical systems were developed, for example, for long-term hydrological impact assessment, LTHIA (Engel, 2000), watershed-based hydrology and pollution modeling, BASIN-2 (Environmental Protection Agency, 2000), or terrain analysis–based hydrologic modeling, TOPOG (CSIRO, 2000).

## 8.4 CASE STUDIES

To illustrate the presented concepts, examples of models, modeling systems, and applications with different levels of complexity and GIS coupling are described. The case studies are focused on modeling of surface water flow and its impact on soil erosion and deposition. Case studies that include global climate, ocean, fresh water, and related land use and ecological modeling are presented in Chapter 10.

### 8.4.1 Modeling Erosion and Deposition Using Standard GIS Tools

Distributed models of water and sediment flow provide valuable tools for assessing the current situation and predicting erosion and deposition patterns for various land management conditions. In this case study, the erosion deposition pattern was estimated using GRASS GIS and its interpolation, flow tracing, topographic analysis, visualization, and map algebra tools. The erosion and deposition pattern was computed using the Unit Stream Power Based Erosion/Deposition model unit stream power-based erosion deposition (USPED) (Mitasova and Mitas, 1999; Mitasova and Mitas, in press). It is a simple model that predicts the spatial distribution of erosion and deposition rates for a steady-state overland flow, with uniform rainfall excess, for transport capacity limited case of erosion process. The model is based on the theory originally outlined by Moore and Burch (1986) with numerous improvements. For the transport capacity limted case, we assume that the sediment flow rate is at the sediment transport capacity $T(\mathbf{r})$, $\mathbf{r} = (x,y)$, which is approximated as a function of upslope area $A(\mathbf{r})$ and slope $b(\mathbf{r})$:

$$T(\mathbf{r}) = R(\mathbf{r})K(\mathbf{r})C(\mathbf{r})P(\mathbf{r})A(\mathbf{r})^m \, (sinb(\mathbf{r}))^n \tag{8.7}$$

where $R(\mathbf{r})$, $K(\mathbf{r})$, $C(\mathbf{r})$ are the spatially variable revised universal soil loss equation (RUSLE) rainfall, soil, and cover factors (Haan et al., 1994), used as weights to incorporate the impact of rainfall intensity, soil, and cover; while and $m$ and $n$ are constants, with $m = 1.6$, $n = 1.3$ used for prevailing rill erosion and $m = n = 1$ for prevailing sheet erosion. Then the net erosion/deposition $E(\mathbf{r})$ is estimated as

$$E(r) = \nabla \cdot [T(r) \cdot s(r)] = \partial(T(r)_* \cos a(r)) / \partial x + \partial(T(r)_* \sin a(r))/\partial y \qquad (8.8)$$

where $s(r)$ is the unit vector in the steepest slope direction and $a(r)$ (deg) is the aspect of the terrain surface. Caution should be used when interpreting the results because the RUSLE parameters were developed for simple plane fields and detachment limited erosion, and to obtain accurate quantitative predictions for complex terrain conditions they need to be re-calibrated.

The model can be implemented using a sequence of GIS commands and map algebra operations. Assume that the elevation, $K$, $C$ is given as raster data and $R = 120$ and resolution res $= 10$ are constants. Then the model can be computed in GRASS5.0 as follows (see a more detailed description of implementation in both ArcView Spatial Analyst and GRASS5.0 by Mitasova and Mitas, 1999):

1. **r.flow** *elevation dsout=flow (computes upslope area)*
2. **r.slope.aspect** *elevation slope=slope aspect=aspect (derive slope/aspect)* 3. **r.mapcalc** *sflow=exp(flow\*res,1.6)\*exp(sin(slope),1.3) (transport capacity term)* *qsx = 120 \* K \* C \* sflow \* cos(aspect) (sediment flow vector)* *qsy = 120 \* K \* C \* sflow \* sin(aspect)*
4. **r.slope.aspect** *qsx dx = qsx.dx (components of sediment flow divergence)*
5. **r.slope.aspect** *qsy dy = qsy.dy*
6. **r.mapcalc** *erdep = qsx.dx + qsy.dy (net erosion/deposition = div of sediment flow)*
7. **r.colors** *erdep rast=erdep.color (assign nonlinear color table)*

Slope, aspect, and partial derivatives *dx, dy* are computed using Horn's formula (Horn, 1981; Shapiro and Westervelt, 1992) implemented in **r.slope.aspect**, but they can be also computed using the neighborhood syntax of map algebra (Shapiro and Westervelt, 1992). The implementation will be similar in any raster GIS; however, the results of the model will greatly depend on the quality of the DEM (Mitas and Mitasova, 1999) and the slope and flow algorithms used in the specific GIS (Figure 8.1b, c). As a one-time application, this model is easy to run as a sequence of commands. For routine, repeated applications for different locations or different land use patterns, the model can be implemented as a script with suitable interface for selecting the input data and viewing the results.

The results of the application of this model to a military installation with combination of high-intensity use and well-preserved areas are shown in Figure 8.3. The model predicts severe erosion on the upper convex parts of hillslopes with bare soil and in areas with concentrated flow in both disturbed and forested areas. Significant portion of soil eroded from hill slopes has a potential for being deposited before entering streams; however, the potential for deposition of soil eroded by concentrated flow is much smaller. This type of analysis is being used for designing the training areas and implementation of conservation and mitigation measures to prevent negative impacts of erosion on soil and water quality.

## 8.4.2 Process-Based Modeling of Water Depth and Erosion/Deposition by a Model with Loose GIS Coupling

Modeling of spatial distribution of water depth provides valuable information for identifying locations that require drainage to prevent negative impact of standing water on yields. Using a high-accuracy DEM, interpolated from rapid kinematic survey data by the regularized spline with tension (RST) method (Mitas and Mitasova, 1999) within GRASS5.0, the water depth distribution was simulated for a typical rainfall for Midwestern agricultural fields (9 mm/h) under saturated conditions. The simplified water flow approximation by kinematic wave (e.g., by using the **r.watershed** or **r.flow** command in GRASS5.0; (Figure 8.1b, c) was not sufficient for flat terrain with depressions, and a two-dimensional approximate diffusive wave solution (Figure 8.1d) implemented in a GIS-independent model SIMWE (Mitas and Mitasova, 1998) had to be used. The gradual accumulation of water in depressions is illustrated by three snapshots from the simulation of water depth during a uniform, steady rainfall (Figure 8.4 a, b, c). Locations where water will stand several hours after the rainfall, taking into account a simplified infiltration, are shown in Figure 8.4d. The resulting water depth maps were used to evaluate location of current drainage and to plan location of a new drainage network in the negatively affected field. While the model was very useful for evaluating and designing the spatial pattern of the drainage network, detailed soil data and more complex dynamic simulations with coupled surface and subsurface flow (e.g., Badiger and Cooke, 2000) are necessary to optimize size, depth, structure, and other drainage network parameters.

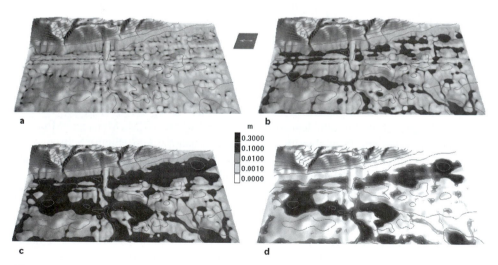

**Figure 8.4** Snapshots from a 2D dynamic model of water depth distribution in a relatively flat field (*0.8 by 1.5km*) during steady uniform rainfall: (a) 3 minutes, (b) 15 minutes, (c) 1 hour, (d) 8 hours after the rainfall. Water depth, represented by color, is draped over exaggerated DEM. Time is approximate, modeling was performed by SIMWE.

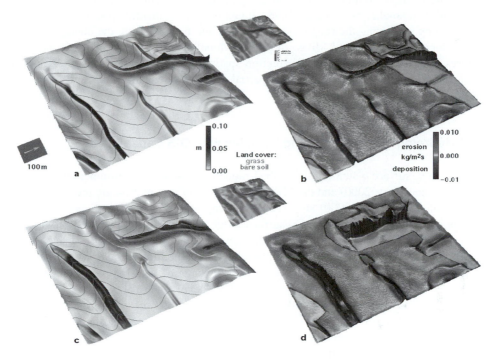

**Figure 8.5** Modeling water depth, sediment flow, and net erosion/deposition for different land use designs using SIMWE. Land use without conservation measures: (a) low water depth due to fast runoff; (b) potential for gullies in areas of concentrated water flow. Land use with extended grass cover including a grassway: (c) increased water depth in grass covered areas; (d) deposition in grassway with erosion along its edges. Sediment flow is represented as a surface with erosion/deposition draped over it as color.

Water depth, sediment flow, and net erosion and deposition pattern can be very complex in landscapes with spatially variable land cover. Borders between different land covers, such as bare tilled soil in a field and dense grass in a meadow, cause abrupt changes in flow velocities, creating effects important for erosion prevention and soil conservation. Process-based, two-dimensional flow models must be used to capture the complex interactions among the terrain shape, land cover, and soil properties. Therefore, the GIS-independent SIMWE model was used to evaluate the impact of various land use patterns on erosion and sediment transport (Mitas and Mitasova, 1998) in an experimental field. Figure 8.5 shows distribution of water depth and net erosion deposition, draped over sediment flow surface, for two different land use alternatives (Auerswald et al., 1996). The original land use design (Figure 8.5a inset) leads to gully erosion in valleys and severe rill erosion on the exposed part of the steeper hillslope (Figure 8.5b). New design (Figure 8.5c inset) reduces the risk of gullies by implementation of a grassway and transformation of the steepest tilled hilllsope to pasture. However, this new design, which increases the protective grass cover from 20% to 40% can cause problems if the roughness in the tilled fields is reduced (e.g., by compaction). While the grassway is very effective in holding water and reducing runoff rate (Figure 8.5c), rapid change in the flow velocity (from smooth field to dense grass) could cause erosion along the grassway, effectively replacing one big gully with two smaller ones (Figure 8.5d). It can be shown, that this phenomenon can be prevented by increasing the roughness in the field and thus reducing the change in water flow velocity (Mitasova and Mitas, in press). Modeling of this type of complex effect requires both the support for high-resolution spatial data provided by GIS and sophisticated numerical methods for solution of partial differential equations currently available only outside GIS.

### 8.4.3 Hydrologic Modeling Using Systems with GIS Capabilities

Groundwater, surface water, and watershed modeling systems (Environmental Modeling Research Laboratory, 2000) and the MIKE/MOUSE family of hydrologic programs (Danish Hydrologic Institute, 2000) are excellent examples of systems that incorporate their own set of GIS tools. For example, a surface water modeling system (SMS) provides an interface and tools for performing simulations with several hydrodynamic surface water models. Specialized GIS capabilities are provided within the Map module, which uses data model similar to ArcInfo, facilitating efficient exchange of data between the SMS and ArcInfo/ArcView. The Map module is designed for development of a high-level conceptual model of the studied site—a task that would be possible, but more time consuming, with general GIS tools. A 2D Mesh module is used for automated mesh generation and editing needed for numerical computations. While it is still necessary to use GIS for such tasks  as data projection, shifting some of the specific data preparation tasks from GIS to

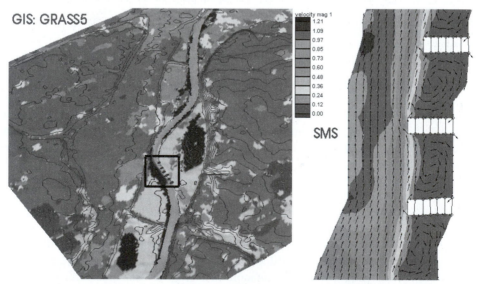

**Figure 8.6** Modeling surface water flow on a section of the Mississippi River. Surface water modeling system (SMS) was used to predict flow by the finite element two-dimensional hydrodynamic flow simulation program FESWMS http://www.ems-i.com/sms/. 3D visualization of terrain model and the river was done in GRASS5 using NVIZ tool. Figures courtesy Dr. Mingshi Chen, William M. Brown, RiverWeb Museum Consortium, http://www.ncsa.uiuc.edu/Cyberia/RiverWeb/Projects/RWMuseum/.

SMS makes modification of model configuration much faster and simpler, so the model can be run for different scenarios efficiently. On the other hand, modifications of the mesh independent from the GIS can cause inconsistencies between the representation of the studied object (river, lake) in SMS and GIS. Application of the SMS is illustrated by a model of surface water flow velocity on a section of Mississipi River with weirs, prepared for the RiverWeb Museum Consortium (Figure 8.6).

## 8.5 CONCLUSIONS AND LESSONS LEARNED

The current research and development in physical systems modeling is focused on distributed, process-based models, often dynamic in 3D space. This trend has been stimulated by the availability of geospatial data and supporting GIS tools. GIS has greatly reduced the time for preparation of inputs; however, this task can still be rather tedious and time consuming. Also, the support for temporal and multidimensional data is not adequate. Therefore, coupling of GIS and models is done at various levels and incorporation of GIS functionality within the modeling systems is now quite common.

The current research in the area of coupled GIS and physical systems modeling focuses on object-oriented model development (e.g., Band, 2000; Helleweger and Maidment, 1999; Naumov, 2000), real-time simulations (e.g., Cantin and Fortin, 2000), multiscale simulations and modeling with heterogeneous data (e.g., Mitasova and Mitas, in press; Green et al., 2000), as well as distributed online modeling (Engel, 2000). At the same time, GIS as a single general system is disappearing and GIS is melting into the general computing infrastructure (GIS Industry Outlook, 1999). We can see increasing assimilation of GIS into end-user applications, including the models of physical systems, rather than integrating applications into the GIS.

Just a few years ago, there were numerous efforts to built comprehensive modeling and problem-solving environments that would provide essentially everything for doing both the routine processing and advanced modeling as well as development of new methods and technologies. The practice, however, seems to be going in other directions as well. The large, universal, thought-through, all-powerful systems that were expected to support almost every possible research or development need ("cathedral," Raymond, 1999) are, in fact, not practical. The maintenance of large software packages is expensive, rigid, and inefficient. A concept of cooperation between a number of smaller software units and tools, environments, and program packages seems to be more successful. This concept enables creation of more independent smaller pieces of software with simplified interdependencies. It enables a number of groups or individuals to contribute and work on various parts simultaneously. If some branch of development proves to be uninteresting or unproductive, it rapidly dies out without going through decision hierarchies usually present in the "cathedral" paradigm. In contrast to the "cathedral," such a framework creates a "bazaar" that offers variety of combinations and provides, in effect, a market of tools that can be combined or used for data processing, modeling, method development, and their combination. An excellent example of this approach is the open source movement, which has created entire infrastructure supporting the bazaar-type development structures. The most widely used product developed within this framework is probably Linux, and among the 10 largest systems is also GRASS5 (Code Catalog, 2000; Neteler, 2000). A similar trend in the commercial GIS sector is reflected by a wave of high-level spatial component products designed for software developers. In many cases the bazaar-type approach is useful for new advances in scientific exploration as the most exciting and influential research breakthroughs happen through stepping outside the established routes.

## 8.6 ACKNOWLEDGMENTS

We would like to acknowledge the long-term support for environmental modeling research from Geographic Modeling Systems Laboratory director Douglas M. Johnston as well as GIS assistance by William M. Brown. Funding was provided by the U.S. Army Construction Engineering Research Laboratories, Strategic Environmental Research and Development Program, and the Illinois Council on Food and Agricultural Research (CFAR). We greatly appreciate the sharing of data by S. Warren, K. Drackett, and K. Auerswald. .

## 8.7 REFERENCES

Auerswald, K., Eicher, A., Filser, J., Kammerer, A., Kainz, M., Rackwitz, B., Schulein, J., Wommer, H., Weigland, S., and Weinfurtner, K. 1996. Development and Implementation of Soil Conservation Strategies for Sustainable Land Use—the Scheyern Project of the FAM. In H. Stanjek (ed.), *Development and Implementation of Soil Conservation Strategies for Sustainable Land Use,* Int. Congress of ESSC, Tour Guide, II, pp. 25–68, Freising-Weihenstephan: Technische Universitat Munchen.

Badiger, S. M., and Cooke, R. A. C. 2000. Application of Integrated GIS and Numerical Models in Subsurface Drainage Studies. *Proceedings of 4th conference on Environmental Modeling and GIS* [CD-ROM]. Banff, Canada.

Band, L. E. 2000. Urban Watersheds as Spatial Object Hierarchies. *Proceedings of 4th Conference on Environmental Modeling and GIS* [CD-ROM]. Banff, Canada.

Burnett, D. S. 1987. *Finite Element Analysis: From Concepts to Applications,* Reading, MA: Addison-Wesley.

Burrough, P. A. 1998. Dynamic Modelling and GIS. In P. Longley et al. (eds.), *Geocomputation: A Primer*, pp. 165–192. New York: Wiley.

Burrough, P. A., da Costa, J. R., Haurie, A., Fedra, K., Salvemini, M., and Hauska, H. 2000. MUTATE: A Web-Based Distance Learning Programme for Environmental Modeling with GIS. *Proceedings of 4th Conference on Environmental Modeling and GIS* [CD-ROM]. Banff, Canada.

Cantin, J. F., and Fortin, P. 2000. Integration of Numerical Models and Field Characterization into a Georeferenced System for Oil Spill Emergency Response in the St. Lawrence River, *Proceedings of 4th Conference on Environmental Modeling and GIS* [CD-ROM]. Banff, Canada.

Ceperley, D. M. 1999. Microscopic Simulations in Physics. *Reviews of Modern Physics*, 71, 438.

Ceperley, D.M., and Mitas, L. 1996. Monte Carlo Methods in Quantum Chemistry. In I. Prigogine and S.A. Rice (eds.), *Advances in Chemical Physics*, Vol. XCIII, pp. 1–38, New York: Wiley.

Code Catalog, 2000. Open Source Code Search Engine. http://www.codecatalog.com/topten.

CSIRO. 2000. TOPOG Online. http://www.csiro.au/topog

Danish Hydrologic Institute. 2000. MIKE11, 21. http://www.dhi.dk/

Doe, W. W., Julien, P. Y., and Saghafian, B. 1996. Land Use Impact on Watershed Response: The Integration of Two-dimensional Hydrological Modeling and Geographical Information Systems. *Hydrological Processes*, 10, 1503–1511.

Environmental Modeling Research Laboratory. 2000. GMS, SMS, WMS. http://www.emrl.byu.edu/.

Environmental Protection Agency. 2000: BASINS-2, http://www.epa.gov/OST/BASINS/.

ESRI. 1996. *Avenue: Customization and Application Development for ArcView.* Redlands.

ESRI. 1999/2000. Four Software Products Feature Web-Mapping Functionality. *Arc News,* 21, 7.

Engel. 2000. Development of a GIS- and WWW-Based Runoff and Nonpoint Source Pollution Modeling System. *Proceedings of 4th Conference on Environmental Modeling and GIS* [CD-ROM]. Banff, Canada.

Favis-Mortlock, D., Boardman, J., Parsons, T., and Lascelles, B. 1998. Emergence and Erosion: AModel for Rill Initiation and Development. *Proceedings of the 3rd Conference on GeoComputation* [CD-ROM]. University of Bristol, England.

Gardiner, C. W. 1985. *Handbook of Stochastic Methods for Physics, Chemistry, and the Natural Sciences*, Berlin: Springer-Verlag.

GIS Industry Outlook 2000. 1999, December. *GeoWorld*, 12, pp. 36, 42.

Goodchild, M. F., Parks, B. O. and Steyaert, L. T. (eds.). 1993. *Geographic Information Systems and Environmental Modeling,* New York: Oxford University Press.

Goodchild, M. F., Parks, B. O. and Steyaert, L. T. (eds.). 1996. *GIS and Environmental Modeling: Progress and Research Issues,* Fort Colins, CO: GIS World, Inc.

Goodchild, M. F., Parks, B. O. and Steyaert, L. T. 1997. GIS and Environmental Modeling. *Proceedings of the Third Conference on GIS and Environmental Modeling,* (Santa Fe) [CD-ROM]. NCGIA.

Green, T. R., Ascough, J. C. and Erskine, R. H. 2000. AgSimGIS for Integrated GIS and Agricultural System Modeling: II. Application to Soil-Water Dynamics in an Undulating Landscape in Colorado. *Proceedings of 4th Conference on GIS and Environmental Modeling* . [CD-ROM]. Banff, Canada.

Haan, C. T.,  Barfield, B. J., and Hayes, J. C. 1994. *Design Hydrology and Sedimentology for Small Catchments.* Academic Press.

Heinzer, T. J., Sebhat, M., and Feinberg, B. 2000. Case Studies Using Geographic Information Systems to Facilitate Data Integration with MIKE21 when Modeling Flood Inundation Scenarios. *Proceedings of 4th Conference on Environmental Modeling and GIS* [CD-ROM]. Banff, Canada.

Helleweger, F. L. and Maidment, D. R. 1999. Definition and Connection of Hydrologic Elements Using Geographic Data. *Journal of Hydrologic Engineering*, 4, 10–18.

Holland, J., and Goran, W. 2000. Development of a Land Management System in Support of Natural Resources Management. *Proceedings of 4th Conference on Environmental Modeling and GIS* [CD-ROM]. Banff, Canada.

Horn, B. K. P. 1981. Hill Shading and the Reflectance Map. *Proc. IEEE*, 69, 14–46.

Jensen, H. J. 1998. *Self-Organized Criticality: Emergent Complex Behavior in Physical and Biological Systems.* Cambridge Lecture Notes in Physics, 10, Cambridge, England: Cambridge University Press.

Johnston, D. M., and Srivastava, A. 1999. Decision Support Systems for Design and Planning: The Development of HydroPEDDS (Hydrologic Performance Evaluation and Design Decision Support ) System for Urban Watershed Planning. *6th International Conference on Computers in Urban Planning and Urban Management (CUPUMS'99),* Venice, Italy

Khairy, W., Hannoura, A. P., and Cothren, G. M. 2000. Application of Information Systems in Modelling of Non–Point Source Pollution, *Proceedings of 4th Conference on Environmental Modeling and GIS* [CD-ROM]. Banff, Canada.

Krysanova V., and Wechsung F. 2000. Analysis of Global Change Impacts at the Regional Scale by Means of Integrated Ecohydrological Modelling. *Proceedings of 4th Conference on Environmental Modeling and GIS* [CD-ROM]. Banff, Canada.

Mitas, L., Brown, W. M., and Mitasova, H. 1997. Role of Dynamic Cartography in Simulations of Landscape Processes Based on Multi-Variate Fields. *Computers and Geosciences,* 23, 437–446. See http://www2.gis.uiuc.edu:2280/modviz/lcgfin/cg-mitas.html.

Mitas, L., and Mitasova, H. 1999. Spatial Interpolation. In P. Longley, M. F. Goodchild, D. J. Maguire, and D. W. Rhind (eds.), *Geographical Information Systems: Principles, Techniques, Management and Applications,* 481–492. New York: Wiley.

Mitas, L., and Mitasova, H. 1998. Distributed Erosion Modeling for Effective Erosion Prevention. *Water Resources Research,* 34, 505–516.

Mitasova, H., and Mitas, L. (In press) Multiscale Erosion Modeling for Land Use Management. In: R. Harmon, W. Doe (eds.), *Landscape erosion and evolution modeling.* Kluwer.

Mitasova, H., and Mitas, L. 1999. Erosion/Deposition Modeling with USPED. World Wide Web tutorial. See http://www2.gis.uiuc.edu:2280/modviz/erosion/usped.html.

Mitasova, H., Mitas, L., Brown, W. M., Gerdes, D. P., and Kosinovsky, I. 1995. Modeling Spatially and Temporally Distributed Phenomena: New Methods and Tools for GRASS GIS. *International Journal of GIS,* 9, 443–446.

Moglestue, C. 1993. *Monte Carlo Simulation of Semiconducting Devices.* London: Chapman and Hall.

Moore, I. D., and Foster, G. R. 1990. Hydraulics and Overland Flow. In M. G. Anderson and T. P. Burt (eds.), *Process Studies in Hillslope Hydrology,* pp. 215–254. New York: Wiley.

Moore, I. D., Turner, A. K., Wilson, J. P., Jensen, S. K., amd Band, L. E. 1993. GIS and Land Surface-Subsurface Process Modeling. In M. F. Goodchild, L. T. Steyaert, and B. O. Parks (eds.), *Geographic Information Systems and Environmental Modeling,* pp.196–230. New York: Oxford University Press.

Murray, S., Miller, W., and Breslin, P. 2000. Visual Framework for Spatial Modeling. *Proceedings from the 4th Conference on GIS and Environmental Modeling,* [CD-ROM]. Banff, Canada.

Naumov, A. 2000. Design of High-Level GIS Data Types for Hydroecological Modeling in GRASS. *Proceedings from the 4th Conference on GIS and Environmental Modeling,* [CD-ROM]. Banff, Canada.

Neteler, M. 2000. Advances in Open Source GIS Software. *Proceedings from the 1st National Geoinformatics Conference of Thailand,* Bangkok: Thailand.

Nuclear Energy Agency. 1995. *Advanced Monte Carlo Computer Programs for Radiation Transport.* Paris: NEA.

Park, S., and Wagner, D. F. 1997. Incorporating Cellular Automata Simulators as Analytical Engines in GIS. *Transactions in GIS,* 2, no. 3, 213–232.

Parks, B., Crane, M., and Clarke, K. (eds.). 2000. *Proceedings from the 4th Conference on GIS and Environmental Modeling,* [CD-ROM]. Banff, Canada.

Raymond, E. S. 1999. *The Cathedral and the Bazaar.* Sebastopol, CA: O'Reilley.

Rewerts, C. C., and Engel, B. A. 1991. ANSWERS in GRASS: Integrating A Watershed Simulation with a GIS. ASAE Paper No.91-2621, 1-8. St. Joseph, MO: American Society of Agricultural Engineers.

Saghafian, B. 1996. Implementation of a Distributed Hydrologic Model within GRASS, In M. F Goodchild, L. T. Steyaert, and B. O. Parks (eds.), *GIS and Environmental Modeling: Progress and Research Issues,* pp. 205–208. GIS World, Inc.

Shapiro, M. and Westervelt, J. 1992. *R.MAPCALC, an Algebra for GIS and Image Processing*, pp. 422–425. Champaign, IL: U.S. Army Corps of Engineers, Construction Engineering Research Laboratories.

Sluiter, R., Karssenberg, D., Burrough, P. A., Wesseling, C., de Jong, K., Van der Meer, M., van Steijn, H., and Jetten, V. 2000. GMOR: Interactive Computer Models for Teaching Dynamic Geomorphological Processes, *Proceedings of 4th Conference on Environmental Modeling and GIS* [CD-ROM]. Banff, Canada.

Srinivasan, R., and Arnold J. G. 1994. Integration of a Basin Scale Water Quality Model with GIS. *Water Resources Bulletin,* 30, 453–462.

Srinivasan, R., and Engel, B. A. 1991. A Knowledge Based Approach to Extract Input Data from GIS. ASAE Paper No. 91-7045, 1–8. St. Joseph, MO: American Society of Agricultural Engineers.

Stakgold, I. 1997. *Green's Functions and Boundary Value Problems.* New York: Wiley.

Steyaert, L. T. 1993. A Perspective on the State of Environmental Simulation Modeling. *Geographic Information Systems and Environmental Modeling,* pp. 16–30. New York: Oxford University Press.

Tucker, G., Gasparini, N., Bras, R., and Rybarczyk, P. 1999. An Object-Oriented Framework for Distributed Hydrologic and Geomorphic Modeling using Triangulated Irregular Networks. *4th International Conference on GeoComputation.* Fredericksburg, MD: Mary Washington College.

Vieux, B. E., Farajalla, N. S., and Gaur, N. 1996. Integrated GIS and Distributed Storm Water Runoff Modeling. In M. F. Goodchild, L. T. Steyaert, and B. O. Parks (eds.), *GIS and Environmental Modeling: Progress and Research Issues*, pp. 199–205. GIS World, Inc.

Wesseling, C. G., Karssenberg, D., Burrough, P. A., and van Deursen, W. 1996. Integrating Dynamic Environmental Models in GIS: the Development of a Dynamic Modeling Language. *Transactions in GIS*, 1, 40–48.

## 8.8 INFORMATION RESOURCES

Danish Hydrologic Institute MIKE system http://www.dhi.dk/ (July 8, 2000)

Environmental Protection Agency (1999)  BASIN-2 watershed and water quality analysis models http://www.epa.gov/OST/BASINS/ (July 8, 2000)

Geographic Information Science and Technology Group, Oak Ridge National Laboratory (Department of Energy) http://www.cad.ornl.gov/cad_ica/text/ica_gcm.html (July 8, 2000)

Geographic Modeling Systems Laboratory http://www.gis.uiuc.edu/ (July 8, 2000)

GRASS5.0(2000) http://www.geog.uni-hannover.de/grass/ (July 8, 2000)

IDRISI–raster-based GIS and its applications http://www.clarklabs.org/ (July 15, 2000)

IDOR2D,3D: Hydronamic and pollutant transport simulations http://water.eng.mcmaster.ca/pages/hydro.htm (July 15, 2000)

Environmental Modeling Research Laboratory (2000) GMS, SMS, WMS linked to Arc-View http://www.ems-i.com/ (July 8, 2000)

Hydrologic models linked to GIS http://soils.ecn.purdue.edu/~aggrass/models/hydrology.html (July 15, 2000)

LISEM Erosion model http://www.geog.uu.nl/lisem/  (July 15, 2000)

LTHIA - on-line, long term hydrologic impact assessment http://danpatch.ecn.purdue.edu/~sprawl/LTHIA5/ (June 5, 2000)

ESRI-GRID, Avenue, Spatial Analyst http://www.esri.com/ (July 15, 2000)

Soil Erosion/Deposition Modeling with GIS on WWW http://www.ex.ac.uk/~yszhang/erosion.htm (July 15, 2000)

Spatial Modeling and Visualization http://www2.gis.uiuc.edu:2280/modviz/ (July 15, 2000)

SWAT: Soil and Water Assessment Tool http://www.brc.tamus.edu/swat/ (July 15, 2000)

## 8.9 ABOUT THE CHAPTER AUTHORS

Dr. Helena Mitasova is a Research Associate at the Geographic Modeling Systems Laboratory, Department of Geography,  University of Illinois at Urbana–Champaign. Her research has been focused on modeling of landscape processes, spatial interpolation, and multidimensional visualization for GIS. She has been partcipating in the development of Open Source GRASS GIS since 1991.

Dr. Lubos Mitas is a Research Scientist at the National Center for Supercomputing Applications, University of Illinois at Urbana–Champaign. His research involves computational materials research, quantum Monte Carlo methods and development of methods for spatial interpolation and simulation of landscape processes. He recently joined the faculty in the Department of Physics at North Carolina State University, Raleigh, NC.

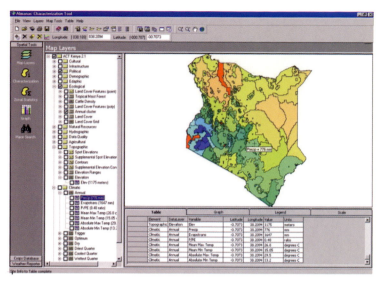

**Plate 1: Figure 3.2**

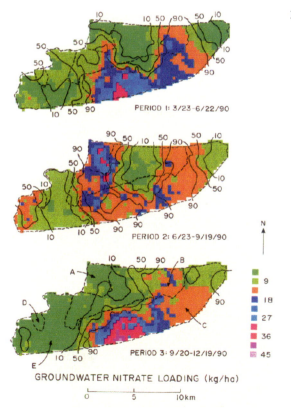

**Plate 2: Figure 6.5.**

PERIOD 1: 3/23-6/22/90

PERIOD 2: 6/23-9/19/90

PERIOD 3: 9/20-12/19/90

GROUNDWATER NITRATE LOADING (kg/ha)

9
18
27
36
45

N

0      5      10km

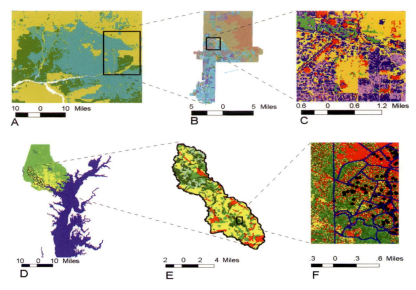

**Plate 3: Figure 7.3**

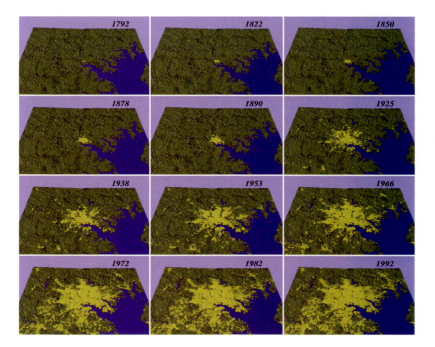

**Plate 4: Figure 7.6**

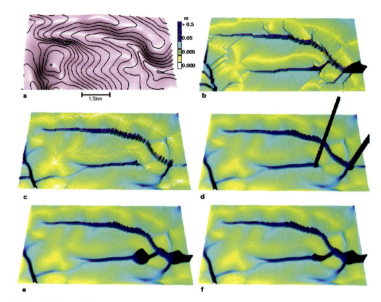

**Plate 5: Figure 8.1**

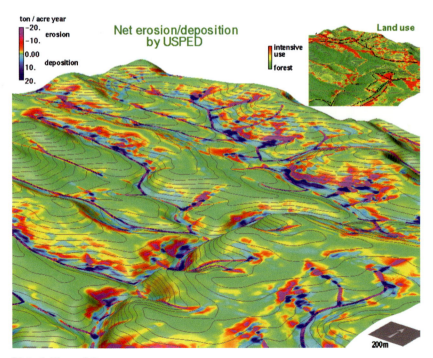

**Plate 6: Figure 8.3**

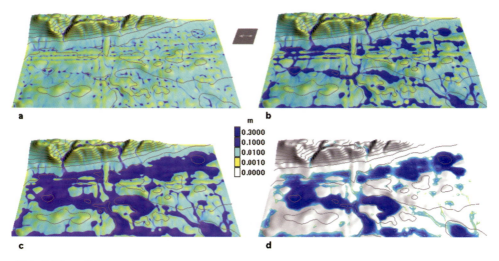

**Plate 7: Figure 8.4**

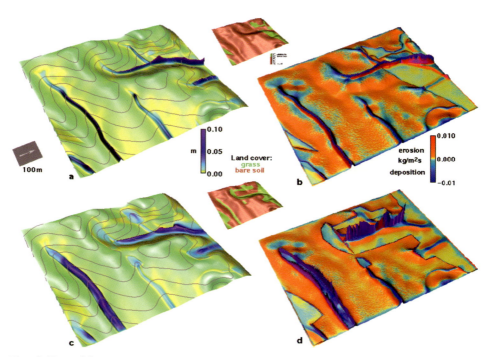

**Plate 8: Figure 8.5**

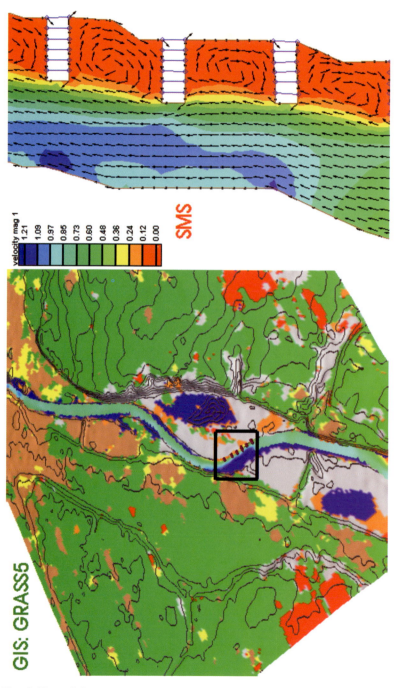

**Plate 9: Figure 8.6**

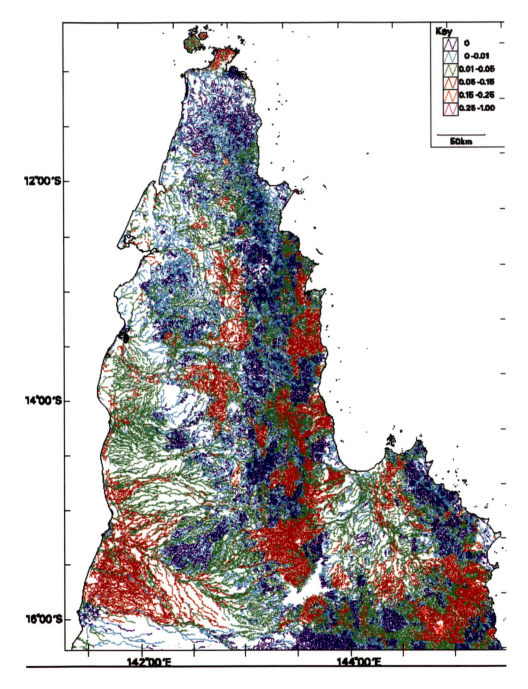

**Plate 10: Figure 10.4**

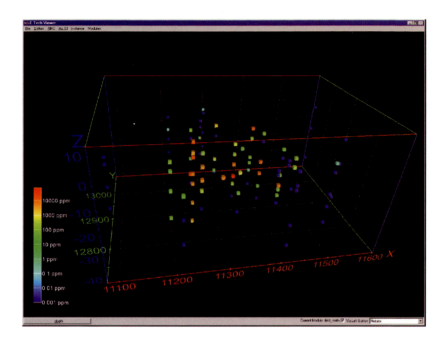

**Plate 11: Figure 11.5**

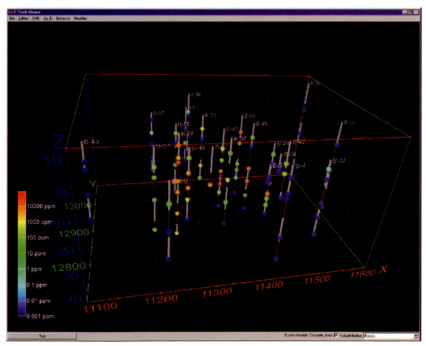

**Plate 12: Figure 11.6**

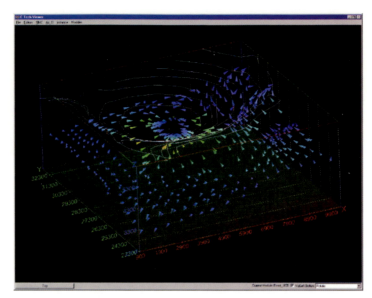

**Plate 13: Figure 11.7**

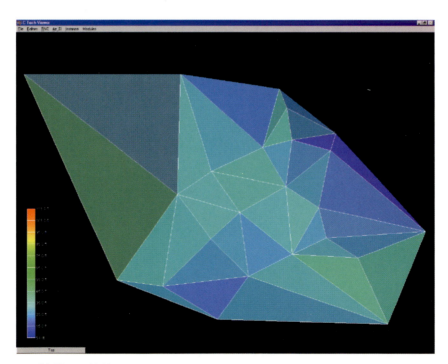

**Plate 14: Figure 11.18**

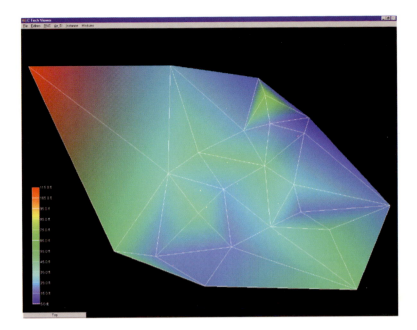

**Plate 15: Figure 11.19**

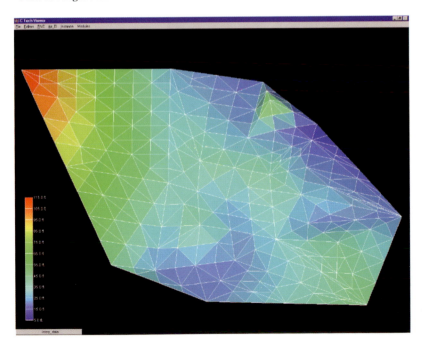

**Plate 16: Figure 11.20**

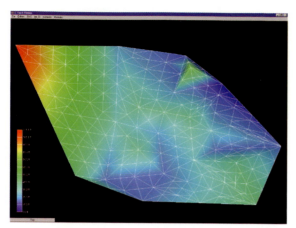

**Plate 17: Figure 11.21**

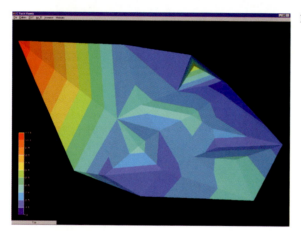

**Plate 18: Figure 11.22**

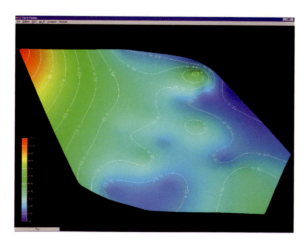

**Plate 19: Figure 11.23**

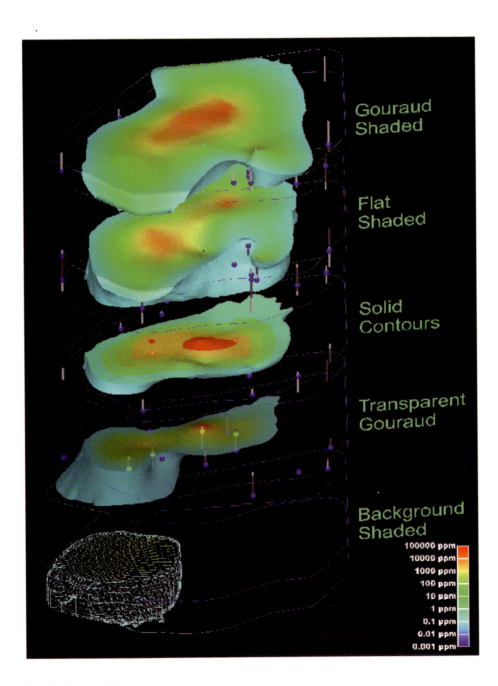

**Plate 20: Figure 11.25**

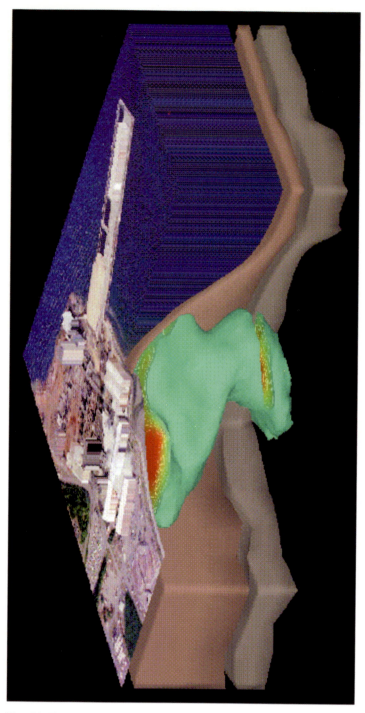

**Plate 21: Figure 11.26**

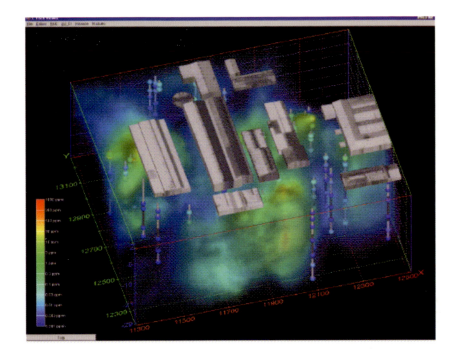

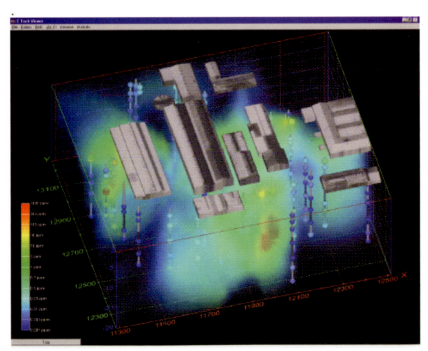

**Plate 22: Figure 11.27**

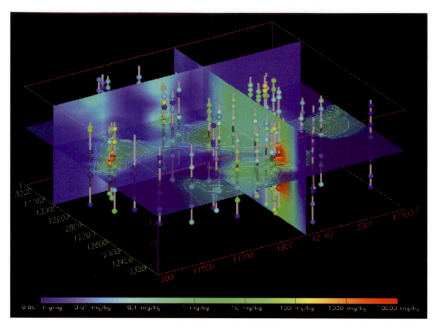

**Plate 23: Figure 11.29**

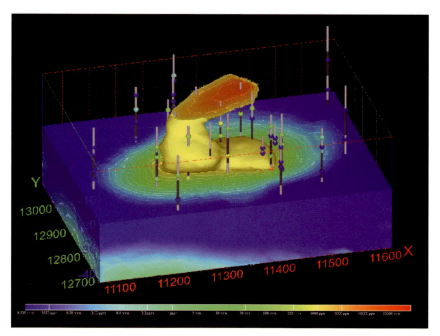

**Plate 24: Figure 11.30**

# 9

# *Integrative Environmental Modeling*

## 9.0 SUMMARY

Environmental models are a critical element of environmental decision support systems. Yet to be of practical value to the decision maker, such models, which are typically quite constrained in their scope, need to be integrated into modeling systems in order to address the space of the decision problem. In carrying out environmental model integration, one must attend to such issues as the implicit and explicit interfaces between the models, including modeling the physical interface and reconciling assumptions and boundary conditions. Uncertainty must also be managed in the integrated modeling system, typically through use of simulation techniques. Finally, such practical concerns as software integration of the modeling codes and development of the integrated user interface are crucial to the success of the environmental decision support system.

## 9.1 INTRODUCTION

This chapter addresses the important topic of environmental model integration. The need for integration as an element of environmental decision support is established, along with the dominant associated issues. Uncertainty management in the integrated system is addressed by advocating the use of simulation methods. Practical issues in integration of environmental models are discussed, with an eye toward helping readers plan and carry out their own integration efforts.

### 9.1.1 What Is Meant by Model Integration?

We use models to help us understand the real world by trying to explain the past and predict the future. But whether these models are of a rigorous, mathematical nature or are more conceptual, mental models, they are limited in their scope. Each model attempts to represent accurately a limited subset of the physical reality we are modeling. This is not only because it is hard to develop more comprehensive models (which is true), but also, and perhaps more important, because comprehensive models would be harder for us to grasp. Since we're using models to help us understand a reality too comprehensive for us to apprehend directly in the first place, it is natural for us to use the simplest models possible to accomplish the task at hand.

There are at least two ways to simplify a model. The first way is to constrain the problem space addressed by the model, limiting the scope of the system the model attempts to recreate. For environmental models, this often corresponds to physical space partitioning. For example, a groundwater flow model might be designed to address only the saturated zone of the soil. The second way to simplify a model is to apply simplifying assumptions within the selected problem space. To continue the example, within the groundwater model we might choose to assume that there are no chemical interactions occurring during movement of water through the saturated zone. This simplifies the model both conceptually and mathematically.

These simplification techniques are useful insofar as they allow us to develop a better understanding of the constrained systems they emulate. But in the context of environmental management, such models alone are not useful to the decision maker because the decision problem space is rarely as constrained as these models. While one can attempt to remedy this by developing more comprehensive and complex models, a more pragmatic approach is to integrate existing models to form a system that can be used to support decision making.

### 9.1.2 Integration in Environmental Decision Support Systems

Environmental decision support systems (EDSS) are computer systems that help humans make environmental management decisions. They facilitate natural intelligence by making information available to humans in a form that maximizes the effectiveness of their cognitive decision processes, and they can take a number of forms (see, for example, Guariso and Werthner, 1989). As the term is used here, though, EDSSs are focused on specific problems and decision makers, and toward this end they incorporate a variety of technologies, including geographic information systems (GIS), mathematical models, simulation, optimization, and expert systems (Frysinger, 1995).

In order to use modeling to address a particular decision problem, an EDSS needs to include modeling support for the entire problem space, not the constrained spaces used to develop simpler models. This requires integration of models into a cooperative system. For example, if decision makers need to predict the fate of a contaminant plume from a hazardous waste site, they will need to consider the movement of the material through (at least) the vadose zone and the saturated zone of the soil. Since these two zones are typically modeled separately, the EDSS will need to integrate these two models in such a way that the interaction is transparent, giving the effect of a single, comprehensive model. However, the second form of simplification, assumption making, cannot be forgotten; if the models are not consistent in their underlying assumptions, the integrated modeling system will not be valid.

To explore environmental model integration, we first examine the nature of interactions between the types of models to be involved, paying special attention to their interfaces and their respective assumptions. One particular area of interaction that requires special attention involves the role of uncertainty in systems of models, and this is addressed separately. We then look toward the implementation of integrated systems of models, followed by a treatment of the important topic of the user interface to the resulting systems.

## 9.2 INTERACTION BETWEEN MODELS

Any meaningful integration of models must focus on the interaction between the models involved. Interaction between models can take several forms, including data interfaces, such as input and output data; user interfaces, such as displays produced by the modeling codes; boundary conditions and constraints; and interactions between conceptual models, such as conflicting assumptions. The first two of these—data interfaces and user interfaces—are addressed separately in Sections 9.2 and 9.3, respectively. The remainder are addressed here.

### 9.2.1 Interaction in the Real World

When considering interactions between models to be coupled, it is useful to look at the interactions between the systems those models emulate in the real world. One should attempt to understand the *real* interface in order to gain insight into the model interface. In some cases the physical reality will present a relatively clean connection. For example, an outlet pipe from a sewage treatment plant (STP) into a river suggests a fairly clean interface between an STP model and a stream dissolved oxygen model. On the other hand, returning to the previous groundwater example, the interface between the vadose and saturated zones is less straightforward, and can be downright complex if high-resolution modeling at either side of the interface is involved. In fact, *both* of these interfaces could be complex at the right resolution (imagine, if you will, an attempt to capture the mixing dynamics associated with the pipe outfall).

To some extent, EDSS architects seeking to simplify the model integration task can stack the deck in their favor by selecting models whose interfaces (inputs and outputs) fall on relatively well-behaved portions of the real system. Many modelers strive for this goal in any case because it can help them to calibrate and validate their models. Thus, the integration process ideally starts with an examination of the *entire* system to be modeled, identifying the natural interface segmentation points and then seeking models that appropriately address these segments.

### 9.2.2 Modeling at the Interface

Building an interface between two models is, in effect, modeling the interface between the two systems represented by the two models. The foregoing section leads us to conclude that the complexity of the modeled interface will depend on both the complexity of the real interface and the resolution of the models involved. For this reason, particular attention has to be paid to the information flow across this interface. Since most, or perhaps all, models are structured in the typical inputs/processing/outputs fashion, representation of the interfaces between models will occur through the judicious matching of one model's outputs with a successor model's inputs.

This "plumbing" need not be a passive pipe. In many instances some data manipulation must be performed in order to reconcile format issues, often using scripting or shell languages for the purpose (more about this later). Therefore, one has the opportunity to effect more substantial change in the interface model at this point. For example, when integrating the STP model with a stream dissolved oxygen (DO) model, one may choose to reduce the level of temporal variability in the STP outflow in order to be consistent with the time scale assumed by the DO model. This can be done through connective scripting between the two models.

It is no accident that this modification is generally of the sort that *reduces* information—in no case can information content at the interface increase without a third source (which must then have its own interfaces to the other models in the system). The typical approach to increasing data resolution is some form of interpolation, which is itself a modeling process (though it is not often described as such). This suggests a simple rule of thumb: The temporal or spatial resolution of an integrated modeling system will not be greater than that of the coarsest model in the system.

### 9.2.3 Conflicts in Model Assumptions

The preceding section addressed the direct interface between two coupled models, represented by the data flowing from one to the other, perhaps with feedback. But special attention must be paid to an indirect, implicit interface. The assumptions underlying any model are well understood to be key to an understanding of the model and its potential applicability to a given situation. Likewise, the collective set of assumptions is crucial to the integrated system of models, and harmony among the assumptions of the integrated models is therefore essential. Some violations of this might be obvious; for example, if we coupled a steady-state saturated zone model with a transient vadose zone model we have clearly compromised the validity of the resulting integrated system.

But other cases might not be so obvious. Modeling assumptions are often embodied in the boundary conditions, and these assumptions are not always easy to detect. Let's suppose we want to integrate a stream flow model with an agricultural runoff model in order to evaluate the effects of agricultural best management practices on surface water quality. Some stream flow models might embed in their boundary conditions a simplifying assumption that there is no flow across the streambed interface. If we couple such a model with an agricultural runoff model that considers root and vadose zone infiltration and transport, we will inadvertently eliminate this source of contaminants in the stream because only over-bank flow was permitted in the stream model.

It is therefore essential that each model to be integrated into a system be thoroughly analyzed with respect to its assumptions. These assumptions must then be compared with those of the other models to ascertain their compatibility. Of course, the combined set of assumptions must be consistent with the goals of the integrated modeling system and the decisions one wishes to make.

# 9.3 UNCERTAINTY ANALYSIS AND MODEL INTEGRATION

An additional indirect interface between models deserves separate and detailed attention. Uncertainty is implicit in environmental decision making. Complex technical decisions must be made regarding events—both in the past and the present—that depend on many different variables. Solutions to such problems often depend on the use of a variety of mathematical modeling techniques. These techniques, in the main, attempt to predict the future performance of a complex system on the basis of relatively sparse empirical data. The predictions drawn from these modeling studies form the basis for the management action to follow, including such expensive decisions as the design of a treatment system, or even of a national policy. Ultimately, the environmental effectiveness of the outcome, in terms of protection of human health and reduction of environmental risk, depends on these results. But the accuracy (or indeed validity) of predictions based on these models is invariably influenced by uncertainty in the models.

### 9.3.1 Uncertainty in Environmental Models

Environmental modeling studies are unavoidably visited by uncertainty of various types, ranging from conceptual model uncertainty to parameter uncertainty. Uncertainty in such environmental management problems exists in part because of a lack of empirical data, errors in the data, incorrect models, and the general nondeterminism of nature.

The first of these, a lack of empirical data, is easy to understand. We routinely live with imperfect knowledge of the current state of systems, due to lack of data (in a usable form). This and the second (errors in the data) are the issues typically addressed in scientific and engineering studies when the goal is to reduce uncertainty. The usual approach is to collect more data and to attempt to reduce the measurement error in the data collected.

The third reason, the use of incorrect models has recently received more attention in environmental management. As decision makers come to accept that model building (whether mental or mathematical) is an essential part of problem solving, the disagreements as to which models are correct become more apparent. Some would argue that a model is correct to the extent that it accurately predicts the future behavior of the system; the limiting factor for environmental problems is the complexity of the system in question. Here is where an interesting human factor emerges.

As mathematical models are expanded to account for more of the details of the natural system under study, the mental models of the analyst become inadequate. While humans are capable of recognizing and apprehending in a gestalt sense the breadth of complex systems, they are ill equipped mentally to manage the myriad simultaneous details attending such systems. It can be argued that we build mathematical models precisely because we cannot manage such details mentally. Yet, as we build these models, they too become more complex than we can fully grasp, resulting in a great deal of effort and controversy associated with the development of the mathematical models. Many environmental modelers spend more time studying their models than studying the natural systems they emulate. In integrated modeling systems this model complexity issue is quite real and must be addressed both in the user interface (more about this later) and in the metadata provided with the modeling system to those who would consider its applicability to a particular decision problem.

This problem becomes especially acute when the decision maker is not the developer of the mathematical model, because an opportunity exists for mismatch between the analyst's mental model and the quantitative mathematical model the analyst is attempting to use. This results in uncertainty, both subjective (i.e., lack of confidence on the part of the analyst) and objective (i.e., a measurable variability in the decisions made by several analysts or by one analyst on several occasions). Ultimately, this uncertainty finds its way into public perception, causing the public at large to wonder how to interpret the products of science and engineering (the public's awareness of the modeling debate surrounding global warming is a good example of this).

Finally, the fourth cause of uncertainty in environmental problems arises from the nondeterministic character of the natural environment, at least as it is currently understood. We should not expect to eliminate uncertainty entirely in solving environmental problems. Like the other three, this cause of uncertainty applies to both spatial and aspatial data, and some adaptive approaches have been proposed to help analysts arrive at accurate descriptions of the uncertain natural parameters (e.g., Heger et al., 1992). Unfortunately, humans have some difficulty in reliably making probabilistic judgements (Hogarth, 1987). There is a tendency toward a "fish-eye" view of uncertainty—perception of unfamiliar issues or events is related to familiar ones—resulting in distortion not unlike the familiar cartoon maps of "The New Yorker's view of the World." This is evident in studies examining human perception of risk, and it applies to probabilistic judgments more generally.

Quantification of uncertainty has been widely acknowledged as a critical issue in risk assessment (see, for example, National Research Council, 1993). A variety of methods for managing uncertainty have been studied (Morgan and Henrion, 1990), most of which are beyond the scope of the present chapter. One of these, which figures prominently in EDSS and is especially valuable for integrated modeling systems, involves the use of computer simulation methods to quantify the uncertainty associated with a model result, conditioned on the correctness and appropriateness of the model for the problem at hand.

### 9.3.2 Accumulation of Uncertainty in Cascaded Models

One common approach to dealing with parameter uncertainty is to select input parameters that represent in some sense an upper boundary value. For example, a source concentration value might be chosen so as to assure with 95% confidence that the actual source concentration is no higher. It is particularly tempting to make such so-called conservative choices in environmental and other health-related fields in the interests of playing it safe. But is it possible to be too conservative?

When we choose such upper bound input parameters for the first model in a cascade and then feed the results of that model into a subsequent model (whose input parameters are similarly chosen), we arrive at conclusions considerably more dire than the actual distributions for the input parameters would suggest (McKone and Bogen, 1991).

### 9.3.3 Avoiding "Creeping Conservatism"

In considering the uncertainty of quantitative models, one considers the output of the model to be some function of one or more input coefficients. These coefficients become the parameters of a numerical representation of the model. The quantitative uncertainty in the modeling solution, then, results from the combined uncertainties of the input parameters. Stochastic analysis can be used to avoid the "creeping conservatism" that results from the use of upper bound parameter values alone to model risk.

Stochastic analysis of uncertainty is predicated on the ability to articulate the probability distributions of each uncertain parameter and then iteratively solve one or more model equations involving these parameters. To accomplish this, samples are drawn from the parameter distributions, most often employing Monte Carlo or Latin hypercube sampling methods.

To generate $N$ Monte Carlo samples from a given probability distribution, one first produces the corresponding cumulative distribution function (CDF). The ordinate of the CDF, which ranges from zero to one, is then sampled uniformly, and the corresponding abscissa values are taken as pseudorandom samples of the target distribution.

Latin hypercube sampling, a variation on the Monte Carlo method, forces the uniform samples drawn on the ordinate to cover the entire range (zero to one) by dividing the axis into N equal-width bins. From each bin a sample is drawn, with uniform sampling within each bin. This modification helps to ensure that the tails of the target distribution are sampled, and therefore can result in convergence on the target distribution in fewer samples than the unmodified Monte Carlo method (Iman and Helton, 1988; Iman and Shortencarier, 1984).

To solve environmental models using such stochastic methods, one solves the model equation iteratively, each time using parameter values drawn from the uncertain parameter distributions by the methods just described. The results of these calculations form, themselves, a distribution that aggregates the uncertainty of each of the parameters and whose characteristics can be used to describe the model. The moments and upper and lower quantile bounds of such a calculated distribution can be directly employed in decision making based on the model. For example, if one calculates individual exposure to radionuclides using such an approach, the CDF of the distribution of results can be used to find the probability that exposure will exceed 25 mrem/year. While such a quantitative probabilistic approach to environmental risk is not yet widely embraced within the regulatory community, the trend toward risk/cost/benefit analyses is driving regulators in that direction.

### 9.3.4 Calibrating and Validating the Integrated Modeling System

Any modeling exercise must include two important steps: calibration and validation. Calibration refers to the use of real-world data to arrive at model coefficients (or, as discussed previously, their distributions). Validation is the process by which a model's performance is compared to real-world data (different from those used for calibration) to assess the fidelity of the model with respect to the modeled system. Ideally, these steps would be applied to integrated systems of models as well, since these really represent models in their own right. But because of the complexity of these systems, the opportunities to do this may be limited. It must be noted that integrating a set of validated models does not yield a validated modeling system. While there is some advantage to starting with validated models, one should look for the chance to calibrate and validate the integrated system in its own right. Among other things, failure to do so substantially reduces the chance of discovering problems at the interfaces. After all, as has already been pointed out, these interfaces are themselves models.

## 9.4 APPROACHES TO MODEL INTEGRATION

So far we have identified several issues requiring attention when one is considering the integration of multiple environmental models in the interest of environmental decision support. We now turn our attention to the implementation of such a modeling system.

### 9.4.1 Scripting Languages

There are a variety of languages or model development systems in varying levels of use. Some of the more modern packages take advantage of the graphical user interfaces (GUI) common in entry-level programming support packages. For example, Visual Basic programs can be written to integrate simple models into a GUI incorporating GIS tools. Similarly, the STELLA systems modeling package has been applied successfully to environmental problems and is especially effective at allowing students and practitioners alike to explore and understand the modeled systems through use of an intuitive visual interface (Deaton and Winebrake, 1999), though integration with GIS software is clumsy at best. JAVA also provides the potential for platform independence for those systems that are amenable to a Web-like implementation. With recent improvements in Web-based GIS tools, JAVA can be expected to play an increasingly important role in environmental model integration.

But the most widely used (and extensively tested) environmental models have been written in compiled programming languages such as FORTRAN and C/C++. These model codes typically use data files for input and output and sometimes respond to command line parameters. One of the early drivers of environmental model integration was the desire to provide access to such legendary (and frequently coupled) codes as the U.S. Geological Survey's MODFLOW and MODPATH programs without requiring the user to interact directly with the programs or manipulate the data files by hand (e.g., Frysinger *et al.*, 1993).

For such model codes, scripting languages provide the most flexible means of achieving software integration. One widely used scripting language, Tcl/Tk, was used to develop a Windows-based version of the open-source GRASS GIS (which was developed in C by the U.S. Army Corp of Engineers). The result, called GRASSLANDS, still provides an open environment permitting end users and application developers familiar with the language to integrate environmental models further and customize user interfaces. And programs such as the MODFLOW model code can still be readily integrated into modeling systems and Unix-based GIS platforms (such as GRASS) using any of the shell command interpreters available for Unix systems (sh, csh, ksh, &c).

### 9.4.2 GIS Platforms

As has been intimated, the models available and the tools accessible will depend on the ultimate delivery platform, which, in turn, interacts with the choice of GIS. The GRASS GIS, already mentioned, was developed in the C programming language to be run on the Unix operating system with windowing support provided by X Windows. This combination is powerful and essentially public domain, making it a natural projects that require extensive customization of the model suite and user interface. While the U.S. Army Corps of Engineers has ceased development and oversight of the GRASS system, it is still available from and supported by Baylor University's Center for Applied Geographic and Spatial Research and the University of Hannover's Institute of Physical Geography-Landscape Ecology. Given the widespread availability and growth of Linux (a public domain implementation of Unix), this combination remains one of the most powerful platforms for environmental decision support system development.

Model integration can be achieved using any combination of shell languages (such as ksh), as well as such GUI scripting languages such as PERL and Tcl/Tk. For modeling systems that need to run on Microsoft operating systems, one must look to commercially available GIS tools, such as ArcView or GRASSLANDS. The first of these is a closed source system but still permits some flexibility with respect to customization. For example, one can use Visual Basic to develop custom interfaces (to both users and modeling codes) for ArcView. GRASSLANDS provides somewhat more flexibility because source code (in C) is provided for the base GRASS modules, and the GUI implementation, in Tcl/Tk, is similarly open for customization.

### 9.4.3 Modular Designs

The foregoing has assumed that the integrated modeling system is to be developed from two or more stand-alone model codes. If this is the case, then the input/output interfaces inherent to these programs will naturally lead to a modular design for the modeling system. While one could modify the software to achieve tighter coupling than that provided by an input/output file interface (assuming source code is available), doing this raises issues about code veracity. For example, the MODFLOW program is widely used in its present form and is accepted as a standard by many organizations. But modifications to the program to integrate other modeling codes would necessarily lead to questions about the quality of the result.

However, there may be circumstances in which tighter coupling, at the source code level, is justified. This situation might occur when the models to be integrated are not readily available in existing codes and one is therefore compelled to implement them. It will still behoove the EDSS developer to enforce strictly the software engineering principles of modularity and information hiding. In addition to representing good software development practice, this will facilitate individual model calibration and validation which, though not sufficient to validate the entire system, will be a useful step along that path.

# 9.5 USER INTERFACE ISSUES

It is well to remember that the point of environmental model integration is to provide support for the human decision-making process. Therefore, the user is a part of the system, and the user interface is a critical element of the integrated modeling system. Apart from conventional human factors engineering principles (which are themselves very important), there are some user interface issues that are peculiar to integrated modeling systems and deserve special consideration.

### 9.5.1 Developing the Integrated Conceptual Model

Mathematical modeling software carries with it an implicit user interface: the conceptual model. This is essentially the user's view of the model and is generally seen from the perspective of adjustable parameters and input data. Naturally, there is also strong coupling to the set of assumptions underlying the model. In particular, the assumptions built into the mathematical solution must be consistent with the assumptions understood by the user.

This aspect of the user interface is complicated enough with a single model but becomes much more challenging when two or more models are to be integrated. In part, this is because the conceptual model meets the mathematical model(s) through the input parameters. On the one hand, a parsimonious interface to these initial parameters is highly desirable and would lead one to develop a dialogue that does not distinguish between the models within the system. That is, given some likely redundancy of parameters between incorporated models, the user should only be asked for that information once, both to provide a friendlier interface and to avoid the opportunity for the same parameter to be given different values for different models.

On the other hand, this unified user interface effectively hides from the user the existence of the multiple models. The user does not necessarily have any explicit interface with the separate models and therefore cannot be held directly accountable for agreement between their assumptions and those of the various models. Therefore, in addition to ensuring that the models to be integrated make consistent assumptions, the EDSS architect must also ensure compatibility between this composite set of assumptions and those of the user.

A first step, of course, is to express the underlying assumptions to the user explicitly.

This is fairly straightforward for assumptions that do not vary (i.e., that are not parametric). For example, if all models in the system assume steady-state conditions, this can be explained to the user in whatever terms are appropriate to the user's training and experience.

But not all assumptions are fixed. A groundwater model might allow the user to specify whether a contaminant reacts with soil particles. It may be sensible for some applications to allow this assumption to be different for the unsaturated versus the saturated zones, but the user needs at least to be made aware of a potential inconsistency so that he or she can make this judgment intentionally. When the set of such choices becomes large and complex enough, it can be difficult to deal with this issue. The matter is further complicated when the user lacks sufficient technical training to make these judgments in a consistent and sound manner (as will often be the case when real decision makers use the EDSS). Lacking a friendly expert in the room, this may be an application for an expert system that could essentially guide users as they edit their conceptual model before initiating the simulation.

### 9.5.2 Data Manipulation and Management

The other major challenging user interface issue that is somewhat unique to integrated modeling systems and EDSS revolves around the management of data. In many cases, both empirical and modeling data will be involved in the analysis of a given scenario, and the handling of these data can be cumbersome at best. This is not especially unique to integrated systems, though having more models and therefore modeling a more complex system may exacerbate it. What is unique, though, is the possible need to represent and evaluate intermediate data produced by one model and normally consumed by another. Remembering that the user's view of the system may not include an awareness of the various models, any critical intermediate modeling results must be captured and made available (as necessary) in a manner that is independent of the particular models involved. This is not an especially complicated task, assuming data representation is systematically handled in the system, but it must be considered in order to produce an effective EDSS with integrated environmental models.

## 9.6 CONCLUSIONS AND LESSONS LEARNED

The integration of multiple environmental models into a GIS-based environmental decision support system requires careful attention to a variety of issues, including explicit and implicit model interaction, uncertainty analysis, calibration and validation, developmental platform, and user interfaces. The potential payback for "getting it right" is that real decision makers (who are not necessarily technical experts) can take advantage of the state of the science by using accessible and robust decision support tools.

## 9.7 REFERENCES

Deaton, M. L., and Winebrake, J. J. 1999. *Dynamic Modeling of Environmental Systems.* New York: Springer-Verlag.

Frysinger, S. P., Thomas, R. P., and Parson, A. M. 1993. Hydrological Modeling and GIS: The Sandia Environmental Decision Support System. In K. Kovar and H. P. Nachtnebel (eds.), *Applications of Geographic Information Systems in Hydrology and Water Resources Management.* pp. 45-50. Proceedings of HydroGIS 93, April 1993. Institute of Hydrology, Wallingford, England: IAHS Press.

Frysinger, S. P. 1995. An Open Architecture for Environmental Decision Support. *International Journal of Microcomputers in Civil Engineering,* 10, no. 2, 119–126.

Guariso, G. and Werthner, H. 1989. *Environmental Decision Support Systems.* Chichester, England: Ellis Horwood Books.

Heger, A. S., Duran, F. A., Frysinger, S. P., and Cox, R. G. 1992. Treatment of Human-Computer Interface in a Decision Support System. In *IEEE International Conference on Systems, Man, and Cybernetics,* pp. 837–841.

Hogarth, R. M. 1987. *Judgement and Choice.* New York: Wiley.

Iman, R. L., and Shortencarier, M. J. 1984. A FORTRAN 77 Program and User's Guide for the Generation of Latin Hypercube and Random Samples for Use with Computer Models. *SAND83-3165* and *NUREG/CR-3624.* Albuquerque, NM: Sandia National Laboratories.

Iman, R. L., and Helton, J. C. 1988 An Investigation of Uncertainty and Sensitivity Analysis Techniques for Computer Models. *Risk Analysis,* 8, 71–90.

McKone, T. E., and Bogen, K. T. 1991. Predicting the Uncertainties in Risk Assessment. *Environmental Science and Technology,* 25, no. 10, 1674–1681.

Morgan, M. G., and Henrion, M. 1990. *Uncertainty.* Cambridge, England: Cambridge University Press.

National Research Council. 1993. *Issues in Risk Assessment.* Washington, DC: National Academy Press.

## 9.8 ABOUT THE CHAPTER AUTHOR

Dr. Steven P. Frysinger is Professor and Environment Group Coordinator in the Integrated Science and Technology (ISAT) program at James Madison University. This follows a varied career in environmental systems engineering and management, information systems engineering, and computer/human interface research at Exxon Research and Bell Laboratories. Dr. Frysinger has authored numerous papers and presentations in these areas. His research in the last ten years has focused on environmental decision support systems, with particular emphasis on user interfaces.

# 10

# Case Studies in Geographic Information Systems and Environmental Modeling

## 10.0 SUMMARY

This chapter examines five case studies illustrating the application of geographic information systems (GIS) to terrestrial environmental modeling. The themes of the case studies are predictive vegetation mapping, landscape-level modeling of forest ecosystem processes, continental-wide modeling of wild rivers, integrated modeling of global climate change, and integrated land use evaluation and planning. For each case study, the aims and objectives of the modeling exercise are reviewed, followed by an account of the targeted system components and fluxes. The spatial data and analyses undertaken by the modeling exercise are then detailed. Each case study review concludes with a brief discussion of model outcomes, benefits, and limitations.

To provide a context for considering the case studies presented here, we outline a typology of models using five axes to describe model characteristics. It is hoped that this will assist the reader in understanding not only the case studies presented but also the many other applications that exist in the literature. These can be found in an ever-increasing range of published literature as more disciplines become aware of the analytical capabilities of GIS and attempt to use it to help integrate their disciplinary expertise into a broader environmental context. The axes used are the system drivers (deterministic/stochastic), the functional mode (quantitative/qualitative), geographic context (implicitly spatial/explicitly spatial), time and model state (static/dynamic), and the data model (entity/field).

We conclude with some observations about the difficulties that can be encountered when applying GIS to environmental models. These include three considerations: (1), the need to recognize the different kinds of processes that may be relevant to an environmental problem; (2), the difficulty in environmental systems of defining system boundaries in a nonarbitrary way; and (3), the inescapable fact that the single most important factor limiting the application of GIS to environmental modeling is the availability of data. Model development and spatial database development must go hand in hand.

In this chapter we present case studies that illustrate the diversity of approaches that have been taken in applying GIS to environmental modeling. As previous chapters have detailed, GIS and environmental modeling are now extensive fields of investigation with substantial histories and bodies of supporting knowledge and documentation. As in all modeling, environmental modeling attempts to simplify systems so that our understanding of key behaviors can be improved. GIS enhances our capacity to represent and analyze the space/time characteristics of environmental phenomena. The main published models discussed here as follows:

- ANUCLIM; climate interpolation model
- ANUDEM; interpolation model for elevation surfaces
- BIOME-BGC; terrestrial ecosystem process model
- BIOME; predictive mapping model for plant functional types
- Decision tree models; used to generate predictive mapping models
- EOS; spatially distributed erosion model
- FOREST-BGC; stand-level ecosystem process model
- Generalized linear models; used to generate predictive mapping models
- IMAGE 2.0; integrated global change model
- McHarg's overlay model; manual land use evaluation model
- Models for monotonic functions; used to generate predictive mapping models
- Multicriteria evaluation; land use evaluation model
- Multiple objective linear analysis; land use allocation model
- PRISM; climate interpolation model
- RHESSys; regional hydroecological simulation model
- TAPES; digital terrain analysis model
- Wild Rivers model; assessment and inventory of human perturbation of rivers and catchments model
- TOPOG; quasispatially distributed hydrological model

# 10.1 A TYPOLOGY OF ENVIRONMENTAL MODELING

Our typology of environmental models with GIS is defined by a set of five axes. These are nonexclusive, and any given model may be described by locating it somewhere in this continuous modeling space. The axes of the typology are system drivers, the functional mode, geographic context, time and model state, and the data model (Figure 10.1). The typology describes only the approach taken to the act of modeling itself in a GIS context. As such it does not consider whether a model is predictive or exploratory, or the type of system being modeled. The model characteristics better describe the objective of the model or the process being modeled. They remain important considerations for any modeling project, but not for describing the model type as considered here.

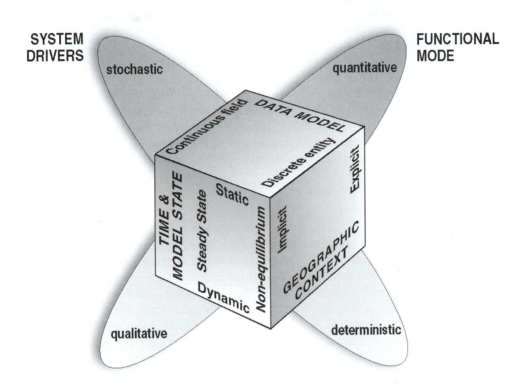

**Figure 10.1** A typology of models defined by five characteristics or dimensions. A GIS-based model will exist somewhere in this system.

### 10.1.1 System Drivers

System drivers are the parameters that drive the modeled processes. These can be internal, such as properties of the model components, or can represent external environmental conditions. In a statistical sense these drivers can be treated stochastically or deterministically. Stochastic models allow the drivers to vary randomly, while deterministic models impose order on the system response based on theoretically or empirically defined relationships. Stochastic drivers can be used where a process is, for practical purposes, impossible to simulate deterministically, such as modeling the spatial distributions of lightning strikes in a landscape.

### 10.1.2 Functional Mode

The functional mode refers to the analytical or functional relationships defining a model. In the typology these are either quantitative or qualitative. Environmental modeling using GIS has generally concentrated on quantitative functions. Qualitative functions are actually very common, although people may not envisage relationships defined in this way as constituting a model. Indeed, some phenomena defy quantitative analysis and are best handled qualitatively (for example, the interpretation of aerial photography). Approaches are becoming available where qualitative models may be implemented in a computational environment (for example, Fraser, 1999; Richards, 1999), and these may become 'spatially enabled' in the future. Expert and knowledge-based systems (for example, Coughlan and Running, 1996) are examples of qualitative approaches. However, in many cases these are derived by converting qualitative statements to nominal and ordinal scales or by defining quantitative equivalents to qualitative statements.

### 10.1.3 Geographic Context

By definition, a GIS-based environmental model is a spatially distributed analysis. However, models vary in terms of how they make use of spatial information. The spatial coordinates that define the location or distribution of an entity may be used implicitly or explicitly. Implicitly spatial models do not use the spatial relationships between locations in the modeling process. Rather, these locations are only used to map the data and to spatially colocate model variables. Consequently, each spatial location (such as a grid cell) is analyzed by calculating a function based only on values estimated for that location. A classic example of this is McHarg's (1969) overlay analysis. Explicitly spatial models use the relationship between a location and its neighbors as inputs to an analysis. Examples here include modeling hydrological fluxes across a landscape (Moore et al., 1991), space/time variation in the fire regime experienced by a landscape, and geostatistical approaches (for example, Hutchinson, 1991; Iasaaks and Srivastava, 1989; Hutchinson, 1991a) that estimate values at a location based on interpolation between a network of data points.

### 10.1.4 Time and Model State

Models can be static or dynamic, representing a system in an equilibrium or nonequilibrium state. A static model predicts system behavior at one point of time only, while dynamic models simulate system response over multiple time periods. In effect, static models usually assume that the system has reached some kind of steady state or equilibrium condition. Dynamic models frequently represent changes through nonequilibrium transitional phases in response to internal or external system drivers, which may eventually converge on an equilibrium condition as a model solution. Dynamic modeling is discussed in Chapter 6.

### 10.1.5 The Data Model

The data model refers to whether spatial variation is accounted for as discrete entities or continuous fields (Burrough and McDonnell, 1998; Chrisman, 1997). The discrete entity approach is useful for modeling individual units, such as individual animals or populations of a tree species. Continuous fields, as the name suggests, are useful for dealing with variation in continuously varying environmental variables or conditions. Chrisman (1997) notes that the data structures used in GIS imply that fields are normally divided into connected surfaces of entities, such as the pixels of a digital elevation model (DEM). It is possible to use both these data types in a single model (for example, the movement of a population of pathogens across a ground surface).

## 10.2 CASE STUDIES

Five case studies in GIS and environmental modeling are discussed using a range of spatial scales and comprising different aspects of the model typology. The models discussed involve applications of GIS to predictive vegetation mapping, forest ecosystem processes, assessment of wild river status, integrated modeling of global change, and integrated land use assessment as a GIS/environmental modeling learning tool. In considering the following case studies, and others you find in the literature, use the typology of Section 10.2 to help identify the kinds of models being presented and the use being made of GIS to extend the models' capacities to deal with the space/time dimensions of environmental systems.

### 10.2.1 Predictive Vegetation Mapping

**Aims.** Vegetation mapping has made pioneering use of GIS and environmental modeling. The vegetation cover at a location can be described in terms of floristics (for example, the taxonomic composition of the plants), community assemblages (that is, the dominant plant species), or structure and physiognomy (that is, the vegetation's height, density, biomass, or growth form). The extensive use made of GIS and environmental modeling by vegetation studies stems from the fact that the spatial distribution of vegetation cover at the landscape scale is often highly heterogeneous, particularly in forested landscapes. Field observation is restricted to a relatively small number (tens to thousands) of plot-based or transect-based observations ranging in size from tens of meters to kilometers. Yet somehow an estimate must be derived of the total distribution of the vegetation characteristics in a landscape.

Remotely sensed data (in particular from satellite-based sensors) have the advantage of being in a spatially continuous form. However, the spectral, thermal, or radar values must somehow be interpreted, and this has proven to be no easy task. Remotely sensed data can be used in a manner where patterns are visually discerned by air-photo interpretation (see Gunn et al., 1988). This traditional method can take two approaches. First, vegetation units are classified based on analysis of field survey plot data, and then mapped patterns are assigned to these predetermined vegetation classes. The second approach involves an initial landscape classification based on interpretation of the remotely sensed image, and then vegetation labeling based on field survey. In practice, some combination of the two approaches usually prevails.

As a general approach to dealing with the spatial distribution of vegetation, the traditional method has a number of limitations. First it assumes that there is a single classification of the vegetation that both exists and can be mapped using remotely sensed images. However, neither is necessarily true. Particular attributes of vegetation composition, structure, and productivity will not necessarily spatially covary, forming discrete spatial units. Regarding the second point, key attributes may simply not be recognized by remotely sensed images. For example, in forests where the signal is dominated by canopy features, below-canopy attributes of interest may simply not be sensed and hence escape pattern recognition.

A further limitation of conventional vegetation mapping is that it only provides static and descriptive information about vegetation pattern, whereas we know that vegetation ecosystems are dynamic. At best conventional mapping provides only a static snapshot in time. From this arises questions of how temporally stable these captured landscape patterns are, and uncertainty as to their ecological significance. To answer such questions requires the capacity to examine potential causal processes and to develop a predictive, as well as descriptive, capacity.

**Methods.** GIS-based approaches have provided new analytical tools and methodologies for addressing the vegetation mapping problem. One approach is for the potential distribution of vegetation to be modeled based on statistical correlations between field-based vegetation observations and environmental data, including remotely sensed data.

A variety of empirically based techniques have been developed to apply GIS to the problem of spatially predicting vegetation based on deriving correlations between the spatial variation in sampled vegetation characteristics and values or indices of prevailing environmental conditions. The environmental conditions may be observed in the field (in situ) or estimated from a GIS database. Usually the environmental data indicate levels of the primary environmental resources that are system drivers of plant response, such as heat, light, water, and mineral nutrients (see discussions in Mackey,1996; Barnes et al.,1998), or disturbance factors such as fire, logging, or grazing history.

A variety of statistical and algorithmic techniques have been used to derive correlation functions between observed vegetation attributes and environmental variables. The distribution of the sampled vegetation is then predicted as an implicitly spatial function of measures or indices of the environmental and disturbance variables. It follows that a spatial prediction of the species distribution can be generated by coupling the correlation function to a GIS database containing spatial estimates of the predictors, in this case the environmental and disturbance variables.

The general field of predicting potential vegetation distributions in relation to environmental gradients is reviewed by Franklin (1995; also see Franklin, 1998). Statistical modeling approaches include the use of generalized linear models (for example, Austin et al., 1984) to the distribution of Australian *Eucalyptus* tree species. Algorithmic approaches include the use of decision trees, such as by Lees and Ritman (1991), who mapped the vegetation of a 20 * 20 km study area at Kioloa State Forest in south east Australia. This was done using a GIS database comprising a 30 m resolution digital elevation model, a digitized geology map, Landsat TM data, and an extensive field surveyed vegetation dataset. The resultant rules were used to predict the spatial distribution over the study area of the dominant canopy tree species using the indices contained in the GIS database.

Mackey (1993; 1994) used a GIS-based approach in modeling the spatial distribution of rainforest structure in the Wet Tropics of North East Queensland, Australia. A statistical modeling procedure called Algorithms for Monotonic Functions (AMF) (Bayes and Mackey, 1991) was used to derive empirical correlations between field surveyed samples of vegetation structure, climatic averages generated for each survey plot using the ANU-CLIM procedure of Hutchinson (1991a) and field-based observations of geological substrate and topographic attributes, including slope, aspect, topographic position, and up-slope contributing area. These correlations were then coupled to gridded estimates of the significant environmental variables using a GIS database developed by Mackey et al. (1988, 1989). Predictions were then generated of the potential distribution of mature rainforest structural types as a function of the environmental correlates. In addition, predictive models of potential distribution were generated for individual structural and physiognomic attributes such as dominant canopy leaf size, stand complexity, and canopy height. These potential distributions were masked by land cover data derived from Landsat TM imagery so that predictions were restricted to within the current extent of rainforest in the region.

**Spatial Data.** The spatial scale over which many plant species and vegetation characteristics vary means predictive vegetation mapping studies now make regular use of spatial models of mesoscaled climate, such as those derived from the ANUCLIM method of Hutchinson (1991a, 1996) or the PRISM procedure of Daly et al. (1994). This is understandable given that the climatic variables of rainfall, temperature, radiation, and potential evaporation are fundamental system drivers for the levels of the primary environmental resources available to plants.

For similar reasons there is considerable application by predictive vegetation mapping exercises of terrain-based indices derived from digital elevation models, including radiation values that incorporate the effect of slope, aspect, and horizon shading, and indices that relate to topographic controls on water flow. As discussed in previous chapters, these can be calculated, for example, by the TAPES suite of models (see Gallant and Wilson, 1996; Moore et al., 1991), or various functions available within the Arc/Info GIS (ESRI, 1996).

GIS-based predictive vegetation mapping studies operate at a range of spatial resolutions from around 20 m grid resolution to coarse global grids (see the pioneering work of Box, 1981). Studies that focus on mesoscaled climatic effects generally operate at grid resolutions greater than 250 m, while mapping exercises that incorporate terrain indices usually utilize DEMs with resolutions in the range of 10 to 40 m.

**Output.** Empirically derived spatial models of vegetation are of interest and value in that they quantify plant-environment relations and provide a spatially explicit framework for examining a range of plant response, including biomass productivity and various aspects of biodiversity. However, models that correlate plant distributions with prevailing environmental conditions are limited in that they, in effect, represent steady-state conditions. Hence it is uncertain to what extent they can be used to predict dynamic phenomena such as transitions following recovery from disturbance or from other significant perturbations such as global climate change.

It is also possible that current patterns of vegetation distribution may reflect past disturbances, extreme climatic events (rather than climatic averages), or historical biogeography. Consequently, empirical correlations between vegetation and prevailing environmental conditions may not provide a reliable basis for predicting how these patterns might change in the future. There is also a danger in uncritically assuming that a spatial correlation found through correlation analysis indicates causality.

For example, any phenomenon that has a geographic distribution occupies a climatic envelope, even if climate is not a system driver of its distribution. So, while spatially distributed but temporally static predictive vegetation models are useful, it is important to recognize their limitations. Furthermore, the longer the time period we examine, the more important become biological processes such as the horizontal dispersal of plant propagules and disturbance processes such as fire regimes, and hence the need to incorporate more explicitly spatial functions.

In terms of the typology used in this chapter (Figure 10.1), the case studies described for predictive vegetation mapping generally use deterministic drivers and a quantitative functional mode, are spatially implicit, and temporally static, and use both entity and field data models. The key characteristics needing improvement are the inclusion of dynamic system states and explicitly spatial processes.

### 10.2.2 Modeling Forest Ecosystem Processes

**Aims.** The Boreal Ecosystem-Atmosphere Study (BOREAS; see Hall, 1999) is a large-scale international, interdisciplinary experiment in the northern boreal forests of Canada involving 85 science teams from across North America. Its goal is to improve understanding of the boreal forests through analyzing how they interact with the atmosphere, how much $CO_2$ they can store, and how climate change will affect them. BOREAS aims to utilize satellite data to monitor the forests and to improve computer simulation and weather models so scientists can anticipate the effects of global change.

BOREAS has had a focus on conducting detailed studies of ecosystem processes in individual stands of Boreal forest ecosystems. The kinds of ecosystem processes being studied include carbon and water fluxes between the stand canopy and the atmosphere, primary productivity, and carbon sequestration. A major challenge for the project has been how to extrapolate the results of these fine-scale studies to characterize ecosystem processes over large areas. As the dynamics of carbon sequestration processes are spatially and temporally complex, the integrated effects of ecosystem processes on a regional scale are poorly understood.

**Methods.** In addressing the regional extrapolation problem, the BOREAS research team has developed a methodology based on coupling an ecosystem process model to a GIS spatial database. The ecosystem process model, BIOME-BGC, simulates biogeochemical and hydrological variables within a variety of so-called biomes or forest ecosystem types. The model assumes that the main system drivers are climate and certain life form characteristics of the vegetation. BIOME-BGC is an extension of an earlier forest simulation model called FOREST-BGC. Further descriptions of these models can be found in papers by Running and Coughlan (1988) and Running and Hunt (1993). While FOREST-BGC was originally conceived as a stand-level model, various researchers have worked to evolve its spatial capabilities. For example, Band et al. (1993) developed a model called RHESSys (Regional HydroEcological Simulation System) that made use of FOREST-BGC and the quasidistributed hydrological model TOPOG (Bevan and Kirby, 1979) to simulate the influence of potential lateral water fluxes on forest ecosystem processes.

A regional study based on BIOME-BGC is described by Kimball et al. (2000). The analyses reported in that paper focused on assessing the relative magnitude and spatial complexity of annual net primary productivity across a boreal landscape and the sensitivity of the system to short-term, interannual weather variations. The spatial capacities of BIOME-BGC were further extended in this study by incorporating spatially distributed daily meteorological fields derived from a microclimate simulator, and remote sensing–based surfaces describing the spatial distribution of important landscape characteristics. The study area covered 1205 km$^2$, and the spatial unit of analysis was a regular grid with a 30 m cell size.

**Spatial Data.** The landscape simulation of Kimball et al. (2000) required development of a spatial database comprising a digital elevation model and various vegetation and daily meteorological attributes. In addition, for each vegetation type (for example, deciduous broad leaf, evergreen conifer) an array had to be defined of critical physiological constraints to define environmental response curves.

Remotely sensed data from airborne synthetic aperture radar (AIRSAR) was used to derive basic land cover classes describing the spatial distribution of the major vegetation types and estimates of vegetation crown cover, leaf area index, and stem biomass. The following land cover classes were recognized: dry conifer, wet conifer, open water, disturbed, deciduous, mixed-deciduous-conifer, and wetland.

Values for some of the model parameters, such as foliar leaf nitrogen content (which in the model influences photosynthesis), were taken from the BOREAS stand-level monitoring programs. Soil rooting depth and water holding capacity were taken from the $1 : 10^6$ scale soil inventory database for Canada, complemented by some field site data. Other parameters, such as some of the vegetation type environmental responses, were taken from the published literature. The model calculates daily carbon and water fluxes and hence requires daily estimates of maximum and minimum air temperatures, solar radiation, and precipitation. These were generated by interpolating, onto a 1-km resolution grid using a DEM, daily records from 60 weather stations in the region.

In summary, the exercise coupled spatially distributed estimates of land cover, biomass, soil, terrain, and daily climatic data to an ecological process model that estimated spatial and temporal behavior in net primary productivity across a boreal landscape.

**Outcomes.** While an impressive exercise, this analysis, like all models, has limitations. For example, there was no modeling of the effect of overstory/understory interactions such as shading on stand dynamics. Also, at a landscape scale, the model did not simulate the lateral transfer of matter and energy, such as the effects of runoff on soil moisture. Thus, this modeling exercise did not incorporate the TOPOG-based procedures used by Band and colleagues noted previously. Furthermore, the climatic analysis was limited to a 1 km resolution, and so topographically scaled climatic effects were ignored. Finally, the vegetation-type data did not include the attribute of stand age, which affects the rate of net primary productivity and the amount of standing biomass.

A critical question for any modeling exercise is the total error associated with the final model output. This is particularly difficult to estimate for a complex landscape process model such as BIOME-BGC. However, the authors did note that the difference between the remotely sensed AIRSAR data and field measurements ranged from 34 to 110%. Similarly, it would be useful to quantify how much the final model output would be influenced by including lateral fluxes of materials and water or using soil data from sources mapped at finer scales. Nonetheless, BIOME-BGC, and its related family of models, represents a bold advance in the application of GIS to environmental modeling at the landscape scale, as it addresses the critical interface between the land surface and the atmosphere. This provides the potential for examining the bidirectional impacts of land cover disturbance and future climate.

In terms of the model typology used here, BIOME-BGC is based on a quantitative functional mode and uses deterministic system drivers and implicitly spatial analyses but does simulate selected system dynamics through time, such as the flux of carbon in response to seasonal variation in weather conditions.

### 10.2.3 Continental Wild Rivers Analysis

**Aim.** The Wild Rivers project was an initiative of the Australian government that aimed to assist agencies in identifying near-pristine rivers and to encourage protection and proper management of their total catchment. The Australian Heritage Commission was given the lead role in developing the Wild Rivers project. This required development of a national (continental) database, analytical models, and project specific GIS-based software as described in Stein et al. (2000).

A wild river is defined by the project as a channel or a connected network of water bodies of natural origin and exhibiting overland flow (perennial, intermittent, or episodic).

In this channel the biological, hydrological, and geomorphological processes associated with river flows, and the biological, hydrological, and geomorphological processes in those parts of the catchment with which the river is intimately linked, must not have not been significantly altered by modern or colonial society (the latter refers to the settlement of Australia by Europeans in 1788; prior to this the continent had been inhabited by its indigenous populations for around 50,000 years).

**Methods.** A Wild Rivers model was developed that distinguishes direct changes to the flow regime, such as a dam, from indirect impacts to the hydrological, geomorphological, or ecological functions of the stream resulting from human-induced changes in the catchment. It was decided not to try to describe the individual physical processes associated with pollutant generation or transport to provide, for example, precise measures of pollutants or stream loads. Rather, an index approach was taken that would enable a preliminary screening of rivers. However, the model does link the impact of human disturbance with the stream via simulation of overland flow within the catchment and via stream topology.

The model is underpinned by three basic assumptions. First, the degree to which the biological, hydrological, and geomorphological processes of a stream have been altered forms a continuum of alteration from severely degraded to near-pristine. Therefore, the indices derived from the model must be based on a continuous measurement scale. Second, surrogate measures of the extent to which natural river processes have been degraded can be derived from data about the intensity and extent of human activities within the catchment and data indicating changes to the flow regime. Third, all river systems need to be assessed in order to identify the full range of wildness.

Figure 10.2 shows the main analytical stages of the Wild Rivers model. For each stream section a contributing catchment area was determined. This subcatchment (the immediate catchment of a stream section) became the primary unit of spatial analysis.

The model first derives a rating for every stream section (each tributary and each section of a mainstream between junctions), based on disturbances to that section within its immediate subcatchment. Scores are calculated for four major disturbance sources: land use activity, settlements and structures, infrastructure, extractive industries, and other point sources of pollution. The scores reflect the intensity and magnitude of the disturbance and the proximity to the stream of point pollution sources. Beyond a threshold, the values are reduced by a simple decay function. This index is called the subcatchment disturbance index (SCDI) and is assigned to all cells in each stream section's subcatchment.

A catchment disturbance index (CDI) is then calculated for each subcatchment by first multiplying the simulated runoff for each grid cell by the SCDI value for that sucatchment. Second, these values are accumulated over the subcatchment into a single value. Third, this value is divided by the subcatchment accumulated flow. Thus the CDI is a function of subcatchment area, potential runoff, and disturbance. The accumulated flow calculations are based on flow paths calculated from the DEM. The relative runoff is estimated as the overflow from a cell-based water balance calculation.

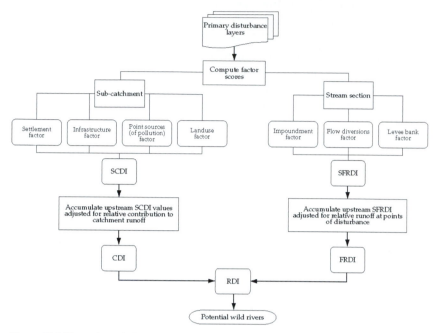

**Figure 10.2** The main analytical steps in the Wild Rivers model, leading to the calculation of the river disturbance index.

A second set of index values is calculated to reflect direct alterations to the flow regime from impoundments, flow diversions, or discharges and levee banks. This index is called the section flow regime disturbance index (SFRDI). A flow regime index (FRI) is then calculated by accumulating upstream SFRDI values adjusted by an estimate of the relative runoff accumulated at the points of disturbance.

These two scores (CDI and FRI) are then combined to give an overall score that is standardized to a dimensionless index (0.0–1.0) called the river disturbance index, with zero indicating least disturbed rivers.

**Spatial Data.** The primary input data used by the Wild Rivers model were from three sources: a National Wilderness Inventory developed by Lesslie and Maslen (1995) with recent updates from state government agencies, the TOPO 250K GEODATA digital data set of the Australian Survey and Land Information Group (AUSLIG, 1992), and the MIN-LOC database (Bureau of Resource Sciences, 1996). These databases contain information on settlements, infrastructure, point sources of pollution and land use, and some instream structures such as weirs, dams, flow diversions, and levee banks. They included a large number of categories that were combined into a smaller set, where each category was assumed to have a similar magnitude of impact on a stream. A distance threshold was also set for each point source category to allow the values to be modified according to their proximity to the stream (that is, point sources were assumed to have a greater impact the closer their location to a stream).

A critical component of the Wild Rivers assessment procedure was the development and application of a new continental, 250-m resolution DEM. A major effort was invested to ensure that the DEM accurately captured the topographic controls on surface hydrological flows. The DEM was used for several key analyses, including delineation of subcatchments for each stream section, modeling of drainage paths, calculation of accumulated flow, identification of stream lines, calculation of distance to streams, and determination of direction of water balance surplus. The DEM was interpolated using the ANUDEM procedure of Hutchinson (1991b; 1996) with the assistance of new, complementary terrain analysis functions. Flow directions were calculated using the ANUDEM program, while Arc/Info functions were applied to delineate stream section subcatchments and calculate surface distance to streams.

A simple 'bucket' water balance model (Nix, 1981) was used to estimate soil water surplus on a grid cell basis. Inputs to this calculation were long-term mean monthly values of precipitation and evaporation estimated from the Australia-wide interpolated climate surfaces generated by the ANUCLIM procedure (Hutchinson, 1991a). These hydrological calculations were coupled to the various continent-wide geographical information relating to human activity, as noted previously.

**Outcomes.** A Wild Rivers database was produced that included the spatial estimates of the final river disturbance index, together with the values for the component indices, and the data layers required for their computation (including the disturbance information, and catchment and surface flow details). A user-friendly interface was written in Arc/Info Arc Macro Language (AML) called the Wild Rivers Software Toolkit, which enables users to access readily the entire spatial database of primary and modeled data.

Figure 10.3 shows river disturbance values for the continent of Australia aggregated on a catchment or water basin basis. The gray scale indicates the percentage of streams in each water basin with an RDI value greater than or equal to 0.01. The geographic extent of human perturbation is apparent, with relatively unperturbed river systems being largely restricted to northern tropical Australia, the western arid country, and the south west of the island state of Tasmania. Figure 10.4 shows the river disturbance values for Cape York Peninsula (the northeast corner of Australia). The relatively high values illustrate the extent to which Cape York contains among the least disturbed river systems on the continent—indeed they are amongst the least disturbed of any tropical savannah and seasonal tropical forest landscape ecosystems in the world.

The Wild Rivers model provides a quantitative assessment of river and catchment disturbance assuming, in effect, equilibrium conditions. It incorporates explicitly spatial analyses but it is temporally static. The system drivers are deterministically defined, while most of the environmental entities are treated using a continuous data model. The continental scale and use of digital terrain analysis, together with the focus on human-environment interactions, makes this a particularly interesting application of GIS.

### 10.2.4 Integrated Modeling of Global Change

**Aims.** The greenhouse problem is a significant challenge facing humanity. Atmospheric concentrations of carbon dioxide are increasing due to a combination of the use of fossil fuel by modern methods of industrial production and transportation, together with deforestation and unsustainable land use management. IMAGE 2.0 is an advanced attempt to integrate models of the key global subsystems of climate change, namely the climate system, the global biogeochemical cycles, and the human societal components. Thus, IMAGE 2.0 attempts to simulate the dynamics of the global society-biosphere-climate systems. This research is being undertaken as part of the International Geosphere-Biosphere Program (IGBP; Walker and Steffen, 1997). The case study description given here is a brief summary of the detailed account of IMAGE 2.0 given in the book edited by Alcamo (1994)..

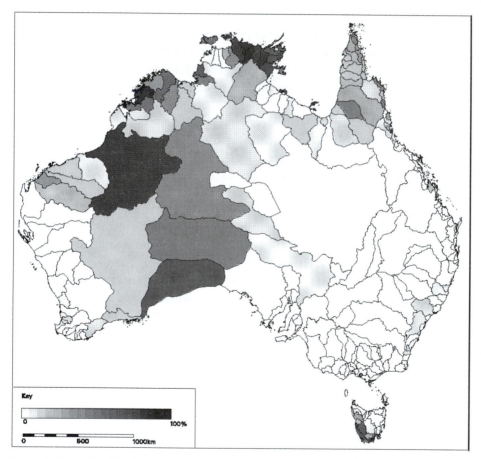

**Figure 10.3** The river disturbance index for continental Australia. The scale indicates the percentage of total stream length in each major water basin (catchment) with a river disturbance index greater than or equal to 0.01.

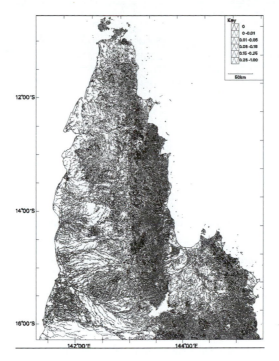

**Figure 10.4** The river disturbance index for Cape York Peninsula, Australia.

The four scientific goals of the IMAGE 2.0 model are to provide insight into the relative importance of different linkages in the society-biosphere-climate system, investigate the relative strengths of different feedbacks in the systems, estimate the most important sources of uncertainty in such a linked system, and identify gaps in knowledge about the system in order to help set the agenda for climate change research. In addition, the project has a number of policy-related goals, including linking scientific and policy aspects of global climate change in a spatially explicit manner to assist decision makers and providing a quantitative basis for analyzing the costs and benefits of various measures to address climate change

**Methods.** IMAGE 2.0 is a dynamic simulation model. Dynamic calculations are performed to year 2100. The time steps used vary between submodels but range from one day to five years. The main components of the IMAGE 2.0 model and their relationships are illustrated in Figure 10.5. The model consists of three fully linked subsystems: energy-industry, terrestrial environment, and atmosphere-ocean.

The energy-industry models compute the emissions of greenhouse gases in 13 world regions as a function of energy consumption and industrial production. End use energy consumption is computed from various economic and demographic driving forces.

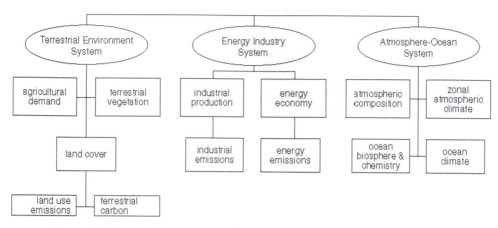

**Figure 10.5** The main components of the Global Change Model IMAGE 2.0 (Alcamo, 1994).

The submodels included are energy economy, energy emissions, industrial production, and industrial emissions.

The terrestrial environment models simulate the changes in global land cover on a grid based on climate and economic factors. The roles of land cover and other factors are then taken into account to compute the flux of $CO_2$ and other greenhouses gases from the biosphere to the atmosphere. This subsystem includes the submodels of agricultural demand, terrestrial vegetation, land cover, terrestrial carbon, and land use emission.

The atmospheric-ocean models compute the buildup of greenhouse gases in the atmosphere and the resulting zonal-average temperatures and precipitation regimes. Submodels are incorporated that deal with atmospheric composition, zonal atmospheric climate, ocean climate, and oceanic biosphere and chemistry.

**Spatial Data.** The model aims to provide as much information as possible on a grid of half a degree of latitude by half a degree of longitude in order to capture the spatial variability in the potential impacts of climate change, such as coastal flooding and changes to agricultural production and biodiversity. Furthermore, land use—related greenhouse emission (such as $NH_4$ from agricultural activity) depend on local environmental conditions and human activity. The policy implications also demand a geographically distributed approach as their implementation tends to be location specific. Another advantage of a grid-based GIS approach is that the output is more readily tested against observational data.

However, not all aspects of the integrated model could be analyzed on the half-degree-cell basis. The economic calculations were undertaken on a regional basis, where countries were grouped into one of 13 economically similar regions. The required economic, technological, and demographic data were aggregated on this regional basis. The zonal atmosphere climate model divides the atmosphere into 18 layers for calculating radiative forcing. Earth's heat balance is computed in eight vertical layers and along 10-degree latitudinal bands. The main output here is annual average surface temperature in 10-degree latitudinal bands covering the entire globe, both oceans and land.

In its entirety the model requires extensive spatial data sets. For example, the terrestrial biosphere component uses land cover change to compute the greenhouse gas fluxes between the terrestrial biosphere and the atmosphere. Potential land cover for both natural ecosystems and agrosystems is determined by the terrestrial vegetation model. This model consists of separate submodels for a water balance calculation, the distribution of potential natural vegetation, and crop distributions and yield. The global datebase of Olson et al. (1985) is used to describe the current land cover patterns required by IMAGE 2.0. The potential vegetation cover is derived from the BIOME model of Prentice et al. (1992), which predicts plant functional types as a function of climate and soils data. Climate estimates were derived from a gridded database developed by Leemans and Cramer (1991), whose values were interpolated from a global network of meteorological stations. These long-term monthly means were further interpolated to generate daily values. Soils data are also needed to calculate the water balance component. A soil properties data set was developed based on the database of Zobler (1986).

**Outcomes.** The complex, interconnected set of models represented by IMAGE 2.0 generates a suite of predicted outputs, including emissions of greenhouse and ozone-related gases, atmospheric concentrations of greenhouse and ozone-related gases, changed climate, new land use patterns, agricultural impacts, and perturbation and risks to ecosystems. Some of these are used as system drivers to alter the inputs to the terrestrial environment and energy-industry systems.

IMAGE 2.0 presents some interesting scale issues. For example, we can ask if it is really feasible to derive average soil conditions for a half-degree-grid given that key characteristics of the soil profile, such as depth and texture, vary at spatial scales that traditionally have demanded mapping at cartographic scales of 1 : 5,000 to 1 : 20,000. The grid resolution needed to capture the appropriate level of spatial variability is around 2 to 20m. Should models such as IMAGE 2.0 aim to increase their spatial resolution so that the minimum grid cell size matches the spatial variability of the attribute with the finest grain? The data implications of increasing the grid resolution are substantial, as doubling the resolution will quadruple the number of cells. So increasing the resolution of IMAGE from half a degree (approximately 50 x 50 km) to 0.0167 degrees (about 1.76 x 1.76km) would quadruple the number of grid cells five times.

Does a global model need to operate at a spatial resolution commensurate with topographic sequences of soil catena? The answer is no as long as we do not expect IMAGE 2.0 (and subsequent versions of this and other systems) to provide a representation of the earth system that can form the basis of detailed policy and management prescriptions. Perhaps we should be satisfied if such a model could simply provide some idea of the direction (rather than the magnitude) of the potential impact of global change. It may be that the spatial, and thus data, requirements could be relaxed.

In terms of our modeling typology, the calculations of IMAGE 2.0 are largely implicitly spatial, in particular, the terrestrial functions that are calculated on a half-degree-cell basis. However, various explicitly spatial calculations are undertaken. For example, in the land cover model, the set of rules that guide the prediction of new agricultural areas includes an adjacency constraint where new areas are only permitted if they are adjacent to existing agricultural areas. Also, in the ocean climatic model the world's oceans are divided into vertical layers of 400 m thickness, and horizontal segments of 10 degrees of latitude. The simulation of advective circulation in the world's oceans includes horizontal flow as well as upwelling and downwelling estimates. IMAGE 2.0 clearly uses a quantitative functional mode, largely deterministic functions, both discrete and continuous data models, and is able to simulate dynamically a range of nonequilibrium system responses.

### 10.2.5 Integrated Land Use Assessment as a GIS/EM Learning Exercise

The final case study examines land use conflict by applying land use evaluation models that integrate biophysical and human related data. This exercise has been developed and implemented by Brian Lees as the laboratory component of the GIS course in the Department of Geography and Human Ecology at the Australian National University (see http://geography.anu.edu.au).

**Aims.** The laboratory exercise is structured as an environmental impact assessment (EIA) for a hypothetical urban subdivision at the Australian National University's field station at Kioloa, New South Wales, Australia (see Lees, 1995; Lees and Ritman, 1991). EIAs require that various potential land uses and constraints be evaluated and their interactions considered. By using this problem as the focus for a GIS and environmental modeling learning exercise, students gain experience in a wide variety of environmental modeling applications and methods and an understanding of the benefits and limitations of using GIS-based approaches to examining environment/human interactions.

**Methods**. The exercise comprises three main components: derivation of the required spatial attribute data, calculation of suitability or hazard ratings for the selected land uses and constraints, and combination of the output from the various submodels to generate a final land use model. The submodels include the land uses of agriculture, forestry, and conservation; and the constraints, or hazards, of fire, erosion, and flooding. Land uses are merged with relevant hazards before they are passed through the final stage of analysis. Figures 10.6 and 10.7 show the flow of data and modeling used in the exercise. Actual modeling methods are not shown, as these change to reflect advances in GIS and modeling.

**Spatial Data**. To implement the EIA requires a detailed landscape-scale data set. Here a subset of the NASA Pathfinder data set for Australian Eucalyptus forests is used (Lees, 1995). The source data sets used in the modeling process are a Landsat TM image, DEM, geological coverage, and cadastral information about roads, fire trails, and property boundaries. Thus both discrete entity and continuously varying environmental datasets are used. The raster data has a resolution of 30 m to be consistent with the Landsat data.

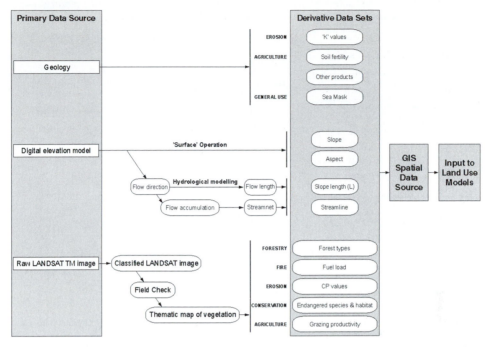

**Figure 10.6** Flow chart indicating the main source data and the derived attributes that together constitute the Kioloa spatial database. These data drive the land use evaluation models for the Kioloa EIA exercise.

The source data sets are used to generate the attribute values required as input to the land use and hazard submodels (Figure 10.6). In conjunction with field work, the Landsat data are used to generate a thematic map of vegetation, from which are derived coverages of forest types, fuel loads, vegetation erosion values, endangered species distribution and habitat, and grazing productivity values. The geology coverage and the DEM are used as a surrogate for soil properties, and from these are derived indices of soil erodibility and fertility. Geology is also used to generate a mask to exclude grid cells in the ocean from further analysis. The DEM is used to derive topographic indices, including slope, aspect, streams, and slope lengths. Depending on the background and interests of the student class, other data sets can be included, such as sites of archaeological, cultural, and natural significance.

**Submodels.** The submodels used are as noted previously, and Figure 10.7 depicts a standard modeling combination (although details of the erosion and fire hazard models are not shown for reasons of brevity). These submodels can be implemented using a range of algorithms and predictive functions, allowing for a flexible learning experience adaptable to student abilities. However, the modeling approaches used are generally simple, as the law of diminishing returns suggests that increasingly complex models may not greatly improve model results. Two example models are described here, namely, the flood model and the conservation model.

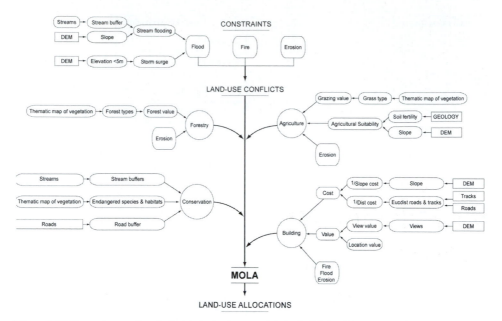

**Figure 10.7** The main steps in calculating the land use models in the Kioloa EIA exercise. Optimal land use allocation is then explored by applying multi objective linear analysis (MOLA) procedures.

The flood model is one of the hazards to a building subdivision (Figure 10.7). It is devised in two parts: the threat from a storm surge over a high tide and the threat of over-bank flow from local streams. All the necessary parameters to calculate these are derived from the DEM (for example, the 5-m storm surge is calculated as all cells below 5 m elevation). The stream flooding is a combination of distance from streamlines and those locations with a low slope value.

The conservation model requires greater use of the source data sets than does the flood model. The streamlines, derived from the DEM, are buffered to provide a riparian protection zone. Endangered species and habitat locations are derived from the Landsat vegetation map and buffered. Finally, roads and fire trails (tracks) are also buffered.

The modeling approaches used generally fall within the steady-state, temporally static category for simplicity and are applied as a combination of entity and field, implicitly and explicitly spatial models. Most of the model functions use deterministic approaches; for example, the Universal Soil Loss Equation (USLE; Foster et al., 1975a, b; Wischmeier and Smith, 1965; ) is used to assess erosion hazard.

Suitability scores for the land use submodels are presently derived using multicriteria evaluation (MCE; Carver, 1991; Voogd, 1983), where various criteria affecting a land use are combined according to a weighting system. Those factors considered more important to the submodel value are given higher weighting and therefore importance. Hazard models are generally applied to the MCE models as limiting factors.

**Final Combination.** As illustrated in Figure 10.7, the final land use allocation is accomplished using multiple objective linear analysis (MOLA; Carver, 1991). As with the MCE, this is a weighted combination of the different model inputs. MOLA is used to generate a decision as to what land use (if any) should be allocated to a given location. The weights used in the MOLA analysis are adjusted until there is sufficient area for each land use to be economically viable, or until some other criterion (perhaps a more integrative sustainability goal) is reached. For example, if there is insufficient area on which to build, then the weights of the building component of the model are increased and the new land use allocation is calculated. If, at the same time, there is too much of the conserved land being impinged on by the other three land uses, the relative weights may again be adjusted. Such an approach is a good representation of what occurs in the real world, as most results require that some kind of compromise is reached between the conflicting demands of stakeholders.

**Outcomes.** A key learning outcome of the exercise is that students are made aware of many of the factors that need to be considered when implementing an environmental model. The exercise is more complex and demanding than one where students merely follow a "recipe". Furthermore, students develop an understanding of the many different modeling approaches that can be implemented in a GIS framework. The laboratory exercise is complemented by a field trip where the students attempt to validate the derived attribute data and model predictions, such as the vegetation classification derived for the Landsat image. Asa result, they gain first-hand experience in both the benefits and limitations of spatial modeling.

As noted previously, the exercise has evolved to make use of improved modeling approaches as GIS and environmental modeling develop. For example, the application has evolved from using essentially boolean modeling methods to fuzzy modeling approaches. The course exercise also illustrates that the application of GIS to environmental modeling is not restricted to any particular software platform, as it has been implemented using several GIS systems, including, in the last ten years, Idrisi (Clark Labs, 2000), GRASS (Byars, 2000), Arc/Info (ESRI, 1996), and Arcview (ESRI, 2000).

## 10.3 CONCLUSIONS AND LESSONS LEARNED

GIS provides tools for "spatially enabling" environmental models. GIS-based environmental models allow for the representation of multiscaled phenomena and the aggregate affects of human-induced environmental impact through space and time. They have the potential to output information in a format that can be readily incorporated into environmental policy and strategic planning. From both a research and implementation perspective, it is now taken as axiomatic that addressing environmental problems demands a multidisciplinary approach. The development of GIS provides a much needed platform that facilitates applications at this critical junction between traditional disciplines.

The case studies presented here illustrate three problem areas that are common to all applications of GIS to environmental modeling—namely, determining the multitude of patterns and processes that comprise what we call the environment; delineating system boundaries in the environmental continuum; and obtaining the spatial data needed to drive a model. Generally, the more complex the model, the more these considerations become constraining factors. IMAGE 2.0 can be considered a complex model as it focuses on the dynamics and nonequilibrium conditions of a large-scale system, inclusive of environment-human interactions, and incorporating a mix of deterministic and stochastic drivers and implicit and explicit spatial analyses. Therefore, critical decisions must be made in terms of determining which biogeochemical-human societal processes to include/exclude, how to define boundaries between submodels, and how to integrate spatial data from differently scaled sources.

### 10.3.1 Defining the Environment

The case studies reviewed illustrate the complex suite of phenomena that can be incorporated within environmental modeling, and clearly there is no simple definition of the environment. Our case studies suggest that a comprehensive definition of the environment must recognize the roles played by physical, biological, ecological, and human-societal processes.

Earth is unlike any other planetary system we know because it is a living planet. The term *Biosphere* was first defined by Vernadsky in 1926 (see Vernadsky, 1998) as those zones of earth whose environmental conditions are to some degree modified by biological processes. Hence, the Biosphere includes the atmosphere, the land surface, the soil mantle, and the oceans.

Earth's environmental conditions are not the product of physical processes alone as biological processes mediate physical fluxes, particularly through gaseous exchanges between living organisms and the atmosphere and the transformation of solar energy to biomass by photosynthesizing plants and microorganisms (see discussion in Gorshkov et al., 2000). The term ecological is used to describe such interactions between the biota and physical processes.

Of course, not all of earth is under the influence of these ecological processes—earth's core is a phenomenon that is under the control of geophysical processes. Similarly, there are physical processes that affect the Biosphere that are not under biotic mediation. Examples are changes in the flux of solar energy reaching the top of the atmosphere, which can change earth's energy balance, and volcanic activity leading to the discharge of carbon from the lithosphere into the atmosphere.

We must also recognize that humans have now become a major driver of global environmental change. Therefore, in an increasing number of contexts, environmental models must now incorporate the influence of processes associated with human activity, such as the industrial use of fossil fuel and land clearance. As the IMAGE 2.0 model illustrates, data concerning such aspects as human economies, urbanization, and demographics are now essential inputs to comprehensive global change modeling.

Many researchers choose to ignore the biological or human factors when defining environmental systems (for example, by studying atmospheric conditions in terms of only physical processes). This may be done for a number of reasons, such as to identify the mechanisms surrounding a particular physical process, or because researchers consider the biological processes to be only a minor component of the subsystem under investigation. The number and type of environmental components considered in a model will thus vary according to the context and end use.

The predictive vegetation models discussed in Case Study 1 often aim to generate spatial estimates of natural vegetation, exclusive of the effects of human activity. However, this is difficult to achieve empirically given the long periods (10,000 to 60,000 years) that humans have inhabited many landscapes and the increasing impact of modern, technological society. The results of the Wild Rivers project given in Case Study 3 illustrate the extraordinary environmental impact modern society has had on much of the Australian environment. But even excluding the human dimension, the scope of most predictive vegetation studies must still consider the effects of a complex array of biophysical processes, including mesoscale climate, topographic controls on surface radiation and hydrological fluxes, soil/regolith physical and chemical properties, and landscape-wide disturbances such as fire regimes.

### 10.3.2 Model Boundaries

A related problem in the application of GIS to environmental modeling is identifying the boundary of the environmental system being investigated. Engineered systems (that is, systems that have been designed and constructed by humans) have readily defined components, flows, and boundaries. This is not the case with environmental systems. While we can often readily identify the components and flows of interest, defining system boundaries in a nonarbitrary way is fraught with difficulty.

If we again consider Earth in toto, then boundary definition is not a problem. Earth represents a system that is essentially closed to all flows except for solar energy (and the occasional meteorite). Earth therefore represents a well-bounded closed system in terms of its major biogeochemical characteristics (for example, water, carbon, nutrients) and biota. However, within the Biosphere, which is where humans mainly operate, system boundaries are generally poorly defined and fuzzy. For example, even the boundary between terrestrial and oceanic systems is open to the flux of water and chemical substances, such as carbon, and populations of living organisms.

So how can systems, and in particular their spatial boundaries, be defined in ways that are nonarbitrary rather than merely reflecting the biases of the investigator? To answer to this question we must evoke the critical concept of *scale* (see Allan and Hoekstra, 1992). All biophysical elements within the Biosphere are ultimately connected, however they are related (and separated) by space and time. Generally, we define a system by identifying a set of phenomena operating within specified space/time scales. The spatial dimensions include the extent of the study (that is, the maximum geographic area of the study), the resolution (the geographic unit of spatial analysis), and the grain (the minimum spatial unit of observation). Temporal scales can often best be considered in terms of the frequency or return time of an event (for example, the mean number of years between fires experienced by a landscape). It follows that while the boundaries of environmental systems may be inevitably fuzzy, they do not have to be arbitrary. Rather, the onus is on the analyst to make explicit the spatial and temporal dimensions that define the phenomena being investigated.

The full spatial extent of the IMAGE 2.0 model is defined in a nonarbitrary manner as it encompasses the total earth system. Similarly, given that the focus of the Wild Rivers project discussed in Case Study 3 was on terrestrial catchment quality and hydrology, the use of the Australian continental coastline for boundary definition was environmentally justifiable. Conversely, the predictive vegetation models discussed in Case Study 1 can suffer from the application of arbitrarily defined system boundaries if, for example, the plant species under study have much broader geographic ranges and more complex environmental relations than contained in the defined study area.

### 10.3.3 Data! Data! Data!

The spatial dimension that GIS brings to environmental models generates spectacular geographic images at landscape, regional, and continental scales. However, we must be careful not to let these intrinsically appealing visuals prevent us from carefully evaluating the underpinning model assumptions and errors and the accuracy of the output. In particular, we must always consider carefully the quality of the spatial data driving the modeling exercise. The major factor limiting the application of GIS to environmental modeling is not modeling capacity but rather the lack of reliable data that account for the spatial distribution and density of key biophysical entities and process flux rates.

Humans are acquiring an ever-increasing volume of data, particularly from remotely sensed sources. However, these data do not necessarily directly measure the entities and processes of interest. Rather, direct quantitative observation of environmental characteristics remains dominated by fine-scaled field survey or instrumented field locations. Interpreting remotely sensed data and spatially extending field data remain critical steps in the application of GIS to environmental modeling. This partly explains the ongoing interest in predictive vegetation modeling noted in our first case study, as there is an ever-increasing demand from models such as BIOME-BGC and IMAGE 2.0 for reliable, spatially distributed data about the structure and taxonomic composition of vegetation ecosystems.

The literature is littered with examples of technically good models that can never be applied in real-world situations due to the large number of parameters and the time and expense needed to generate the required data. A review of water projects published in the journal *Water Resources Research* between 1965 and 1985 (Biswas, 1990, p. 5) found that 723 papers were published on systems analysis from the United States. Only 38 of these papers were found to deal with identifiable (that is, practical) water projects. Of these, only four were built, with just one considered to have been implemented according to the optimization procedures described.

Herein lies the greatest challenge in the application of GIS—to ensure that spatial database development occurs in parallel and is not outstripped by, model development.

## 10.4 REFERENCES

Alcamo, J. 1994. *IMAGE 2.0. Integrated Modeling of Global Climate Change.* Dordrecht, The Netherlands: Kluwer Academic Publishers.

Allen, F. T. H. and Hoekstra, T. W. 1992. *Towards a Unified Ecology.* New York: Columbia University Press.

AUSLIG. 1992. *GEODATA TOPO-250k User Guide.* Canberra, Australia: Australian Survey and Land Information Group.

Austin, M. P., Cunningham, R. B., and Fleming, O. M. 1984. New Approaches to Direct Gradient Analysis Using Environmental Scalars and Statistical Curve-Fitting Procedures. *Vegetation,* 55, 11–27.

Band, L. E., Patterson, P., Nemani, R., and Running, S. 1993. Forest Ecosystem Processes at the Watershed Scale: Incorporating Hillslope Hydrology. *Agricultural and Forest Meteorology,* 63, 93–126.

Barnes, B. V., Zak, D. R., Denton, S. R., and Spurr, S. R. 1998. *Forest Ecology,* 4th ed. New York: Wiley.

Bayes, A. J., and Mackey, B. G. 1991. Algorithms for Monotonic Functions and Their Application to Ecological Studies in Vegetation Science. *Ecological Modeling,* 56, 135–159.

Bevan, K. J., and Kirby, M. J. 1979. A Physically Based, Variable Contributing Area Model of Basin Hydrology. *Hydrological Science Bulletin,* 24, 43–69.

Biswas, A. K. 1990. Environmental Modeling for Developing Countries: Problems and Prospects. In A. K. Biswas, T. N. Khoshoo, and A. Khosla (eds.). *Environmental Modeling for Developing Countries,* pp. 1–12. London: Tycooly.

Byars, B. 2000. GRASS GIS (Geographic Resources Analysis Support System). http://www.baylor.edu/~grass/index2.html (July 5, 2000).

Box, E. O. 1981. *Macroclimate and Plant Form. Tasks for Vegetation Science 1.* The Hagve: Dr. W. Junk Publishers.

Bureau of Resource Sciences. 1996. Mineral Occurrence Locations Database (MINLOC). Technical Manual. Canberra, Australia.

Burrough, P. B., and McDonnell, R. A. 1998. *Principles of Geographical Information Systems.* Oxford, England: Oxford University Press.

Carver, S. J. 1991. Integrating Multi-Criteria Evaluation with Geographical Information Systems. *International Journal of Geographical Information Systems,* 5, no. 3, 321–339.

Chrisman, N. 1997. *Exploring Geographic Information Systems,* New York: Wiley, New York.

Clark Labs. 2000. *IDRISI Geographical Analysis and Image Processing Software.* http://www.clarklabs.org/01home.htm (July10, 2000).

Coughlan, J. C. and Running, S. W. 1996. Biophysical Aggregations of a Forested Landscape Using an Ecological Diagnostic System. *Transactions in GIS*, 1, no. 1, 25–39.

Daly, C., Nielson, R. P., and Phillips, D. L. 1994. A Statistical-Topographic Model for Mapping Climatological Precipitation over Mountainous Terrain. *Journal of Applied Meteorology.* 33, 140–158.

ESRI. 1996. *Arc/Info Version 7.0.4.* Redlands, CA: Environmental Systems Research Institute, Inc.

ESRI. 2000. Arcview GIS. http://www.esri.com/software/arcview/. (July 10, 2000).

Foster, G. R., Meyer, L. D., and Onstad, C. A. 1975a. A Runoff Erosivity Factor and Variable Slope Length Exponents for Soil Loss Estimates. *Transactions of the American Society of Agricultural Engineers,* 20, 683–687.

Foster, G. R., Meyer, L. D. and Onstad, C. A. 1975b. An Erosion Equation Devised from Basic Erosion Principles. *Transactions of the American Society of Agricultural Engineers,* 20, 678–682.

Franklin, J. 1995. Predictive Vegetation Mapping: Geographic Modeling of Biospatial Patterns in Relation to Environmental Gradients. *Progress in Physical Geography,* 19, 474–499.

Franklin, J. 1998. Predicting the Distribution of Shrub Species in Southern California from Climate and Terrain-Derived Variables. *Journal of Vegetation Science,* 9, 733–748.

Fraser, D. 1999. *QSR Nudist VIVO Reference Guide.* Melborune, Australia: Qualitative Solutions and Research Pty Ltd.

Gallant, J. C., and Wilson, J. P. 1996. TAPES-G: A Terrain Analysis Program for the Environmental Sciences. *Computers and Geosciences,* 22, 713–722

Gorshkov, V. G., Gorshkov, V. V., and Makarieva, A. M. 2000. *Biotic Regulation of the Environment: Key Issue of Global Change.* Chester, England: Springer/Praxis.

Gunn, R. H., Beattie, J. A., Riddler, A. M. H., and Lowrie, R. A. (eds.) 1988. *Australian Soil and Land Survey Handbook: Guidelines for Conducting Surveys.* Melbourne, Australia: Inkata Press.

Hall, F. G. 1999. The Boreal Ecosystem-Atmosphere Study. http://boreas.gsfc.nasa.gov/BOREAS/BOREAS_Home.html (July 15, 2000).

Hutchinson, M. F. 1991a. A Continental Hydrological Assessment of a New Grid-Based Digital Elevation Model of Australia. *Hydrological Processes,* 5, 45–58.

Hutchinson, M. F. 1991b. Climatic Analyses in Data Sparse Regions. In R.C. Muchow and J. A. Bellamy (eds), *Climatic Risk in Crop Production,* pp. 55-71. Wallingford: CAB International.

Hutchinson, M. F. 1996. *ANUDEM Version 4.4 User Guide.* Canberra, Australia: The Centre for Resource and Environmental Studies, The Australian National University.

Isaaks, E. H. and Srivastava, R. M. 1989. *An Introduction to Applied Geostatistics.* Oxford, England: Oxford University Press.

Kimball, J. S., Keyour, A. R., Running, S. W., and Saatchi, S. S. 2000. Regional Assessment of Boreal Forest Productivity Using an Ecological Process Model and Remote Sensing Parameters Maps. *Tree Physiology,* 20, 761–775.

Leemans, R., and Cramer, W. 1991. *The IIASA Database for Mean Monthly Values of Temperature, Precipitation and Cloudiness on a Global Terrestrial Grid.* Research Report RR-91-18, Int. Institute of Applied Systems Analyses, Laxenburg, Austria.

Lees, B. 1995. The Kioloa GLCTS Pathfinder Site. http://geography.anu.edu.au/path-finder/index.html (July 10, 2000).

Lees, B. G., and Ritman, K. 1991. Decision Tree and Rule Induction Approach to Integration of Remotely Sensed and GIS Data in Mapping Vegetation in Disturbed or Hilly Environments. *Environmental Management,* 15, 823–831.

Lesslie, R., and Maslen, M. 1995. *National Wilderness Inventory Australia. Handbook of Procedures, Content and Usage.* Canberra, Australia: Australian Heritage Commission.

Mackey, B. G. 1993. A Spatial Analysis of the Environmental Relations of Rainforest Structural Types. *Journal of Biogeography,* 20, 303–336.

Mackey, B. G. 1994. Predicting the Potential Distribution of Rain-Forest Structural Characteristics. *Journal of Vegetation Science,* 5, 43–54.

Mackey, B. G. 1996. The Application of GIS and Environmental Modeling to the Conservation of Biodiversity. *Third Annual Conference/Workshop on GIS and Environmental Modeling.* National Center for Geographical Information Analysis/Santa Barbara. Conference Proceedings, Santa Fe, January 1996. CD-ROM

Mackey, B. G., Nix, H. A., Hutchinson, M. F., McMahon, J. P., and Flemming, P. M. 1988. Assessing the Representativeness of Places for Conservation Reservation and Heritage Listing. *Environmental Management,* 12, 501–514.

Mackey, B. G., Nix, H. A., Stein, J. A., and Bullen, F. T. 1989. Assessing the Representativeness of the Wet Tropics of Queensland World Heritage Area. Biological Conservation, 30, 279–299.

Moore, I. D., Grayson, R. B., and Ladson, A. R. 1991. Digital Terrain Modelling: A Review of Hydrological, Geomorphological and Biological Applications. *Hydrological Processes,* 5, no. 1, 3–30.

Nix, H. A. 1981. Simplified Simulation Models Based on Specified Minimum Data Sets: the CROPEVAL Concept. In A. Berg (ed.) *Application of Remote Sensing to Agricultural Production Forecasting*, pp. 151–169. Luxenbourg: Commission of the European Communities.

Olson, J., Watts, J. A., and Allison, L. J. 1985. *Major World Ecosystems Complexes Ranked by Carbon in Live Vegetation: A Database.* Report NDP-017, Oak Ridge National Laboratory, Oak Ridge, TN.

Prentice, I. C., Cramer, W., Harrison, S. P., Leemans, R., Monserud, R. A., and Solomon, A. M. 1992. A Global Biome Model Based on Plant Physiology and Dominance, Soil Properties and Climate. *Journal of Biogeography,* 19, 117–134.

Richards, L. 1999. *Using NVIVO in Qualitative Research.* Melbourne, Australia: Qualitative Solutions and Research Pty. Ltd.

Running, S. W., and Coughlan, J. C. 1988. A General Model of Forest Ecosystem Processes for Regional Applications. I. Hydrological Balance, Canopy Gas Exchanges and Primary Production Processes. *Ecological Modeling,* 42, 125–154.

Running, S. W., and Hunt, E. R. 1993. Generalization of a Forest Ecosystem Process Model for Other Biomes, BIOMe-BGC, and an Application for Global-Scale Models. In J. R. Ehliringer and C. Fields (eds.) *Scaling Physiological Processes: Leaf to Globe.* Academic Press.

Stein, J. L., Stein, J. A., and Nix, H. A. 2000. *The Identification of Wild Rivers: Methodology and Database Development.* A Report for the Australian Heritage Commission by the Centre for Resource and Environmental Studies The Australian National University. Education and Communication Section, Australian Heritage Commission. Commonwealth of Australia.

Vernadsky, V. I. 1998. *The Biosphere.* New York: Copernicus/Springer-Verlag (first published in Russian in 1926).

Voogd, H. 1983. *Multicriteria Evaluation for Urban and Regional Planning.* London: Pion.

Walker, B., and Steffen, W. 1997. *The Terrestrial Biosphere and Global Change: Implications of GCTE and Related Research.* IBGBP Science No. 1. Stockholm, Sweden: IGBP.

Wischmeier, W. H., and Smith, D. D. 1965. Predicting Rainfall-Erosion Losses from Cropland East of the Rocky Mountains. *U.S. Department of Agriculture Handbook,* 282, 1–47.

Zobler, L. 1986. *A World Soil File for Global Climate Modeling.* Technical Memorandum, NASA, New York.

## 10.5 INFORMATION RESOURCES

The following references provide further cases studies in the application of GIS to environmental modeling:

Goodchild, M. F., Steyaert, L. T., Parks, B., Johnson, C., Maidment, D., and Glendinning, S. (eds.). 1996. *GIS and Environmental Modeling: Progress and Research Issues.* Fort Collins, CO: GIS world Books.

Heit, M., Dennison Parker, H., and Shortreid, A. (eds.). 1996. *GIS Applications in Natural Resources* 2. Fort Collins, CO: GIS World Books.

McDonald, A. D., and McAleer, M. (eds.). 1997. *MODSIM 97: International Congress on Modeling and Simulation.* Proceedings. Hobart, Tasmania, Australia, December 1997. Volumes 1–4. The Modelling and Simulation Society of Australia, Inc.

NCGIA. 1996. *Third International C0onference/Workshop on Integrating GIS and Environmental Modeling.* Proceedings, January 21–25, Sante Fe, New Mexico. National Center for Geographic Information and Analysis. CD-Rom.

**Journals:**
- *Ecological Modelling*
- *Global Ecology and Biogeography*
- *Journal of Biogeography*
- *Landscape Ecology*
- *Journal of Vegetation Science*
- *Transactions in GIS*
- *International Journal of Geographic Information Science*

## 10.6 ABOUT THE CHAPTER AUTHORS

Dr. Brendan Mackey is a reader in environmental science and ecology at the Department of Geography and Human Ecology, Faculty of Science, The Australian National University, Canberra. He conducts research and teaching in the application of GIS and environmental modeling to ecological problems, with a focus on forest ecosystems and global change.

Shawn Laffan is a Lecturer in GIS at the Department of Geography and Human Ecology, Faculty of Science, The Australian National University, Canberra. He undertakes research and teaching in GIS and environmental modeling, with a focus on landscape hydrological and geomorphological problems.

# 11

# *Visualizing Environmental Data*

## 11.0 SUMMARY

Environmental models and the data they demand can quickly become extremely complex. When dealing with site assessments, environmental remediation design or monitoring, public hearings, or environmental litigation, the quantity of data involved can quickly become overwhelming. Maintaining and organizing that data is insufficient. Visualization is the only means for condensing and effectively communicating vast quantities of spatially referenced data. Whether the data consumer is an environmental engineer, geologist, or the public, visualization provides an invaluable tool to communicate complex data in a form that makes the data universally intelligible.

This chapter addresses issues related to the visualization of environmental data. Compiling the model input data and ensuring that they are consistent, adequate, and sufficiently complete is the only first step. Beyond the data, the chapter addresses by example the myriad of methods that can be applied to visualize, analyze, and model environmental data.

Conscientious data visualization should begin with a direct visualization that displays the data with a minimal amount of interpretation or assumption. The next stage of investigation and display involves gridding the data and interpolating and extrapolating the data to the nodes and/or cells of the grid. This process can be complicated by the incorporation of geologic information, site features (such as roads, buildings, and bodies of water), and aerial photography. Once mapped to the grid, numerous examples are given of how to present this data visually using different representational methods. Unlike standard cartographic visualization (Slocum, 1999), presenting many types of environmental data generally requires finding ways to delve into the three-dimensional properties of the information. Usually this is performed using data subsetting techniques.

Visualization is visual communication. It is the art and science of compressing and presenting information in a useful form that we humans can easily comprehend (Friedhoff and Benzon, 1989). Whether the data collected at an environmental site represent a portion of a page or many bookshelves of reports, visualization can benefit everyone involved by transforming words and numbers into comprehensible images.

# 11.1 VISUALIZING ENVIRONMENTAL DATA

This chapter is dedicated to the complex task of visualizing environmental data. There are many ways of visualizing data ranging from simple graphs, plots, and charts to more representational two- and three-dimensional visualizations and temporal data animations. The focus of this chapter is representational graphics that portray the spatial characteristics of environmental data. To this end, the data requirements associated with this chapter will necessitate that all data be spatially referenced. The chapter uses a sequenced approach, taking the data through a process flow from initial exploration of input data, to the production of finished high-quality graphic outputs.

### 11.1.1 Exploratory Data Analysis

When approaching any new project involving environmental data visualization, the first step is to assess the form and content of the data. Data are commonly delivered in many forms (discussed in Section 11.2.2). A visual assessment of the content of the data is just as useful as evaluating basic summary statistics, such as the number of samples; minimum average, and maximum data values; and the spatial extent.

### 11.1.2 Data Sources

The form of the data is usually governed by the application in which the data were authored and/or stored. Typical applications include database managers, geographic information systems (GIS), spreadsheet programs, and ASCII files.

**Databases and GIS.** Environmental databases can be built in generic database software such as Microsoft Access or Oracle, or products like Earthsoft's EQuIS or Integrate's TerraBase, which are specifically developed to provide environmental data management for both chemistry and geologic data. GIS programs like ESRI's ArcView provide database functionality that is inherently tied to spatial information. For that reason, GIS systems provide an ideal platform for storing and retrieving data to be used for environmental visualization. GIS files may also include maps that contain features such as roads, building outlines, rivers, and other geographic features. GISs also provide some level of visualization capability though it is generally two dimensional or limited three dimensional. Most GISs have added many of the visualization methods discussed here into their newer software releases, while others provide bridges to packages that support these capabilities.

Environmental databases and GIS systems generally contain far more information than is needed for data visualization. Sample/measurement date, laboratory, analytical methods, quality assurance information, well construction details, and name(s) of companies or individuals are just a few examples of information that may not be necessary to the task of visualization. However, this additional data is often crucial to the documentation and defense of visualization tasks.

**Spreadsheets.** Spreadsheet programs like Microsoft's Excel are commonly used as the repository for environmental data. Their ease of use and ability to perform many database-like functions enhance their appeal. Spreadsheets can import and export most common data file formats, including many database and some GIS files.

**ASCII Files.** ASCII files provide the most portable form for environmental data. Spreadsheets and databases can import and export ASCII files and they can be edited with a myriad of programs ranging from simple text editors (like Microsoft Notepad) to spreadsheets and databases. It is this portability, especially across operating systems, that makes ASCII files a common choice as an input file format for environmental visualization.

**CAD Files.** Spatial environmental data is often contained in CAD (computer-assisted drawing) files. Many providers of stereo photography and topographic data deliver the produts in CAD files. CAD files are also used to display roads, buildings, and other site features. These data are most useful when the CAD files are drawn in the same, consistent coordinate system used for all other data.

**Images.** Image files (also known as raster or bitmaps) are digital photographs. These photographs can range from snapshots of features to orthorectified aerial photography. Orthorectification requires that the image axes are parallel to the north-south and east-west coordinate axes and artifacts such as radial and terrain displacement have been removed. Snapshots merely provide documentation of features and usually do not include any quantitative spatial information. Orthorectified photos can be used in visualizations as texture maps. This allows them to be projected onto ground surfaces, geologic layers, and/or buildings and other features in the visualization. In order to use a photo as a texture map it must be georeferenced. Some image file formats include georeferencing information in the form of a similarly named "world" file. The world file consists of the spatial coordinates of one corner of the image and the real-world size (both width and height) of a pixel. Additional information about the rotation of the image is included, but to be used in a GIS, the rotation terms are zero. GeoTIFF (a TIFF image and its associated world file with a TFW extension) is an example of a georeferenced image file format supported by many GIS packages. These file formats simplify the process of registering aerial photography and satellite imagery. Some GIS software, such as ESRI's Arc/Info, are capable of georeferencing raw images.

## 11.1.3 Data Content Requirements

This discussion of environmental data is limited to data that include spatial information. When spatial data are collected with a GPS (global positioning system) receiver, the spatial information is often represented in latitude and longitude. Generally, before these data are visualized or combined with other data, they are projected to a Cartesian coordinate system with a known projection and datum. Converting from geographic coordinates to other coordinate systems is called map projection. Many different projections and coordinate systems can be used (Clarke, 1999). It is most important to maintain consistency. Projecting these data is especially necessary for three-dimensional visualization because we want to maintain consistent units for $x$-, $y$-, and $z$- coordinates. Latitude and longitude angle units (degrees, minutes, and seconds) do not represent equal lengths and there is no equivalent unit for depth. Projections convert the angles into consistent units of feet or meters.

**Chemistry.** Chemistry data files must contain the spatial information ($x$-, $y$-, and optional $z$- coordinates) as well as the measured analytical data. The file should specify the name of the analyst and should include information about the detection limits of the measured parameter. The detection limit is necessary because samples where the analyte was not detected are often reported as zero or "nd." It is generally not adequate (especially when logarithmically transforming this data) to use 0.0 where tha analyte was not detected.

To create a graphical representation of the borings or wells from which the samples were taken, the chemistry data file should also include the boring or well name associated with each sample and the ground surface elevation at the location of that boring.

**Geology.** Geologic information is considerably more difficult to represent in a single, unified data format because of its nature and complexity. Geologic data files can be grouped into one of two classes: those representing interpreted geology and those representing boring logs. By some definitions, boring logs are interpreted data since a geologist was required to assign materials based on core samples or some other quantitative measurements. However, for this discussion interpreted geological data will be defined as data organized into a geologic hierarchy.

C Tech Development Corporation's Environmental Visualization System (EVS)[1] utilizes one of two different ASCII file formats for interpreted geologic information. These two file formats both describe points on each geologic surface (ground surface and bottom of each geologic layer), based on the assumption of a geologic hierarchy. Simply stated, geologic hierarchy requires that all geologic layers throughout the domain be ordered from top to bottom and that a consistent hierarchy be used for all borings. At first, it may not seem possible for a uniform layer hierarchy to be applicable for all borings. Layers often pinch out or exist only as localized lenses. Also, layers may be continuous in one portion of the domain but are split by another layer in other portions of the domain. However, all of these scenarios and many others can be modeled using a hierarchical approach. The easiest way to describe geologic hierarchy is by example. Figure 11.1 shows a clay lens in sand with gravel beneath.The borings (A & C) on the left and right sides of the domain would not detect the clay lens that would be detected in B. On the sides, it appears that there are only two layers in the hierarchy, but in the middle there are three or four layers depending on your point of view.

The hierarchical geologic modeling approach accommodates the clay lens by treating every layer as a sedimentary layer. By "pinching out" layers (making the thickness of those layers zero) most geologic structures can be created. Geologic layer hierarchy requires that we treat this domain as four geologic layers. These layers would be upper sand, clay, lower sand, and gravel, as in Figure 11.2 When this geologic model is visualized in 3D, both upper and lower Sand can have identical colors or hatching patterns.

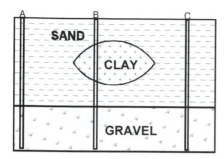

Figure 11.1 Clay lens in sand.

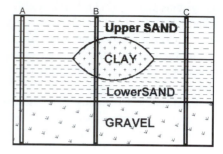

Figure 11.2 Geological hierarchy of clay lens.

Since the layers will fit together seamlessly, dividing a layer will not change the overall appearance (except when layers are exploded). For sites that can be described using the aforementioned method, it is generally the best approach for building a three-dimensional geologic model. Each layer has smooth boundaries, and the layers (by nature of hierarchy) can be exploded apart to reveal the individual layer surface features. An example of a much more complex site is shown in Figure 11.3. Sedimentary layers and lenses are modeled within the confines of a geologic hierarchy.

With EVS, there are two other geology file formats. One of them is a more generic format for interpreted (hierarchical) geologic information. With that format, $x$-, $y$-, and $z$-coordinates are given for each surface in the model. There is no requirement for the points on each surface to have coincident $x$-$y$ coordinates or for each surface to be defined with the same number of points. The borehole geology file format described previously could always be represented with this more generic file format.

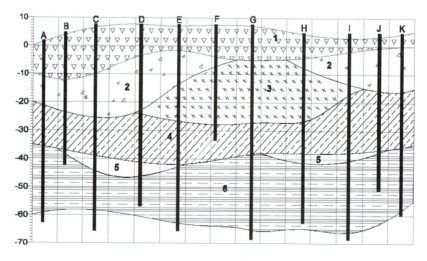

Figure 11.3 Complex geologic hierarchy.

The last file format is used to represent the materials observed in each boring. Borings are not required to be vertical, nor is there any requirement on the operator to determine a geologic hierarchy. C Tech refers to this file format as Pregeology, referring to the fact that it is used to represent raw 3D boring logs. This format is also considered to be "uninterpreted." This is not meant to imply that no form of geologic evaluation or interpretation has occurred. On the contrary, it is required that someone categorizes the materials on the site and in each boring.

In the EVS software, the raw boring data can be used to create complex geologic models directly using a process called geologic indicator kriging (GIK). The GIK process begins by creating a high-resolution grid constrained by ground surface and a constant elevation floor or some other meaningful geologic surface such as bedrock. For each cell in the grid, the most probable geologic material is chosen using the surrounding nearby borings. Cells of common material are grouped together to provide visibility and rendering control over each material.

### 11.1.4 Direct Data Visualization

Many methods of environmental data visualization require mapping (interpolation and/or extrapolation) of sparse measured data onto some type of grid. Whenever this is done, the visualization includes assumptions and uncertainties introduced by both the gridding and interpolation processes. For these reasons, it is crucial to incorporate direct visualization of the data as a part of the entire process. It becomes the operator's responsibility to ensure that the gridding and interpolation methods accurately represent the underlying data.

**Glyphs.** A common means for directly visualizing environmental data is to use glyphs. A "glyph" refers to a graphical object that is used as a symbol to represent an object or some measured data. For the purposes of this chapter, glyphs will be positioned properly in space and may be colored and/or sized according to some data value. For a graphics display, the simplest of all glyphs would be a single pixel. A pixel is a dot that is drawn on the computer screen or rendered to a raster image. The issue of pixel size often creates confusion. Pixels (by definition) do not have a specific size. Their apparent size depends on the display (or printer) characteristics. On a computer screen, the displayed size of a pixel can be determined by dividing the screen width in inches or millimeters by the screen resolution in pixels. For example, a 19" computer monitor has a screen width of about 14.5 inches. If the desktop area is set to 1280 by 1024, the width of a pixel would be approximately 0.011 inches (~0.29 mm). If the desktop area were reduced, the apparent size of a pixel would increase.

There are virtually no limits to the type of glyph objects that may be used. Glyphs can be simple geometric objects (e.g., triangles, spheres, and cubes) or they can be representations of real-world objects like people, trees, or animals.

**Glyphs in 2D.** For two-dimensional displays we generally use glyph objects that are two dimensional (having no depth or $z$-coordinate information). Figure 11.4 is an example of such a display.

**Glyphs in 3D.** It is once we move to the three-dimensional world that glyphs become

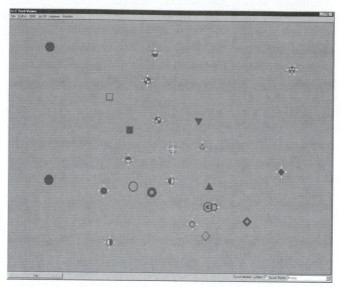

**Figure 11.4** Two-dimensional glyphs.

much more interesting. In Figure 11.5, cubes (hexahedron elements) are positioned, sized, and colored to represent chemical measurements made in soil at a railroad yard in Sacramento, California. Axes were added to provide coordinate references and this picture was rendered with perspective effects turned on. This results in a visualization where parallel-lines do not remain parallel and objects in the foreground appear larger than those in the background.

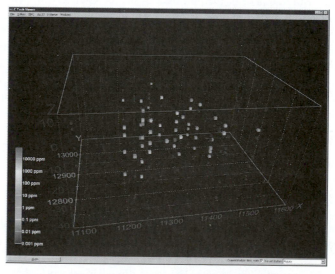

**Figure 11.5** Two-dimensional cubic glyphs.

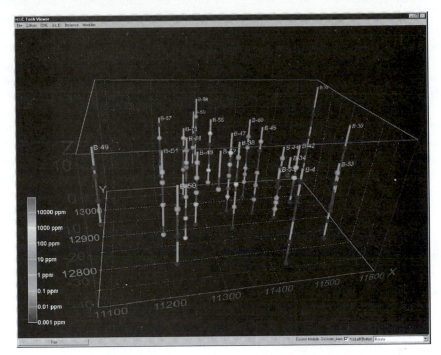

**Figure 11.6** Three-dimensional glyphs with boring tubes.

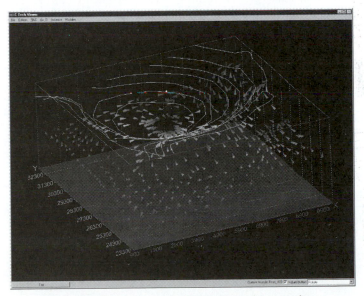

**Figure 11.7** Three-dimensional glyphs representing vector data.

When representations of the borings are added, the figure becomes much more useful. Figure 11.6 shows the samples represented by colored spheres and borings represented by tubes. The tubes are colored alternating dark and light gray, where the color changes on 10-foot intervals. This provides a reference to allow the viewer to determine quickly the approximate depth of the samples. The borings are also labeled with their designation. These last two figures both represent the same data; however, it is clear which one provides the most useful information

Glyphs can also be used to represent vector data. The most commonly encountered vector data represents groundwater flow velocity.  In this case, the glyph is not only colored and sized according to the magnitude of the velocity vector, but the glyph can also be oriented to point in the vector's direction. For this type of application, an asymmetric glyph (as opposed to a sphere or cube) is used. Figure 11.7 uses a glyph that is referred to as "jet." It is an elongated tetrahedron that points in the direction of the vector. The data represented in this figure are predicted velocities output from a MODFLOW simulation to predict the groundwater flow field resulting from the dewatering of a gold mine pit.

## 11.2 GRIDDING AND DIMENSIONALITY

Although there is great value in directly visualizing measured data, it does have many limitations. Without mapping sparse measured data to a grid, computation of contaminant areas or volumes is not possible. Further, the techniques available for visualizing the data are very limited. For these reasons and more, significant attention should be paid to the process of creating a grid onto which the data will be interpolated and extrapolated.

For this chapter, a grid is defined as a collection of nodes and cells. Nodes are points in two or three dimensions with coordinates and one or more data values. The word cell and element are both used as generic terms to refer to geometric objects. The cell type and the nodes that form their vertices define these objects. Commonly used cell types are described in Table 11.1 and Figure 11.8.

Table 11.1 Common Cell Types

| Cell Type | Number of Nodes | Dimensionality |
|---|---|---|
| Point | 1 | 0 |
| Line | 2 | 1 |
| Triangle | 3 | 2 |
| Quadrilateral | 4 | 2 |
| Tetrahedron | 4 | 3 |
| Pyramid | 5 | 3 |
| Prism | 6 | 3 |
| Hexahedron | 8 | 3 |

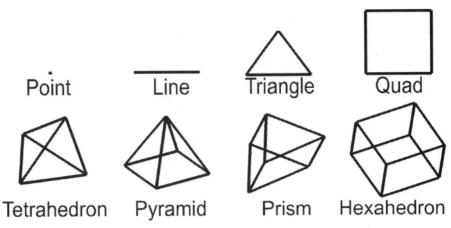

**Figure 11.8** Common cell types.

Dimensionality refers to the space occupied by the cell. Points do not have length, width, or height; therefore, their dimensionality is zero (0). Lines are dimensionality 1 because they have length. Dimensionality 2 objects (areal) such as quadrilaterals (quad) and triangles have area, and dimensionality 3 objects (volumetric) ranging from tetrahedrons (tet) to hexahedrons (hex) are volumetric. When creating a two-dimensional grid, areal cells are used, and for three-dimensional grids, volumetric cells are used.

### 11.2.1 Rectilinear Grids

Rectilinear (a.k.a. uniform) grids are among the simplest type of grid. The grid axes are parallel to the coordinate axes and the cells are always rectangular in cross section. The positions of all the nodes can be computed knowing only the coordinate extents of the grid (minimum and maximum $x$, $y$ and optionally $z$) and the number of nodes in each dimension. Two-dimensional rectilinear grids are comprised of quadrilateral cells. For a 2D grid with $i$ nodes in the $x$- direction and $j$ nodes in the $y$- direction, there will be a total of ($i$ - 1)*($j$ - 1) cells.

The connectivity of the cells or the nodes that define each cell—the topology—can be implicitly determined because the nodes and cells are numbered in an orderly fashion. The advantages of rectilinear grids include the ease of creating them and the uniformity of cell area in 2D and cell volume in 3D. The disadvantages are that grid nodes are generally not coincident with the sample data locations and large areas of the grid may fall outside of the bounds of the data. A two-dimensional rectilinear grid is shown in Figure 11.9. Three-dimensional rectilinear grids offer the simplest method for gridding a volume. They are constrained to rectangular prismatic volumes and have hexahedral cells of constant size (Figure 11.10). For some processes and visualization techniques such as volume rendering, this is advantageous and may even be required. For a grid with $i$ by $j$ by $k$ nodes there will be ($i$ - 1) x ($j$ - 1) x ($k$ - 1) hexahedron cells whose connectivity can be implicitly derived.

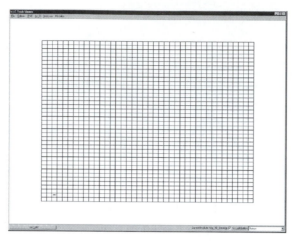

**Figure 11.9** Two-dimensional rectilinear grid.

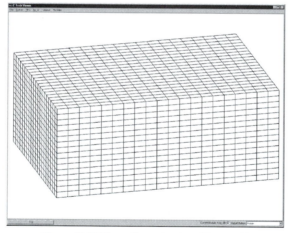

**Figure 11.10** Three-dimensional rectilinear grid.

## 11.2.2 Finite Difference

This type of grid derives its name from the numerical methods that it employs. Simulation software such as the U. S. Geological Survey's MODFLOW utilizes a finite difference numerical method to solve equilibrium and transient groundwater flow problems. This solution method requires a grid that contains only rectangular cells. However, the cells need not be uniform in size. For two-dimensional grids, this results in rectangular cells; however, it is possible that no two cells are precisely the same size. Some simulation software requires that finite difference grids be aligned with the coordinate axes. EVS does not

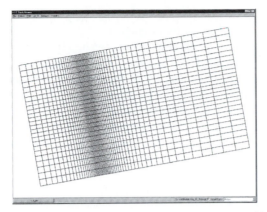

**Figure 11.11** Two-dimensional rotated finite difference grid.

impose this restriction, but it does provide a means to export the grid transformed so that the grid axes are aligned. Figure 11.11 shows a rotated 2D finite difference grid. Smaller cells are concentrated in areas of the model where there are significant gradients in the data. For groundwater simulations this is usually where wells are located. For environmental contamination it should be the location of spills or areas where DNAPL (dense non-aqueous phase liquids) contaminant plumes were detected. The smaller cells provide greater accuracy in estimating the parameter(s) of interest.

Three-dimensional finite difference grids have the same restrictions as 2D grids with respect to their $x$- and $y$- coordinates. Yet the $z$- coordinates of the grid (which define the cell thicknesses) are allowed to vary arbitrarily. This allows for creation of a grid that follows the contours of geologic surfaces. For a grid having $i$ by $j$ by $k$ nodes there will be ($i$-1) x ($j$-1) x ($k$-1) hexahedron cells whose connectivity can be implicitly derived. The coordinates of the nodes for this grid must be explicitly specified. Figure 11.12 shows the grid created to model the migration of a contaminant plume in a tidal basin.

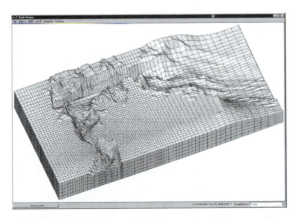

**Figure 11.12** Three-dimensional finite difference grid.

## 11.2.3 Convex Hull

The convex hull of a set of points in two-dimensional space is the smallest convex area containing the set. In the *x-y* plane, the convex hull can be visualized as the shape assumed by a rubber band that has been stretched around the set and released to conform as closely as possible to it. The area defined by the convex hull offers significant advantages. Within the convex hull, all parameter estimates are interpolations. The convex hull best fits the spatial extent of the data. Remember that the convex hull defines an area. That area can be gridded in many ways. EVS grids convex hull regions with quadrilaterals. Smoothing techniques are used to create a grid that has reasonably equal area cells. A two-dimensional example of a convex hull grid is shown in Figure 11.13. In this example, the domain of the model was offset by a constant amount from the theoretical convex hull. This results in rounded corners and a model region that is larger than the convex hull.

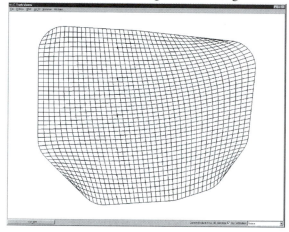

**Figure 11.13** Convex hull grid with offset.

## 11.2.4 Adaptive Gridding

Adaptive gridding is the localized refinement of a grid to provide higher resolution in the areas or volumes surrounding measured sample data. Adaptive gridding or grid refinement can be accomplished in many different ways. In EVS, rectilinear, finite difference, and convex hull grids can all be refined using a similar method. In two dimensions a new node is placed precisely at the measured sample data location. Three additional nodes are placed to create a small quadrilateral cell within the cell to be refined. The corners of the small cell are connected to the corresponding corners of the cell being refined, creating a total of five cells where the one previously was. The resulting nodal locations and grid connectivity must be explicitly defined.

Adaptive gridding offers many advantages. It assures that there will always be nodes at the precise coordinates of the sample data. This ensures that the data minimum and maximum in the gridded model will match sample data and provides more fidelity in defining data trends in regions with high gradients.

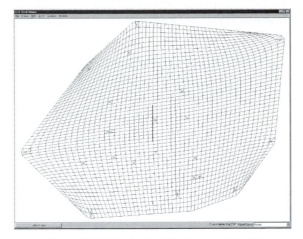

**Figure 11.14** Adaptively gridded convex hull grid.

Figure 11.14 shows a two-dimensional adaptively gridded convex hull model. This model's area was also offset from the convex hull. Since each sample data point results in a refined region, and the sample points define the convex hull, the regions in each corner of the model contain adaptively gridded cells. Figure 11.15 is a close-up view of some refined cells near the lower right in Figure 11.14. It shows one of the special cases. If the point to be refined falls very near an existing cell edge, that edge is refined and the cells on either side of the edge are symmetrically refined. Since the edge must be broken into three segments, the cells on both sides must be affected.

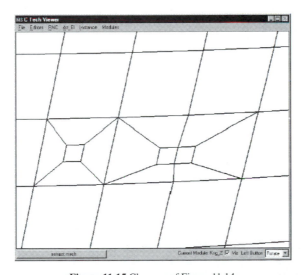

**Figure 11.15** Close-up of Figure 11.14.

**Figure 11.16** 3D adaptively gridded model.

The refinement process can also be applied to all types of 3D grids. When a sample falls in a hexahedron (hex) cell, a new much smaller hex cell is created with one of its corners located precisely at the coordinates of the sample point. The eight corners of the small cell are connected to the corresponding corners of the parent cell. This creates seven hex cells that fully occupy the volume of the original cell. Since the 3D-refinement process occurs internal to the volume of the model, it is more difficult to visualize the process. In order to see the refined cells, we have removed all cells in the grid with any nodes that were below a thresholded concentration level created. By choosing the threshold properly, several of the refined cells become visible in Figure 11.16.

Figure 11.17 is an enlarged view of the upper right-hand corner of Figure 11.16. It reveals the structure, relative sizes, and connectivity resulting from 3D adaptive gridding.

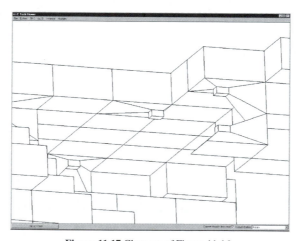

**Figure 11.17** Close-up of Figure 11.16.

## 11.2.5 Triangular Networks

Triangular networks are defined as grids of triangle or tetrahedron cells where all of the nodes in the grid are exclusively those in the sample data. For these types of grids, the cell connectivity must be explicitly defined. In two dimensions, these grids are referred to as triangulated irregular networks, or TINs. The 3D equivalent grids are tetrahedral irregular networks.

**Triangulated Irregular Networks—2D.** Delaunay triangulation is one of the most commonly used methods for creating TINs. By definition, three points form a Delaunay triangle if and only if the circle defined by them contains no other point. Focusing on creating Delaunay triangles produces triangles with fat (large) angles that have preferred rendering characteristics. The boundary edges on the Delaunay network form the convex hull, which is the smallest area convex polygon to contain all of the vertices.

The TIN surface shown in Figure 11.18 has significant variation in the size of the triangles. This is a natural consequence of the grid's being created using only nodes from the input data file. When such a surface is rendered with data, having very large triangles can result in objectionable visualization anomalies. These anomalies result from rendering large triangles that have a range of data values that span a significant fraction of the total data range. There are many methods that could be used to assign color to each triangle. These methods are referred to as surface rendering modes.

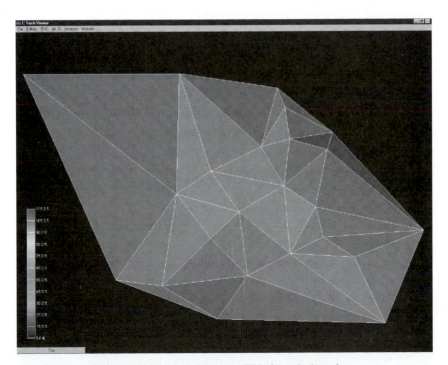

**Figure 11.18** Flat-shaded delaunay TIN of a geologic surface

Two of the most commonly used rendering modes are flat shading and Gouraud shading. Flat shading assigns a single color to the entire triangle. The color is computed based on the average elevation (data value) for that triangle, lighting parameters, and orientation to the viewer camera. In the upper left corner we have a large single triangle that spans a significant range of elevations. When the triangle is assigned a color based on the average color of its nodes, that color will be wrong. More precisely, the color does not fall within the color scale. Note the color of the triangle in the upper-left corner of Figure 11.18 and the one below it, which is is outside the range of the color scale.

ArcView and ArcView 3D Analyst utilize flat shading exclusively in representing 2D and 3D surfaces. 3D Analyst is further restricted to employing perspective effects when performing 3D rendering. This often produces dramatic results, but for scientific visualization the effect can often lead to undesirable distortions.

The problem of large triangles is no better when using Gouraud shading. Gouraud shading assigns colors to each node of the triangle based on the data values. This assures that the colors at the nodes (vertices of the triangles) will be correct. Colors are then interpolated over the area of the triangle based on lighting parameters and orientation to the viewer camera. Consider the triangle in the upper-left-hand corner of Figure 11.19.  The upper right node is assigned the color blue (corresponding to a low value) and the upper-left node is assigned the color red (corresponding to a high value). The color scale for this problem ranges from blue to cyan to green to yellow to red. However, for this anomalous situation the color that will be interpolated between blue and red along the uppermost edge will be magenta, not in the normal range of colors.

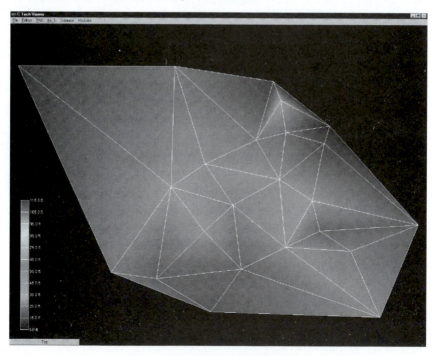

**Figure 11.19** Gouraud-shaded delaunay TIN of geologic surface.

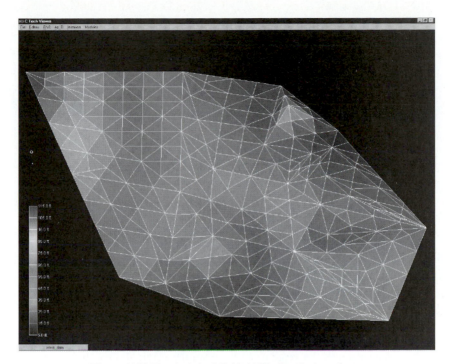

**Figure 11.20** Flat-shaded subdivided TIN of geologic surface.

To overcome the problems caused by large triangles, the triangles can be refined (subdivided) to create a grid that still contains points that honor the original input nodes but has more uniform cell sizes. In Figure 11.20 (which has a spatial extent of 500 feet in x and 380 feet in y), it was specified that no triangle's edge may exceed 45 feet in length. We must interpolate the elevation values (or our data values) to these new nodes created as a result of the triangle subdivision. The simplest means of doing this is bilinear interpolation. The refined TIN grid with bilinear interpolation and flat shaded triangles is shown in Figure 11.21. Note that the all of the triangles have appropriate colors. To avoid the large cell coloring problem (this is a problem with all cell types except points), no single cell should have data values at its nodes that span more than about 20% of the total data range. If Gouraud shading is employed instead of flat shading, the resultant surface has a smoother appearance; however, the fundamental linear interpolation along cell edges is still evident in the colors. If the maximum triangle size were made much smaller, the flat shaded model would approach the appearance of the Gouraud shaded model. However, without using a different interpolation approach the Gouraud-shaded model would not change dramatically.

EVS includes another technique for coloring surfaces. This method, called solid contours, assigns uniform color bands based on the data values. Figure 11.22 demonstrates this method that subdivides cells using bilinear interpolation. As this method inherently includes trianglular subdivision using bilinear interpolation, the figure would be identical

l

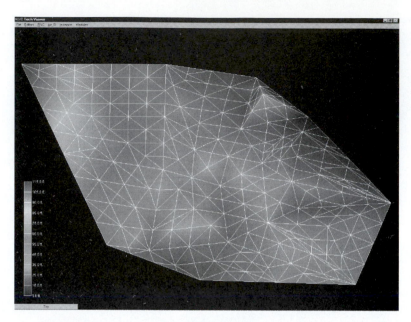

**Figure 11.21** Gouraud-shaded subdivided TIN of geologic surface.

whether the input grid was the large triangles from the original TIN surface or the more refined smaller triangles. The boundaries of the colored bands are effectively lines of constant elevation.

**Figure 11.22** Solid contour TIN of geologic surface.

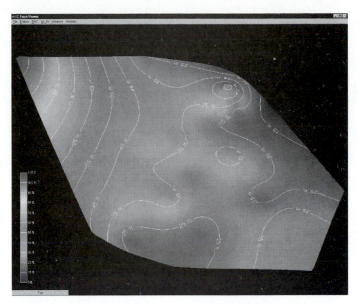

**Figure 11.23** Kriged 2D convex hull grid.

To complete this discussion and comparison of gridding and interpolation methods, the same data file was used to create a convex hull grid and the elevation data was estimated using EVS's two-dimensional kriging software. Kriging will be discussed in more detail in Section 11.4.3. This technique honors all of the original data points but creates much smoother distributions between the values. The result shown in Figure 11.23 is a more realistic and aesthetically superior surface. Labeled isolines on 10-foot intervals were added to this figure. Note that these isolines are similar but much smoother than those in Figure 11.22.

**Tetrahedral Irregular Networks—3D.** Tetrahedral irregular networks (Figure 11.24) provide a method to create a volumetric representation of a three-dimensional set of

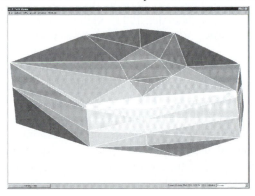

**Figure 11.24** Tetrahedral irregular network.

points. As with a TIN, the nodes in the resulting grid are exclusively those in the original measured sample data. Tetrahedral irregular networks use tetrahedron cells to fill the three-dimensional convex hull of the data as shown in Figure 11.24. The result often contains cells of widely varying volumes having potentially large data variation across individual cells. For this and other reasons, this approach is not often used.

# 11.3 INTERPOLATION METHODS

Spatial interpolation methods are used to estimate measured data to the nodes in grids that do not coincide with measured points. The spatial interpolation methods differ in their assumptions, methodologies, complexity, and deterministic or stochastic nature.

### 11.3.1 Inverse Distance Weighting

Inverse distance weighted averaging (IDWA) is a deterministic estimation method where values at grid nodes are determined by a linear combination of values at known sampled points. IDWA makes the assumption that values closer to the grid nodes are more representative of the value to be estimated than samples farther away. Weights change according to the linear distance of the samples from the grid nodes. The spatial arrangement of the samples does not affect the weights. IDWA has seen extensive implementation in the mining industry due to its ease of use. IDWA has also been shown to work well with noisy data. The choice of power parameter in IDWA can significantly affect the interpolation results. As the power parameter increases, IDWA approaches the nearest neighbor interpolation method, where the interpolated value simply takes on the value of the closest sample point. Optimal inverse distance weighting is a form of IDWA where the power parameter is chosen on the basis of minimum mean absolute error.

### 11.3.2 Splining

Splining is a deterministic technique to represent two-dimensional curves on three-dimensional surfaces (Hutchinson and Gessler, 1994). Splining may be thought of as the mathematical equivalent of fitting a thin sheet of rubber to a series of data points. Like its physical counterpart, the mathematical spline function is constrained at defined points. Splines assume smoothness of variation. Splines have the advantage of creating curves and contour lines that are visually appealing. Some of splining's disadvantages are that no estimates of error are given and that splining may mask uncertainty present in the data. Splining is not constrained to the input data range, as interpolated values can be higher than the maximum observed value and lower than the minimum observed value. Areas outside the convex hull of the data (extrapolations) can exhibit unrealistic behavior as well. Splines are typically used for creating contour lines from dense, regularly spaced data. Splining may, however, be used for interpolation of irregularly spaced data.

### 11.3.3 Geostatistical Methods (Kriging)

Kriging is a stochastic technique similar to inverse distance weighted averaging in that it uses a linear combination of weights at known points to estimate the value at the grid nodes. Kriging is named after D. L. Krige, who used kriging's underlying theory to estimate ore content. Kriging uses a variogram (a.k.a. semivariogram), which is a representation of the spatial and data differences between some or all possible "pairs" of points in the measured data set. The variogram then describes the weighting factors that will be applied for the interpolation. Unlike other estimation procedures investigated, kriging provides a measure of the error and associated confidence in the estimates. Cokriging is similar to kriging except that it uses two correlated measured values. The more intensely sampled data are used to assist in predicting the less sampled data. Cokriging is most effective when the covariates are highly correlated. Both kriging and cokriging assume homogeneity of first differences. While kriging is considered the best linear unbiased spatial predictor (BLUP), there are problems of nonstationarity in real-world data sets.

## 11.4 VISUALIZATION TECHNIQUES

After selecting the combination of direct data visualization, gridding, and interpolation methods to create the geometry and associated data, many different visualization techniques are available to visualize the model. With so many choices available for gridding, interpolation, and rendering of each object in the total model, there are virtually limitless possibilities for the final visualization.

### 11.4.1 Rendering Methods

Rendering methods determine the appearance of objects in the view. The methods that are chosen have a profound effect on the final product. The large number of rendering options and subsetting techniques allows for the same data to be presented in many different ways. This ability to customize visualizations provides a mechanism to emphasize certain aspects or objects in a model.

   **Points and Lines.** Virtually all visualizations contain lines and points. Whether the visualization specifically includes cells that are lines and/or points or has 2D and/or 3D cells (e.g., triangles and hexahedron) that have nodes and edges, line and point rendering must be considered. Points can be rendered as individual pixels or as square blocks of multiple pixels. Points can also be represented by various glyphs, as discussed in Section 11.1.3. Lines can be rendered solid or with patterns like dotted or dashed in any width (measured in pixels). Lines can also be rendered as cylindrical tubes with user-specified radius. Both points and lines can be colored according to data values or their colors can be set to any shade to outline the cells that they represent. Of course, it is always an option not to display the points and/or lines.

   **Surfaces.** The choice of surface rendering technique has a dramatic impact on model visualizations. Figure 11.25 is a dramatization that incorporates many common surface-rendering modes. These include Gouraud shading, flat shading, solid contours, transparency, and background shading. In this figure, a plume is represented in each geologic layer of this model. The geologic layers are exploded and a unique rendering mode is used for each layer. This allows demonstrating five different surface rendering techniques. Section 11.2.5 included some discussion on surface rendering techniques. In the model, a very fine grid (in the *x-y* plane) was used and the flat shaded plume looks similar to the gouraud shaded one. The solid contoured plume provides sharp color discontinuities at specific plume levels; however, it provides no information about the variation of values within each interval.

The transparent plume was Gouraud shaded. Transparency could be applied to any of the surface rendering techniques except background shading. Transparency provides a means to see features or objects inside of the plume while still providing the basic shape of the plume. Objects inside a colored transparent object will have altered colors, and the colors of the transparent object are affected by the color of the background and any other objects inside or behind the plume.

Background shading is a rather different approach. Each cell of the plume is colored the same color as the background. This makes the cell invisible; however, the cell is still opaque. Objects that are behind the background-shaded cells are not visible. In this example, the cell outlines are shown as lines colored by the concentration values. Background shading of the surfaces provides a "hidden line" rendering where the cells behind are not shown.

An example of the rendering mode called "no lighting" has not been included in this chapter. This technique renders cells as a single color (similar to flat shading) but with no lighting or shading effects. This eliminates all three-dimensional depth cues about the surface and usually produces an undesirable affect. Texture mapping is a process of projecting a raster image onto one or more surfaces. The images should be georeferenced to ensure that the image's features are placed in the correct spatial location. In Figure 11.26, a chlorinated hydrocarbon contaminant plume is shown at an industrial facility on the coast. Sand and rock geologic layers are displayed below the ocean layer. A color aerial photograph of the actual site was used to texture map and render the geologic layer that represents the ocean and was also applied to the three-dimensional representations of the site buildings as well as the ground surface.

**Volume Rendering.** There are several volume rendering techniques. In Figures 11.27 and 11.28, two different methods are shown. For both methods, the data must be represented as a rectilinear grid (see Section 11.2.1) and the data range is partitioned into two linear regions. The first region begins at the data minimum value and ends at a user specified value (break point) somewhere between the minimum and maximum. The second region starts at the break point and continues to the data maximum. The opacity and color at each end of these regions is specified. This provides the parameters for mapping the data to colors and associated opacity. The first method uses a ray-tracing technique. It provides smooth distributions but is computationally intensive.

The other volume rendering technique is often referred to as back-to-front (BTF). This technique is supported in many OpenGL graphics cards. The hardware acceleration can make the rendering using BTF over 100 times faster than ray tracing. With back-to-front, the projection plane is behind the grid and the grid is represented as a series of slices. The slice direction through the volumetric grid is chosen to keep the slices as orthogonal as possible to the viewing direction. Each slice of the volume is projected on the projection plane, from the farthest plane to the nearest plane. The slices are colored and made transparent based on the properties described previously.

**Ray Tracing.** Ray tracing is a computationally intensive form of rendering. It is not uncommon that rendering individual images at modest resolution (640 by 480 pixels) may take 30 minutes or more. What ray tracing lacks in speed it makes up for in fidelity and detail. Ray tracing can display details like shadows and reflections (such an entire scene reflected in a shiny spherical object). Ray tracing is often used to produce photo realistic scenes for architectural and landscape renderings. However, another consequence of photo realism is that a substantial amount of information, such as surface textures, reflectivity, specular reflection coefficients, etc., must be supplied for all objects in the view. Because of the complexity and slow speed, ray tracing is not commonly used for environmental visualization.

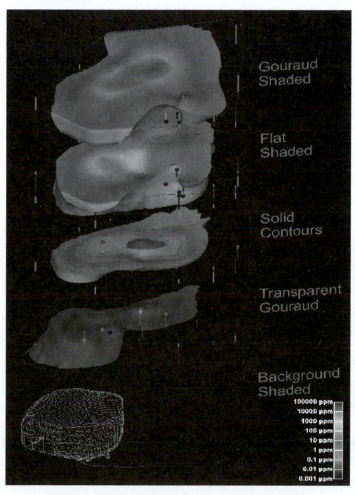

**Figure 11.25** 3D plume showing various shading methods.

**Color.** The choice of color(s) to be used in a visualization affects the scientific utility of the visualization and has a large psychological impact on the audience. Throughout this chapter, a consistent color scale (a.k.a. data map) has been used. This color scale associates low data values with the color blue and high data values with the color red. Values between the data minimum and maximum are mapped to hues that transition from blue to cyan (light blue) to green to yellow to red. People are accustomed to interpreting blue as a "cold" color and red as a "hot" color. For this reason, laypersons more easily understand this color spectrum. It also provides a reasonably high degree of color fidelity, allowing discrimination of small changes in data values.

However, many times color scales with vivid colors like red are deemed too alarming. Since there is not a universally (or even scientifically) accepted standard for color spectrums used for data presentation, the use of softer shades of color and the elimination of red or other garish colors from the spectrum cannot be challenged on a scientific or legal basis. The consequence of this is the distinct possibility of two different visualizations that both communicate the same information with completely different colors. Often the choice of colors is made on aesthetic or political grounds, governed more by the party being represented and their role in the site than by scientific reasons.

**Model Subsetting.** Once the model of the site has been created, visually communicating the information about that site generally requires subsetting the model. Subsetting is a generic term used to convey the process of displaying only a portion of the information based on some criteria. The criteria could be "display all portions of the model with a $y$-coordinate of 12,700. This would result in a slice at $y = 12,700$ through the model orthogonal to the $y$- (or North) axis. As this slice passes through geologic layers and/or contaminated volumes, a cross section of those objects would be visible on the slice. Without subsetting, only the exterior faces of the model will be visible.

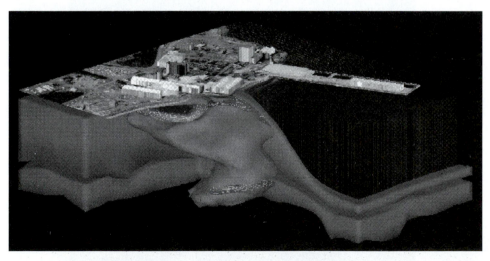

**Figure 11.26** Coast facility showing contaminant plume, geology with texture mapping

When evaluating subsetting operations, the dimensionality of input and output should be considered. For example, consider the slice described previously. If a slice is passed through a volume, the output is a 2D planar surface. If that same slice passes through a surface, the result is a line. Slices reduce the dimensionality of the input by one. The following sections will discuss a few of the more common subsetting techniques.

The first method uses a ray tracing (a.k.a. ray casting or front-to-back) technique. Using ray tracing, the rectilinear grid is considered behind the projection plane. A ray is projected from each point in the projection plane through the volume. The ray accumulates the opacity and color of each cell it passes through.

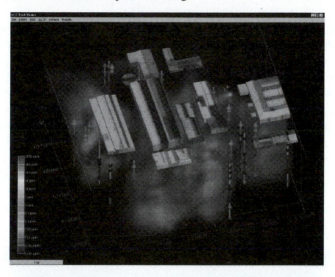

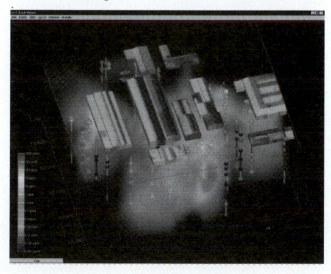

**Figure 11.27** Ray traced volume rendering.

**Plume Visualization.** Contaminant plume visualization employs one of the most frequently used subsetting operations. This is accomplished by taking the subset of all regions of a model where data values are above or below a threshold. This subset is also referred to as an isovolume and its threshold value as the isolevel. When creating the objects that represent the plumes, two fundamentally different approaches can be employed. One approach creates one or more surfaces corresponding to all regions in the volume with data values exactly equal to the isolevel and all portions of the external surfaces of the model where the data values exceed the isolevel. This results in a closed but hollow representation of the plume. This method, which was used in Figure 11.26, has a dimensionality one less than the input dimensionality.

The other approach subsets the volumetric grid outputting all regions of the model (cells or portions thereof) that exceed the isolevel. This method has the same dimensionality output as input. The disadvantages of this approach are the need to compute and deal with the all interior volumetric cells and nodes. The advantages include the ability to perform additional subsetting and to compute volumetric or mass calculations on the subset volume.

**Cutting and Slicing.** Within the EVS software there is a significant distinction between the terms *cut* and *slice*. Slices create objects with dimensionality one less than the input dimensionality. If a volume is sliced, the result is a plane. If a surface is sliced, the result is one or more lines. If a line is sliced, one or more points are created. Figure 11.28 has three slice planes passing through a volume that has total hydrocarbon concentrations on a fine 3D grid. The horizontal slice plane is transparent and has isolines on ½ decade intervals.

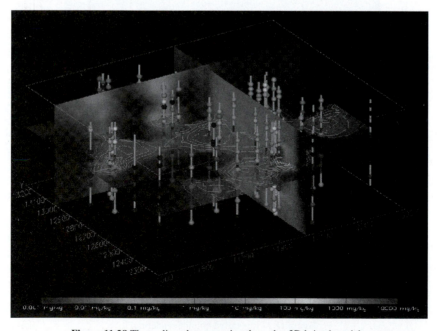

**Figure 11.28** Three slice planes passing through a 3D kriged model.

**Figure 11.29** Cut 3D kriged model with plume and labeled isolines.

By comparison, cutting still uses a plane, but the dimensionality of input and output is the same. Cutting outputs all portions of the objects on one side of the cutting plane. If a volume is cut, a smaller volume is output. In Figure 11.29, the top half of the grid was cut away, but the plume at 1000 ppm is displayed in this portion of the volume. The lower half of the model also has labeled isolines on decade intervals.

**Isolines.** Isolines delineate the model where data values have a constant value. Examples include contour lines (equal elevation) and isopachs (equal thickness). Isolines, sometimes referred to as isopleths, have output dimensionalities that are one less than the input dimensionality. Surfaces with data result in isolines or contour lines that are paths of constant value on the surface(s). Isolines can be labeled or unlabeled. Various labeling techniques can be employed, ranging from values placed beside the lines to labels that are incorporated into a break in the line path and mapped to the three-dimensional contours of the underlying surface. Examples of visualizations using isolines are shown in Figures 11.28 and 11.29.

## 11.5 PUBLISHING VISUALIZATIONS

For the purposes of this chapter, the term *publishing* refers to the creation of a visualization that can be delivered to an end user and does not require that user to have the software that created the visual to view it. With this definition, we expand publishing to include formats used by word processing software, Web publishing, animation, and color printing. In today's world, even when the final use for a graphic is color printing (e.g., magazines, brochures, etc.), the image is usually delivered to the printer in an electronic format.

### 11.5.1 Raster Images

Raster images are the foundational format for visualization publishing. Raster images are the most common method of outputting visualization results. Animations are constructed from a series of raster images, and printing of digital graphics nearly always involves rasterization (conversion to a raster image) at some stage of the process.

**Image Formats and Resolution.** Raster images can be written in one of many different file formats. Common formats include Windows Bitmap (.bmp), TARGA (.tga), TIFF (.tif), JPEG (.jpg), PNG (.png), and GIF (.gif). The critical issues to consider are color depth and whether the format uses lossy or lossless compression. For optimal quality, 24-bit true color should be used. The size of uncompressed true color raster images can be estimated in bytes as three times the product of the pixel resolution (width and height) plus ~30. For example, a 640 x 480 24-bit uncompressed image will be just under 922 Kbytes.

Windows Bitmap (.bmp) files do not use compression and support several color depths from 1 bit monochrome (black and white only) to 24-bit true color. File sizes tend to be large because of the lack of compression, but bmp files generally compress very well in archives (e.g., gzip, pkzip, etc.). An additional advantage of Windows Bitmap files is that all Windows computers include software that can read .bmp files.

TARGA and TIFF formats both use lossless compression algorithms and support a wide range of color depths, including 24-bit. Lossless compression does not compromise image quality. In contrast, lossy compression allows for modifying pixel colors in order to maximize compression. This usually results in the appearance of image noise, which is most pronounced near regions of high contrast such as the pixels adjacent to lines and edges. Lossless compression and uncompressed formats can be saved and converted any number of times without any degradation of image quality.

JPEG and GIF formats are both commonly used for graphics on the Internet. Their advantages are small file sizes and compatibility with virtually all computers and operating systems (provided they have Web browsers or other image viewing software installed).

There is also a copyright issue with some uses of GIF. A commercial company owns the LZW compression algorithm and a charge may be assessed for certain types of use. A replacement for GIF called PNG (Portable Network Graphics) has been proposed (see ftp://ftp.uu.net/graphics/png/README  and  http://www.eps.mcgill.ca/~steve/PNG/png.html). PNG files have the additional advantage of utilizing a lossless compression technique. The proliferation of PNG offers the size advantages of JPG and GIF files with no degradation in quality.

**Color Printing Issues.** The following provides hints and tips for obtaining optimal quality when printing. This assumes you are using a color printer, but it is important to note that the user may print gray scale images with a black and white printer if desired. This would be best implemented by creating gray scale color maps to eliminate ambiguities associated with different colors that have the same gray scale representation.

Optimal printing of a raster image requires taking several factors into consideration. First, you must know the characteristics of the printer and the intended size of the printed image. Printers vary considerably, and no single recommendation can be appropriate. Color printers fall into three primary categories: inkjet, color laser, and dye sublimation.

EVS, for example, produces raster images that are continuous tones with 256 shades each of red, green, and blue for a total of 16.7 million possible colors (256 x 256 x 256). Color printers either produce continuous tones or approximate them using a pattern of primary colored pixels in an *n*-by-*n* grid.

Among these three printer categories there is considerable variation. Inkjet printers are generally capable of producing one of only eight primary colors for each printer pixel (or dot). These colors are white, black, cyan, magenta, yellow, red, green, and blue. Inkjets must therefore use a grid of primary colored pixels to approximate continuous tones. The larger the grid (4 by 4 vs. 2 by 2) the better the color approximation. However, larger grids tend to create artifacts called jaggies that are visually undesirable. The challenge is to balance the need for smoother color rendition with the desire to have higher resolutions.

Dye sublimation printers are at the other end of the spectrum. Their ability to reproduce continuous tones makes the task of choosing a resolution easy. A typical dye-sub printer has a resolution of 300 dots per inch (dpi). If the intended size of the final printed image is 10 inches wide by 7 inches tall, then the optimal image size is 10 x 300 by 7 x 300 or 3000 x 2100 pixels. If quicker image creation and print times are desired, a compromise resolution would be exactly half or 1500 wide by 1050 high.

It is best to have an integer number of printer pixels for each "source" image pixel. When the image size is half of the printer pixel resolution, each source pixel gets a 2-by-2 grid. The *n*-by-*n* grid concept applies to all types of printers. This "rule" is actually a guideline for best results. Other resolutions (noninteger ratios) create banding artifacts that are usually objectionable.

For inkjet printers you should always allow for at least a 2 x 2 grid and usually 3 x 3 to 5 x 5 gives the best results. For an EPSON printer with 720 x 1440-dpi resolution you should use the smaller resolution number (720) for your calculations. The printer uses the additional resolution to better approximate the colors.

For example, for a printer with 720 dpi, to print an image 9 by 7.5 inches (landscape) we recommend that you start at a 4 x 4 grid, which gives an effective printed resolution of 180 dpi. Your image width and height would therefore be

Width = 9.0 x 180 = 9.0 x (720/4) = 1620
Height = 7.5 x 180 = 9.5 x (720/4) = 1350

Finally, color laser printers vary in their abilities to approximate continuous tones. This means that the rules to apply will be somewhere between dye-sub and inkjet properties.

**Stereo Image**s. There are several different types of three dimensional viewing and printing technologies available. Anaglyphs, which present the left and right eye images as red or blue monochromatic images, are among the simplest technique but do not allow the use of full color images. Shutter goggles, which switch the visibility of left and right eyes as in the IMAX Stereo Theaters and Virtual Reality headsets, allow for each eye to view a different image rendered by the computer. These methods require special viewer software and hardware to differentiate the left- and right-eye views. Our binocular vision enables us to view 3D objects because each eye sees a slightly different image of the scene from a slightly different viewpoint. In the brain, these two images are combined into a stereoscopic perception, having the appearance of depth and volume.

Lenticular images utilize a lenticular plastic sheet that has vertical ridges that form narrow cylindrical lenses. These lenses display a slightly different view to each eye depending on the angle between the eye and the lenticular sheet. The resulting lenticular images are autostereoscopic, meaning that they require no special viewers to display 3D imagery.  A significant limitation of lenticular displays is that they display images with an apparent resolution that is governed by the width of the cylindrical lenses. However, the three-dimensional lenticular printouts have all the visual cues of an actual volume of space, including object positioning (foreground and background), parallax and shading, and "look-around". The hologram-like effect is quite remarkable, and the image can have a tremendous impact on the viewer. 3D COM, Inc. provides a commercial service printing stereo image sequences as lenticular printouts (see http://3dhardcopy.com/).

An alternative stereo inkjet printing technology that creates a high resolution, full color stereoscopic hard copy is called StereoJet (Scarpetti 1996; Scarpetti et al., 1998). In the StereoJet image the left and right views are printed onto opposite surfaces of a multi-layer sheet, using inks with polarizing properties. The polarizing axes of the two image-receiving layers are oriented at 90° to one another. By wearing polarized glasses, viewers perceive the composite stereoscopic image in full depth. Each eye sees only the assigned image, and the brain processes the information to provide full binocular stereoscopic perception. Lenticular 3D prints do not have the resolution or clarity produced by this technology; however, StereoJet prints require special glasses. StereoJet printing is quite expensive at about $300 per square foot, but the quality is very high. For additional technical information or a source for StereoJet printing, see http://www.rowland.org/stereo/ and http://www.slidefactory.com/SJmain.html.

### 11.5.2 Postscript

Postscript, developed by Adobe, is a relatively common printer language that can represent both raster images and vector (polygon)–based representations of models. The advantage of vector representations is the elimination of jaggies and other objectionable results in printed output when the raster image being printed is too low resolution. The disadvantage of Postscript's vector format is its limited accommodation of rendering modes and shading methods.

### 11.5.3 VRML Models

Virtual Reality Modeling Language (VRML) is the only common nonraster output format that successfully captures most capabilities and features common in environmental visualizations. VRML is an ASCII file format that provides a vector description of three-dimensional models. It includes information about grids, data, subsetting, rendering properties (e.g., transparency), and raster images that are used as texture maps. Generally, models output as VRML look nearly identical to the rendered model in modeling and visualization software like EVS.

**Capabilities and Limitations.** VRML models provide a great degree of functionality and support nearly all rendering modes and visualization techniques. VRML models are interactive in that the viewer can "walk" or "fly" through the model or "examine" it as if it were a scale model. Ray tracing and all types of volume rendering are not currently supported in VRML.

**Customization.** VRML models can be customized by hand editing or the use of VRML authoring and/or editing software. These packages provide a mechanism to add sounds, interactivity, and functionality to what are normally static VRML models.

**VRML Viewer Browser Plug-Ins.** Viewing of VRML files is possible in most Web browsers, like Internet Explorer and Netscape Navigator. The processing and viewing of VRML models requires a browser plug-in. At the time of writing this chapter, testing performed by C Tech Development Corporation identified the most capable browser plug-in to be Cosmo. Silicon Graphics Inc. originally developed the Cosmo Software, but Computer Associates later acquired it. The Cosmo viewer can be downloaded for free from http://www.cai.com/cosmo/.

### 11.5.4 Animations

Animations provide one of the most powerful means for communicating environmental data. They offer all of the advantages of images and many of the advantages of VRML models. Processing or playing back a sequence of individual raster images creates animations. Animations can be used to clarify three-dimensionality by rotating a model about one or more axes; however, they can do much more. Animations can also include moving objects, changing plume levels, and moving cut or slice planes. They can even be used to present temporal data such as measured water table variation over time. In fact, all of these techniques can be combined into a single animation.

Although animations cannot be manipulated by the user as a VRML model can, animations can incorporate complex model changes that are beyond the capabilities of VRML. Furthermore, VRML's flexibility can sometimes be a problem. The end user can become confused or can miss the point of the model since VRML viewers allow the user to manipulate the model without limitation. Animations follow a script. If the script is properly planned and executed, animations lead the viewer down a path communicating the proper message. Animations are not limited to rendered images. Title images can be used to introduce sequences and sound can be added. The addition of sound allows for background music for dramatic affect as well as narration.

**Creation Process.** The process of creating animations begins with the creation of the image sequences. In C Tech's EVS, a script is generated based on a series of key frames (representing transition points in the animation) and the interframe interpolation method. When the script is played, the software renders each image and writes it as a raster image. Conversion routines then encode the images into an animation file format.

**Formats.** There are many animation file formats. The standard format for Windows computers is Audio-Video Interleaved (AVI). AVI files can employ many different codecs (compression/decompression methods). Virtually all codecs utilized by AVI converters employ lossy compression. So, if an individual frame of an animation were viewed or extracted, it would be noticeably degraded from the original. As with image file formats, the compression artifacts are most pronounced in areas of high contrast, such as lines and edges. Photographs of people and landscapes do not tend to have these high-contrast regions (as compared with digitally created images) and therefore the compression is usually more acceptable. MPEG (developed by the Motion Picture Engineering Group) and Apple's QuickTime are competing animation formats. These also generally employ lossy compression.

A noticeable exception in animation file formats is the proprietary HAV file format developed by Gromada (http://www.gromada.com). HAV uses a lossless compression format that often creates files as small or smaller than lossy codecs. Gromada offers shareware converter software and freeware HAV player. Their player is also capable of playing image sequences without the need to convert them to an animation format. C Tech's EVS software includes an animation converter that creates AVI, MPEG, and HAV files.

The creation of animations intended for playback on televisions requires consideration. The U.S. and foreign video formats (NTSC, PAL, and SECAM) all use interleaved video fields. The first color TV broadcast system was implemented in the United States in 1953. This was based on the NTSC (National Television System Committee) standard. Many countries on the American continent as well as many Asian countries, including Japan, use NTSC. NTSC is capable of displaying 525 horizontal lines/frame. The PAL (phase alternating line) standard was introduced in the early 1960s and implemented in most European countries except for France. The PAL standard utilizes a wider channel bandwidth than NTSC that allows for better picture quality. PAL is capable of displaying 625 horizontal lines/frame. The SECAM (Sequential Couleur Avec Memoire or Sequential Color with Memory) standard was introduced in the early 1960s and implemented primarily in France. SECAM uses the same bandwidth as PAL but transmits the color information sequentially. SECAM is capable of displaying 625 horizontal lines/frame.

Interleaved video displays alternating odd and even scan lines every sixtieth of a second (50 Hz. for some PAL and all SECAM formats). This results in even and odd scan lines flickering at 30 Hz. When lines or edges are drawn only one pixel wide and are nearly horizontal, the lines will exist in only the odd or even fields. This causes the lines to flicker in an objectionable manner. The best solution to this problem is to employ special processing techniques on the images before conversion to broadcast video. The processing reduces the contrast of lines and blurs them slightly to avoid having the entire line being represented in only the odd or even fields. Although this may seem like it would introduce additional negative consequences, the results are usually excellent.

## 11.6 VISUALIZATION APPLICATIONS

There are many applications for visualization of environmental data. Visualization enables technical personnel to better understand their work. It is often the only practical method of communicating complex results to a nontechnical audience, such as in public hearings or litigation support. In these environments, animations are particularly useful and often employed.

Though GIS software often includes some visualization capability, it often falls short as compared with true 3D visualization software. However, GIS software has the ability to store images and animations created with other software in their databases. These images or animations can be viewed by selecting them on a site-by-site basis. In this manner, GIS software can become the repository for detailed visualizations performed on small sites that are a portion of a much larger project. With GIS systems like ESRI's ArcView[2], images and animations can be hot-linked to objects in the GIS project, allowing them to be viewed with a single mouse click.

## 11.7 CONCLUSIONS AND LESSONS LEARNED

This chapter has demonstrated the value of visualization as a scientific and communications tools for virtually all environmental projects. Visualization is a valuable scientific asset to geologists, engineers, and investigators and is often the only effective means for communicating to a lay audience. There are many choices and options that must be considered in developing an effective visualization. These include gridding, interpolating, subsetting, rendering methods, and color selecting. The range of visualization techniques allows for quick first-look evaluations and high-level photorealistic models that incorporate a myriad of data sources.

Footnotes:
   1. Environmental Visualization System and EVS are registered trademarks of C Tech Development Corporation.
   2. ArcView and ARC/INFO are registered trademarks of Environmental Systems research Institute.

## 11.8 REFERENCES

Clarke, K. C. 1999. *Getting Started with Geographic Information Systems.* Upper Saddle River, NJ: Prentice Hall.

Freidhoff, R. M., and Benzon, W. 1989. *Visualization: The Second Computer Revolution.* New York: Abrams.

Hutchinson, M. F., and Gessler, P. E., 1994, Splines—More Than Just a Smooth Interpolator. *Geoderma,* 62, 45–67.

Scarpetti, J. J., Dubois, P. M, Friedhoff, R. M., and Walworth, V. K. 1998. Full-Color 3-D Prints and Transparencies. *Journal of Imaging Science & Technology*, 42, 307–310.

Scarpetti, J. J. 1996. Coating Methods and Compositions for Production of Digitized Stereoscopic Polarizing Images, International Patent W/O 96/23663.

Slocum, T. A. 1999. *Thematic Cartography and Visualization.* Upper Saddle River, NJ: Prentice Hall. .

## 11.9 INFORMATION RESOURCES

All of the figures and graphics in this chapter were created using C Tech Development Corporation's Environmental Visualization System (EVS). Much of the material in this chapter is based on material in the EVS reference manuals and tutorials (http://www.ctech.com).

## 11.10 ABOUT THE CHAPTER AUTHOR

 Reed D. Copsey is the President and CEO of C Tech Development Corporation, a software company exclusively engaged in visualization, modeling, and analysis of environmental, geologic, mining, and archaeological data.

# 12

# GIS/EM
# A Research Agenda

## 12.0 SUMMARY

Like the three influential meetings before it, the Fourth International Conference on Integrating Geographic Information Systems and Environmental Modeling—GIS/EM4—focused not on the specific methods of modeling and GIS, but on the integration of these methods and the benefits that cross-disciplinary work offers to those disciplines that use them for integrated science. Participants came from all quarters of the GIS and EM communities, from biological and physical systems modeling, and from government, academe, and private enterprise. The attendees' unique cross-disciplinary knowledge was exploited by means of work groups to help articulate a research agenda for the next three to five years. These work groups identified a wide range of issues and ideas that are organized and presented here under the headings of stakeholders and users, data, GIS, modeling, complexity and uncertainty, professional practice, education, and communication. A synthesis of these major issues comprises the agenda of proposed research presented in this chapter as The Banff Statement.

## 12.1 INTRODUCTION

Ten years have elapsed since the inception of the GIS/EM conference/workshops series in Boulder, Colorado, in September 1991. The intervening decade has been characterized by significant progress in both the GIS and environmental modeling fields, in the integration of these spatial/analytic tools, in computer technology and communications, in addressing data issues, and in the growing realization that addressing the world's complex environmental problems requires an integrated science approach.

Those familiar with the GIS/EM conference series know that in addition to Boulder, Colorado, subsequent meetings were held at Breckenridge, Colorado; Santa Fe, New Mexico; and most recently, Banff, Alberta, Canada. The selection of meeting sites along the North American continental divide became symbolic of the divide that has existed between the GIS and EM communities. Reasons for the gap are many, but foremost is a lack of communication between the two groups. A goal of the organizers of the GIS/EM conferences has been to narrow and bridge this gap. Although the divide has not been closed, this book attests to continued progress.

To break down barriers to communication and promote discourse across disciplines at GIS/EM4, an innovative program structure was used to organize the paper sessions across 21 themes (Table 12.1). This resulted in a melange of themes being represented at any particular session.

**Table 12.1** GIS/EM4 Paper Sessions
1. Research and development opportunities and constraints
2. Users, advocates, and sponsors
3. Auditing and performance measures
4. Biotic (living) systems
5. Physical (nonliving) systems
6. Managed (agriculture, silviculture, urban) systems
7. Human-environment interactions
8. Problem-solving environments
9. Modeling techniques and tools
10. Statistical techniques and tools
11. Decision support and decision making
12. Integration, coupling, and synergism
13. Complexity and chaos
14. Spatial dimensions
15. Temporal dimensions
16. Uncertainty and predictability
17. Adaptability, feedback, and dynamics
18. Representations and formalisms
19. Measurement (including data) and indicators
20. Information handling and knowledge bases
21. Visualization and explanation

## 12.2 WORK GROUPS

Organizers of the GIS/EM conference series have worked to achieve direct involvement of attendees in the formulation of an articulate and coherent research agenda guiding the integration of spatiotemporal analysis (via GIS) and environmental modeling. Methods used to achieve attendee participation have varied with each conference, and so have the results. For example, the 1991 meeting in Boulder, Colorado, used 10 breakout groups to partition the interests of 600 attendees. Although the sizes of breakout groups exceeded the optimum number for best group dynamics, each of the 10 groups identified pertinent issues and developed recommendations that were presented at the close of the meeting (Crane and Goodchild, 1993).

In contrast, at GIS/EM2 in Breckenridge, Colorado, participants were asked to complete a questionnaire, the results of which were presented on the final day (Goodchild et al, 1996). More time was available for planning GIS/EM4, which enabled organizers to return to the use of work groups to involve attendees in formulating a research agenda that provides a set of goals to guide subsequent research. Unlike conference paper presentation sessions that were formulated along thematic lines (to be provocative and promote interdisciplinary cross fertilization), work groups were created along major disciplinary and environmental management lines—biotic systems, physical (nonliving) systems, managed systems, human-environment interactions—to create four categories of "experience."

A systematic approach was used to subdivide the conference assembly of approximately 400 people and assign them to work groups. This was accomplished by classifying all papers, posters, and other presentations to indicate which of 21 subthemes were addressed by each author's written contribution (conferees who were not contributing authors were asked to self-assign to an already identified work group). A directed graph was then constructed using the resulting database and custom software to cross index four "disciplinary" themes with the remaining seventeen "less-disciplinary" (or more application/tool/technique-oriented) themes. The computer analysis was carried out according to rules formed after inspection of the database and included criteria that aggregated results

**Figure 12.1** Workshop session under way at GIS/EM4. Photo by Keith Clarke.

(using five "major" categories) to identify a limited number of work groups. More important, these rules combined conferees into systematic and logical—but tactically counterintuitive—sets with those they would not otherwise have had reason to regard as colleagues.

Our strategy was the discovery by conferees of previously unknown commonalities or complementarities. This discovery process is a hallmark and performance measure for the organizers of the GIS/EM conferences. The four disciplinary themes used were (1) biotic systems, (2) physical (nonliving) systems, (3) managed (agriculture, silviculture, urban, etc.) systems, and (4) human-environment interactions. The five major themes used (after aggregation) included (1) research prioritization, utilization, and evaluation; (2) natural systems and system relations; (3) analytic frameworks, properties, and applications; (4) system dimensions, characteristics, and behaviors; and (5) language, metrics, monitoring, benchmarks, and interpretation. Certain small work group categories, with only two or three people assigned, were either combined to form a larger group, or people were encouraged to join another group, if their interests warranted it. This resulted in creation of the 12 work groups depicted in Table 12.2.

As many as 30 people participated in some work groups. Facilitators and recorders were assigned to each, and the groups were asked to identify issues and problems and to develop a list of recommendations. Since all work groups acted independently of each other, the issues they identified and their treatment is variable. Most groups covered a broad range of topics, resulting in a fair amount of overlap that serves to reinforce the importance of particular issues and recommendations. For the sake of brevity, results of the work groups have been further integrated by topic, and these are presented in the following subsections.

**Table 12.2** The Twelve GIS/EM4 Working Groups (shaded gray)

| Discipline | | Theme | | | | |
|---|---|---|---|---|---|---|
| | | Research Prioritization | Natural Systems | Analytic Frameworks | System Dimensions, Etc. | Language, Metrics, Etc. |
| | Biotic | | | | | |
| | Physical | | | | | |
| | Managed | | | | | |
| | Human-Env. | | | | | |

## 12.2.1 Users and Stakeholders

The inclusion of users and stakeholders in the model development process was identified as essential to its subsequent success and acceptance. Stakeholders are those people who are affected by the model outputs throughout the modeling process. By their inclusion, stakeholders bring to the table ideas, perspectives, data, opportunities, constraints, and values that may not otherwise be considered in the process. Stakeholders help define matters from a potentially more immediate, practical, or local perspective. Stakeholder buy-in also helps to assure that modeling results will be applied directly to the problem under investigation.

The question was asked, "Whom are we really targeting with models when we say their results will be used by decision makers?" Technical users are normally not decision-makers. Decisions are generally made by legislative groups, agency heads, executive officers, private land owners, and other arbiters. Should end users also be experts, so they understand the processes they decide upon? It would certainly be beneficial if end users were more knowledgeable about the modeling processes they base their decisions on. Perhaps a more realistic scenario would be for technical users and experts to interpret the results of model runs for decision makers. Either way, future GIS/EM conferences need to develop a better understanding of decision makers, their views and needs, and how well current models meet their needs.

### 12.2.2 Data

Interestingly, data issues did not receive as much attention from GIS/EM4 work groups as they previously had. Still, a number of interesting points emerged. There is immediate perceived need for well-documented benchmark data sets that are readily accessible (downloadable from a Web site) to the modeling community. Such data are used to calibrate and validate model runs. These data need to include associated metadata concerning lineage, accuracy, resolution, reliability, and so on. Furthermore, the results of model runs employing these data need to be fed back reciprocally to the Web site, where their posting can benefit subsequent data users.

Work group members also expressed the need for hierarchically organized sets of geospatial data at various scales/resolutions. Similarly, they stated that models and data structures should also be flexible and adaptable to varying scales and resolutions. High-resolution data need to be available for areas of high societal interest/impacts. Such data are being produced in many parts of the world where population growth is creating particularly intense pressure on the environment and resources. For example, high-resolution elevation data is being acquired for certain floodplains in the United States by the Federal Emergency Response Agency (FEMA) in order to depict more accurately the 100- and 500- year flood inundation zones.

Problems of data accessibility also received attention, but no resources were offered. At the federal level of government in the United States, there are increased efforts to make public data available at little (only cost of reproduction) or no cost. In contrast, many local government agencies view data sales as a key revenue source; consequently these data may be relatively expensive.

### 12.2.3 GIS

The decade of the 1990s saw dramatic growth in the use of GIS for the development of integrated, georeferenced spatial and time series data sets, data management, product generation, but to a much lesser extent for spatiotemporal analysis. That GIS is not being used more extensively today for spatiotemporal analysis may, in part, reflect a lack of understanding of spatiotemporal analytic techniques—a deficiency that could be addressed by appropriate training and experience. Several work groups expressed the need for GIS soft-ware tools to include error tracking or propagation functions. Another recognized a need for an expert system interface that provides statistical advice on spatial autocorrelation and geostatistics. Spatial statistics are a vital component for any GIS and should be incorpo-rated in the software as a matter of course. Of course, this will not negate the responsibility of the users to acquire appropriate training and skills in spatial statistics in order to be able to apply the tools properly.

We live in a three-dimensional world and need to develop proper tools that can, at a minimum, effectively handle 3D data sets, models, and visualizations. This capability is particularly important for modeling the physical and natural environment. For example, many critical biological and ecological processes involve flows and interactions within and between marine, terrestrial, aquatic, and atmospheric spheres. True three-dimensional capability is essential to represent these processes. Unfortunately, this has been an identi-fied deficiency in GIS software since before the inception of these conferences and was, in fact, one of the key deficiencies that provoked their organization. It is not clear that this deficiency will be addressed in the foreseeable future. There are special modeling software packages that provide some 3D capability.

### 12.2.4 Modeling

In contrast to the previous GIS/EM conferences, which exhibited a strong GIS emphasis, GIS/EM4 showed greater balance between GIS and environmental modeling, especially in the content of the conference program. So it is interesting that modeling issues dominated responses from the work groups. A common thread in their recommendations is the need for honesty, or disclosure, in modeling so that everyone, from the modeler to the decision maker, has a clear understanding of what the results of a model run may mean. This might entail documentation of error sources, error propagation, accuracy, uncertainty, limita-tions, underpinning theory, data characteristics, and so on.

Nearly all work groups identified the need to develop metadata standards for models. At a minimum, such metadata should include author(s), domain of application, data requirements, predictability, assumptions, listing input/output variables, limitations, uncer-tainty, results of model validation, results of sensitivity, documentation of model features, model lineage, and keywords. A related issue is the need for model documentation to be transparent, current, and complete in order to benefit other potential users of a model. Incomplete or poorly written documentation, and undocumented changes to models, can contribute to the misuse or mistrust of models and their parameters. Other issues relating to standards include the need for a standardized method for documenting knowledge embedded/encapsulated in a model, standards for comparable   landscape/environmental measurements, and a universal spatiotemporal model construction.

The question was raised whether models and digital spatiotemporal data should be peer reviewed and published in professional journals. This would certainly help to elevate the visibility of the modeling community and, in time, might help to strengthen the quality of both models and data and lead to more informed applications by users. Another group called for creation of a Web-based clearinghouse for models (and presumably data) that would employ an intelligent interface capable of guiding users toward the most appropriate model/data combination for an intended application or problem. Such a Web site could also provide online training and/or point to providers of training.

Another intriguing question asked whether there is a role for simple models. For example, should simple nonparameterized models be used as the basis for policy decisions, and should complex models be reserved for parameter development? Simple models entail the risk that decision makers will not fully grasp the complexity of the system being modeled. Understanding system complexity is especially important when applying models for predictive purposes.

Work groups identified a number of desirable tool enhancements, including more effective real-time integration of GIS and EM for simulations of processes and incorporation of feedback effects, such as simulation of flooding/erosion and dynamic modification of the underlying digital elevation model (DEM). Improvements are also needed in the presentation of models and output to nonmodelers. Many of the requested improvements will require a transdisciplinary approach to developing a language common between GIS, EM, and other allied tool sets; user-friendly visualization and animation tools; multiresolution and fuzzy boundary techniques; and the inclusion of spatial statistics and 4D modeling. In addition, frameworks need to be developed to combine models in order to enhance their capability to assess cumulative multi-impacts. Similarly, scientific models need to become better connected to management decision support systems to enhance equally difficult choice-making tasks.

Interesting and specific recommendations were made for the next conference, GIS/EM5, such as holding a competition for the "best modeling application" and awarding a prize. Others would like to see demonstrations of models by real users, as well as examples where modeling has had a known and direct impact on policy and decision making. A less well-defined and ambitious suggestion is to examine development of a modeling theory, an activity that may be suitable for a specialized workshop.

### 12.2.5 Complexity and Uncertainty

A significant issue designed into the conference and examined by the work groups is the need to deal effectively with complexity and nonpredictability in modeling. Nonlinear behavior needs to be accounted for in models of complex systems. Methods are needed to incorporate an understanding of the complex behavior of systems directly into the modeling process from the beginning. However, in order to understand complex behavior, we first have to determine which metrics to use. A related issue is uncertainty—a topic that, curiously, received scant attention at GIS/EM4. There is an urgent need to educate students, stakeholders, and the public about uncertainty and its propagation throughout the modeling process. Unfortunately, practical guidelines and standards for the treatment of uncertainty do not exist. New and better tools are needed to evaluate, quantify, and effectively visualize the nature and degree of uncertainty. Additionally, researchers should be strongly encouraged to study and report on uncertainty in GIS and EM, and this should become one of the program themes for GIS/EM5.

## 12.2.6 Professional Practice

Several work groups suggested the establishment of a professional code of practice that would apply to all aspects of GIS/EM—data, models, verification, validation, and output. This could promote adoption and use of "best practices" by model designers. Another recommendation dealing with professional practice calls for creation of a new professional society—The Association for Spatial Modeling of the Environment—that would be transnational, transdisciplinary, and organized around the broad theme of developing methods of environmental analysis within a spatiotemporal context. This idea was broached at previous GIS/EM conferences but seems to be gaining wider support. It is suggested among participants that establishment of such a society would help to promote the ethical use of scientific methods, as well as a professional code of practice. A professional society might also be able to add impetus for further development of themes that have been the cornerstone of previous GIS/EM meetings. It could serve as a clearinghouse for peer-reviewed models and data, provide guidelines on appropriate use of modeling modules and parameters, and help identify international funding sources to further the aforementioned interests. Most important, it could provide a focal point and home for the diversity of environmental modeling work being conducted. Others oppose creation of a professional society on the grounds that it is inconsistent with the open activities of the conferences and that its basis would be coercive rather than cooperative.

## 12.2.7 Education

A strong need was identified for graduate-level education in environmental modeling. The case could also be made that some exposure to environmental modeling should begin during undergraduate study, when it could create a broader appreciation of this aspect of environmental problem solving. Interestingly, one would be hard-pressed to identify a program at any academic level where theory   and practice of modeling is included as a key part of the curriculum.

Although not mentioned by GIS/EM4 work groups, the editors suggest that a similar need exists for education in spatial analysis techniques using GIS software tools. GIS is being used increasingly for spatial analysis; however, it is still predominantly a tool for integrating geospatial data and generating graphic output. Neither GIS nor EM have reached their separate or joint full potentials at this time.

## 12.2.8 Communication

Appreciation was expressed that GIS/EM4 work groups, because they contained a mix of people from academia, private enterprise, and government, created an environment (microcosm) that was conducive to improving communication among the GIS/EM scientific community. To multiply or expand such communication, the suggestion was made that a Web site be established and run by the editorial board. Such a Web site should include information about current developments in GIS and EM, a catalog of models together with their documentation and metadata, a review section on new models and data sets with feedback from model users, a calendar of significant dates (conferences, due dates for funding proposals, training opportunities, etc.), a listing of experts in specialized disciplines who are involved in environmental modeling, and e-discussion groups on topics such as a metadata standards, error and complexity, theory of modeling, and development of common GIS/EM languages.

Strong recommendations were made to publish spatial data and models in peer-reviewed journals, both digital and traditional hardcopy journals. In addition, authors are encouraged to clean and make available both data and models that underlie the applications they write about. Requiring submission of data and metadata for peer review as part of an expanded publication process may be a useful way to encourage a positive shift in data and model accessibility.

More effective means of exchanging scientific knowledge need to be developed. Complex ideas are not yet presented in an accessible, educative, and communicative manner for stakeholders and nonspecialists. Ecobeaker (a software developer participating at GIS/EM4) is a good example of a system that employs an interactive display of the user's choice of rules and demonstrates how user's choices evolve in time and space. SimCity (not a GIS/EM4 participant) is another example of a software tool that provides an excellent user interface. In general, there is need for improvement in visualization and other communication techniques in order to ensure successful communication of modeling processes and results to decision makers. It should be noted, however, that research on the psychophysical aspects of visualization methods is scant at best.

To reinforce further the transdisciplinary nature of GIS/EM5, it is recommended that keynote speakers should be invited from even more varied disciplines than those already tapped—symbologists, computer specialists (parallel computing), mathematicians (graph, string, category theory), statisticians, economists, and sociologists to name only a few. As anticipated, the idea of a virtual meeting was suggested, and while this might be easier or more economical to organize, it would lack the stimulating spontaneity and intensity of a live meeting. A common feature of virtual meetings has been their unorthodox multidisciplinary settings and programs. Also, because the Internet is still not ubiquitous, some people would be unable to participate.

The foregoing issues, suggestions, and ideas contributed by the 12 working groups have been synthesized into a more cogent expression of an agenda for research—*the Banff Statement.*

## 12.3 THE BANFF STATEMENT

Banff, Alberta, Canada
Friday, September 8, 2000

Four hundred interested scholars and practitioners met at the Banff Centre for Conferences in Banff, Canada, from September 3–8, 2000, at the Fourth International Conference on Integrating GIS and Environmental Modeling. In a plenary meeting of all conferees, the following statement was unanimously confirmed as a summary of the accomplishments at the meeting.

We consider that effective solutions to environmental problems cannot come without significant advances in the understanding of human, engineered, and biophysical environmental systems, and an appreciation that these systems are all components of a single integrated system that includes humankind. We believe that beneficial scientific solutions will derive from building new and better analytical tools by combining the strengths of models and geographic information systems, and from applying these hybrid tools across disciplines, scales, and user communities. We believe that significant progress has occurred in the integration of GIS and Environmental Modeling since the origin of the GIS/EM meetings in 1991. We have identified the following opportunities for eliminating the major barriers to further progress in integrating modeling and GIS:

•Modelers and GIS analysts should seek to develop transdisciplinary behavior, to engage in open sharing of data and methods, and to seek collaboration between the biophysical, socioeconomic, and ecological sciences so that coupled models of holistic systems can be developed. In particular, the role of humans in environmental systems should receive additional attention from science.

•Models must be developed so that they can be applied across spatial scale, and the role of scale variation should be explicitly a component of models themselves.

•Environmental models demand adequate descriptions of the quality, uncertainty, and accuracy inherent in their input data. Similarly, models themselves should be described by metadata so that systems of models can become interoperable. The propagation of error through the model, while perhaps not quantifiable, should be assessed objectively and communicated to the users of model predictions.

•Integrating modeling and GIS requires an increased awareness of new technologies and their potential impacts.

•Integrated modeling demands that stakeholders, policy makers, the marketplace, and alternative strategies be an integral component in modeling.

•Users of spatial models need to develop a new work ethic termed "honesty in modeling." This code of ethics must include accounting for uncertainty and error in predictions, recognizing of alternative management and intellectual strategies and traditions in problem solving, and working to include user communities in the process. The GIS/EM interest group should work actively to engage scholars in the developing world, and encourage them to develop effective solutions to local problems by sharing methods and data, perform educational outreach, and provide intellectual support.

•More effective use needs to be made of education, training, and new technologies to develop and disseminate examples of best practices in GIS/EM. This should include the development of core educational curricula.

•Problems need to be approached using both top-down and bottom-up methods, and integrated assessments and analyses should replace ad hoc solutions.

•The GIS/EM community should seek to provide to regulators and decision makers logical arguments and case studies documenting why modeling integrated with GIS can improve the world in which we live.

Approved. Attendees of GIS/EM4
Banff, Alberta
September 8, 2000

## 12.4 EPILOGUE

One of the objectives of the organizers of the GIS/EM4 Conference was to look back and evaluate the progress and accomplishments that have been achieved since the inception of this series of meetings in 1991. Clearly, significant progress has been made; however, continued efforts are needed to extend further the benefits of integrated science, and promote better integration of tools, techniques, and methods for spatiotemporal modeling.

During the wrap-up session on the final morning of GIS/EM 4, attendees were asked whether there was a need for and interest in having a GIS/EM5 conference. Overwhelming support was voiced for the suggestion, and the decision was made to hold GIS/EM5 in Australia in June 2003, in an effort to bridge the barrier of the Pacific Basin. The many good ideas generated by the work groups will be considered in forming the program for GIS/EM5. The editors hope that readers will keep that date open on their calendars and plan to contribute to the further integration of GIS and environmental modeling through interdisciplinary scientific work.

## 12.5 REFERENCES

Crane, M. P., and Goodchild, M. F. 1993. Epilogue. pp. 481–483 in Goodchild, M. F., Parks, B. O., and Steyaert, L. T. (eds.). 1993. *Environmental Modeling with GIS.* New York: Oxford University Press.

Goodchild, M. F., Steyaert, L. T., Parks, B. O., Johnson, C., Maidment, D., Crane, M., and Glendinning S. 1996. Epilogue. pp. 485–486. In Goodchild, M. F., Steyaert, L. T., Parks, B. O., Johnson, C., Maidment, D., Crane, M. and Glendinning S. (eds.). *GIS and Environmental Modeling: Progress and Research Issues.* New York: Wiley.

## 12.6 ABOUT THE CHAPTER AUTHORS

Keith C. Clarke (upper right) is professor and chair of the Department of Geography, University of California, Santa Barbara and Director of the Santa Barbara site of the NCGIA. Bradley O. Parks (front right) is a senior research scientist with the University of Colorado at Boulder and with NOAA—National Environmental and Space Data Information Systems. He is founder and director of the International Conferences on Integrating GIS and Environmental Modeling (GIS/EM). Michael P. Crane (upper left) is a physical scientist at the U.S. Geological Survey's EROS Data Center, where he serves as the Emergency Response Program manager. Thanks to Yvonne Dixon (front left) of the Banff Center for Conferences, who played a key role in local arrangements of the meeting. The editors also thank Xiaohang Liu for assistance in editing this book.

# Index